Sustainable disease management in a European context

David B. Collinge · Lisa Munk · B.M. Cooke

Sustainable disease management in a European context

Reprinted from European Journal of Plant Pathology, Volume 121, No. 3, 2008

 Springer

David B. Collinge
University of Copenhagen
Denmark

Lisa Munk
University of Copehagen
Denmark

B.M. Cooke
University College Dublin
Ireland

ISBN 978-1-4020-8779-0   e-ISBN 978-1-4020-8780-6

Cover photos:

From top to bottom: A simplified model of defence against pathogens illustrating successful transgenic strategies for making disease resistant plants (Collinge et al., 2008).

Control of leaf blight of tomato by spraying tomato plants with garlic juice 2 h before inoculation. Inoculation was done by spraying whole plants with a sporangial suspension of *P. infestans* (4–5×10^{-4} sporangia ml^{-1}). Top panel, inoculated plants; lower panel, inoculated and sprayed with diluted garlic juice containing 110 μg ml^{-1} allicin (Slusarenko et al., 2008).

Accumulation of H_2O_2 as seen by DAB-staining in (a) & (b) the barley $-B.\ graminis$ f.sp. *hordei* interaction and (c) & (d) the wheat $-Septoria\ tritici$ interaction. In (a), a cell is undergoing HR as a response to penetration and is completely stained with DAB, in (b) only papillae are stained. (c) and (d) show wheat cvs Stakado and Sevin exhibiting resistance and susceptibility to the same isolate of *S. tritici* (Shetty et al., 2008).

Rice cultivar mixtures in China. Single rows of tall traditional varieties are interspersed within fields of high yielding hybrid cultivars (Finckh, 2008).

Schematic representation of domains found in plant LRR Resistance proteins. TIR Toll/interleukin-1 receptor, CC coiled coil, NB nucleotide binding, ARC1/2 APAF1, R protein and CED4, LRR leucine-rich repeat, SD solanaceous domain, BED BEAF/DREAF zinc finger domain, TM transmembrane, Kin kinase, WRKY WRKY transcription factor (Tameling and Takken, 2008).

Printed on acid-free paper

9 8 7 6 5 4 3 2 1

springer.com

European Journal of Plant Pathology

Volume 121 · Number 3 · July 2008

Special Issue: Sustainable disease management in a European context
Edited by: David B. Collinge · Lisa Munk · B.M. Cooke

Instructions for Authors for *Eur J Plant Pathol* are available at http://www.springer.com/10658.

Eur J Plant Pathol (2008) 121:213–216
DOI 10.1007/s10658-008-9316-z

Foreword

David B. Collinge · Lisa Munk · B. M. Cooke

Received: 3 April 2008 / Accepted: 3 April 2008

Sustainable disease management in a European context

The main theme of the book is sustainable disease management in a European context. Of course the issues are global and the papers reflect this. Some of the questions addressed are: How does society benefit from plant pathology research? How can new molecular approaches solve relevant problems in disease management? What other fields can we exploit in plant pathology research? What challenges are associated with free trade across the new borders? How can we contribute to solving problems of developing countries? How does plant pathology contribute to food quality and safety? How does globalization/internationalization affect teaching and extension in plant pathology?

D. B. Collinge (✉) · L. Munk
Department of Plant Biology and Biotechnology,
Faculty of Life Sciences, University of Copenhagen,
Thorvaldsensvej 40, 1871, Frederiksberg C,
Copenhagen, Denmark
e-mail: dbc@life.ku.dk

L. Munk
e-mail: lm@life.ku.dk

B. M. Cooke
School of Biology and Environmental Science,
College of Life Sciences,
University College Dublin, Belfield,
Dublin 4, Ireland
e-mail: Mike.Cooke@ucd.ie

The authors of papers in this special issue of European Journal of Plant Pathology were selected among the invited speakers at the 8th Conference of the European Foundation for Plant Pathology & British Society of Plant Pathology Presidential Meeting 2006[1] that was held at The Royal Veterinary and Agricultural University[2], Copenhagen from 13th–17th of August 2006. This was an intimate conference attended by some 200 largely European delegates from more than 30 countries. The result of the conference and of this volume is an insight into the diversity of problems facing pathologists and the remarkable progress made in recent years. This book is intended to be more than a proceedings volume, and clearly, given the breadth of the subject, it represents a series of readings and not a comprehensive account of the state of research in the field in the middle of 2007 (the deadline for submission of these articles) or even of the excellent research presented at the conference that has not resulted in a paper in this special issue. There are many interesting relevant topics that were not presented at the conference — an obvious example is toxins in our food. Notwithstanding this, we believe the authors have provided a useful series of review articles and case studies of many key areas that we hope can

---

[1] http://www.efpp06.kvl.dk/index.html

[2] Now the Faculty of Life Sciences, the University of Copenhagen.

inspire future research. We have asked the authors to prepare the papers so that they can be used as teaching material for advanced courses and are well satisfied with the result.

We have organised the papers in four sections. As in all classifications of biological material, there is ambiguity as to the correct order and precise classification: alternative models would be entirely appropriate.

The first and largest section **How can biotechnology contribute to sustainable development?** presents an overview of the biological knowledge obtained using molecular biological techniques of the nature of plant-microbe interactions. We start with a review by Collinge et al. of the success stories, progress and challenges associated with developing transgenic disease plants. Weed resistant or rather herbicide and/or insect resistant crops have been grown extensively in various parts of the world, though largely not in Europe, for over a decade, and the area increases annually. Cultivated transgenic disease resistance crops are currently restricted to virus-resistant papaya and courgettes (zucchini) in the USA. The biological — and political — reasons for this are presented and discussed. The techniques of molecular biology and molecular genetics continue to provide an ever deeper understanding of the nature, recognition and regulation of the active and passive defence mechanisms protecting plants from pathogens. Two papers delve into different aspects of the regulation of disease defence mechanisms. Goellner and Conrath review the priming of defences and innate immunity, and in the process cover both the molecular basis of different forms of induced resistance to pathogens and pests and practical experience with the application of the concept. Tameling and Takken consider the downstream signalling associated with race-specificity. Recent knowledge shows how these processes respond to targeting by pathogen effector proteins — of which avirulence gene products represent a special case. The traditional models for and our understanding of Flor's gene-for-gene hypothesis are brought into context. The use of tools of functional genomics to study defence responses are illustrated by Collinge et al. and exemplified by a case study with the NAC transcription protein family of barley in the next paper. Two papers by Shetty et al. and Pruvsky highlight the importance of reactive oxygen species (ROS) in defence and signalling processes. The former reviews the current knowledge of ROS — as directly antimicro-

bial defences, as signals inducing defence responses and in oxidative cross-linking. The latter looks at the role of ROS and other factors in the switch from quiescent to necrotrophic interactions. The implications for control of post-harvest diseases are discussed. Both these papers challenge the simple classification of pathogens into biotrophs, hemibiotrophs and necrotrophs. The final paper in the section, by Ludwig-Müller and Schuller, concerns the study of *Plasmodiophora brassicae*, a fascinating organism quite unlike most of the pathogens we meet, and the use of the model plant *Arabidopsis thaliana* in its study. The major output described in the papers of this section is still fundamental biological knowledge where comparative genomics is an emerging theme. The real and projected impact of this knowledge in combating plant pathogen interactions is discussed.

The second section concerns **Strategies for disease control**. Jørgensen et al. present an analysis of the actual needs and habits of different types of farmer in order to optimise disease control in cereals. The advisory service integrates disease resistance information with fungicide recommendations. Organic agricultural systems are not concerned with the use of fungicides. Two papers consider alternative strategies appropriate for organic growers and provide interesting case studies. Slusarenko reviews the control of plant diseases by natural products and exemplifies this with Allicin from garlic. Finally, Whipps et al. present and review the mycoparasite *Coniothyrium minitans* as a biocontrol agent.

Under the title **Quarantine and diagnostics**, the third section addresses the issues of global pathogen spread. Despite our level of knowledge, new threats from pathogens continue to emerge, resulting in the spread of disease to agricultural systems and natural ecosystems around the world. Both increased free trade and climatic change contribute to these developments. Petter et al. review the progress made to harmonise methods for diagnostics and provide access to the materials developed. Thrane describes the implementation of these tools and provides a case study: potato testing in Denmark. Two papers in this section present two modern methods for diagnostics and identification of pathogens that allow rapid diagnosis of problems without the need for the taxonomic and identification skills developed through a life-time of study of diverse pathogens. Thornton describes the use of monoclonal antibodies for

detecting fungi (*Trichoderma* spp.) in soil and compares their use to other techniques, *e.g.*, nucleic acid-based methods. The use of these methods for following population dynamics and quantities of fungi are discussed. Boonham et al. look at the development and prospects of generic platform technologies, specifically Real-Time PCR, for diagnostics. The final paper of the section by Smith et al. shows how information on new outbreaks can be used globally but, using Africa as a case study, also discusses the global challenges facing agro-industry and quarantine systems.

The final section concerns **Population diversity and dynamics**. Disease resistance is the most effective form of controlling disease, when available. Plant breeding strategies for disease resistance are dependent on an understanding of the diversity of the pathogen population with respect to the frequency of avirulence genes to which the crop is exposed. Hovmøller describes the impact of virulence surveys — mapping virulence specificity (i.e. avirulence genes) in *Puccinia striiformis* populations in Denmark and the use of the data in dissemination to breeders and farmers. Kaur et al. describe how molecular biological methods have developed as essential tools in efficient plant breeding for disease resistance, and cover both the methods now available for identifying allelic variation (eco-TILLING) and the use of molecular markers in the breeding process. Wheat resistance used against powdery mildew is the case study described. Finckh advocates breeding for resistance diversity to provide the rationale for exploiting and implementing resistance.

The future for plant pathology

So what are the most significant advances made in the last few years and what challenges remain?

An increased understanding of the nature of race specificity is currently emerging. It is increasingly clear that resistance genes can function in two ways. They either recognise pathogen molecules directly, or recognise the effect of the pathogen molecule on the host cell. These pathogen molecules are now called effector proteins. A subset of these can be recognised by the host and they are then called specific elicitors, which are coded by the pathogen's avirulence genes. Race specificity is now considered to represent a second level of defence which functions to guard the primary mechanisms of resistance, now commonly known as innate immunity, from being disarmed by the effector molecules produced by the pathogen. At the same time, molecular genetic analyses, especially but not exclusively, in *Arabidopsis*, have vastly expanded our knowledge of the nature and regulation of distinct forms of induced resistance. In other words, several distinct, but interacting (i.e. so-called 'crosstalk') signal transduction pathways have been identified which regulate the defence responses activated by plant pathogens. The challenge now is to put this fascinating knowledge to use.

The increased knowledge of microbial genomes has not only led to major revisions in the taxonomy of plant pathogens but also to the development of molecular diagnostic tools, for instance gene chips — i.e. microarrays which allow specific identifications and PCR-based tools. These tools are useful in some systems but there are still many challenges for other pathogens, especially for asexual organisms where closely related organisms that cause diseases on different crops can be difficult to distinguish. We are getting closer to the idea of being able to put diseased material into one end of a machine and obtain a printout of the probable diagnosis from the other end. This would provide a cost-effective solution and bring diagnostics to technicians rather than letting it remain in the hands of a few specialists. These developments are welcome as increasing free trade can facilitate more rapid spread of pathogens between continents and the more variable climates we seem to be experiencing that are attributed to global warming create new habitats for pathogens in regions where particular pests and pathogens can find new niches.

Major progress has been made in the development of models for forecasting the development of diseases both between and within seasons. This can have considerable impact in the development of decision systems for farmers in terms of which varieties to plant in the next season, and how and when to spray. A major challenge is to devise the means for providing appropriate input data for the models. A challenge for the industry is the increasing costs of the development of new pesticides versus the potential for profit.

Plant pathology can both benefit from and contribute to biotechnology. In industrial countries, plant pathogens are emerging as a resource for new enzyme products for preparing biomass for biofuels. The study of the biochemistry of the complexity of cell walls and the tools that pathogens use to degrade them are one of many opportunities for plant pathologists. We hope you are inspired.

A challenge for the subject of plant pathology lies in the way the universities and research institutions worldwide are funded. In an increasing number of countries, assessment systems and the division of resources is based on impact factors and the like, rather than societal benefit. The field of plant pathology covers both fundamental research and finding solutions for problems facing growers.

The impact of our work is not always measured in citations in the scientific literature but in increased yields of healthy crops. Universities often appoint new lecturers on the former criterion. This is a worrying trend. In addition, although the numbers of university students continue to increase worldwide, the interest in biological sciences and agriculture is waning among the young. As a profession it is vital that we continue to demonstrate that the research and teaching that we perform reaches our targets — society, the industry and students, and that the benefit of our efforts is visible to society. We hope this volume contributes to these aims and needs.

Eur J Plant Pathol (2008) 121:217–231
DOI 10.1007/s10658-007-9229-2

What are the prospects for genetically engineered, disease resistant plants?

**David B. Collinge · Ole Søgaard Lund ·
Hans Thordal-Christensen**

Received: 9 May 2007 / Accepted: 27 September 2007
© KNPV 2007

Abstract Insect and herbicide-resistant plants are the most widely grown transgenics in agricultural production. No strategy using genetically engineered plants for disease resistance has had a comparable impact. Why is this? What are the prospects for introducing transgenic disease resistant plants to agriculture? We review the biological background for strategies used to make disease resistant GM crops, illustrate examples of these different strategies and discuss future prospects.

Keywords Genetically engineering ·
Disease resistant plants · Plant virus ·
Fungal disease · Bacterial disease

D. B. Collinge (✉) · O. S. Lund
Department of Plant Biology,
Faculty of Life Sciences, University of Copenhagen,
Thorvaldsensvej 40,
1871 Frederiksberg C, Denmark
e-mail: dbc@life.ku.dk

O. S. Lund
e-mail: osl@life.ku.dk

H. Thordal-Christensen
Department of Agricultural Sciences,
Faculty of Life Sciences, University of Copenhagen,
Thorvaldsensvej 40, 1871 Frederiksberg C, Denmark
e-mail: htc@life.ku.dk

O. S. Lund
Department of Genetics and Biotechnology,
Faculty of Agricultural Sciences, University of Aarhus,
Thorvaldsensvej 40, Opg. 8, 2. sal,
1871 Frederiksberg C, Denmark

Introduction

Disease resistance is the most effective means of controlling disease. However, there are many pathogens for which no effective sources of disease resistance have been identified. Genetic engineering has been promoted for two decades as a solution for this problem, but to date only very few GM disease resistant cultivars have been introduced to commercial agriculture. This is in stark contrast to the situation for two other key disciplines of plant protection, namely insect pest and weed control where Bt[1] and herbicide-tolerant crops represent well over 90% of all GM crops (James 2006), and have been on the market for more than ten years. The answer to this lies primarily in the complexity of the biology of the traits concerned. Economics has undoubtedly also played a role in that investment in transgenic insect and herbicide resistance was considered safe since the key technologies concerned were well established in agricultural practice prior to their biotechnological application. Furthermore, the implementation of new products is delayed as a result of moratoria resulting from negative public opinion and expense of commercialisation.

Enhanced disease resistance has been achieved using several strategies. These are depicted in Fig. 1 and are described briefly here. The most straightfor-

---

[1] Abbreviations: Bt *Bacillus thuringiensis* toxins; GM genetically modified.

Fig. 1 A simplified model of defence illustrating successful transgenic strategies. See Table 1 for examples. Strategy 1 concerns direct interference with pathogenicity or inhibition of pathogen physiology. Thus 1a involves constitutive expression of antimicrobial factors and 1b involves pathogen-induced expression of one or more genes in the transgenic plant. Strategy 2 concerns the regulation of the natural induced host defences. 2a concerns altering recognition of the pathogen (e.g., R-genes) and 2b concerns downstream regulatory pathways (e.g., SAR), and includes transcription factors. Strategy 3 is pathogen mimicry: the manipulation of the plant to prime recognition of a specific pathogen through pathogen derived gene sequences (genetic vaccination). See Table 1 for examples

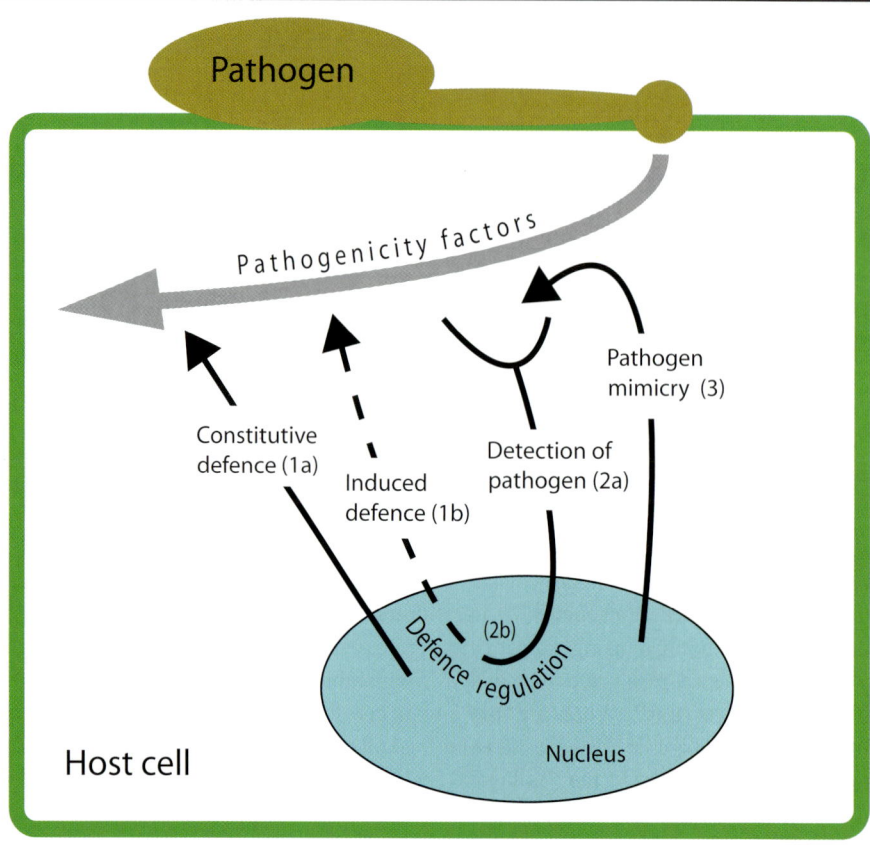

ward approach is to add genes encoding antimicrobial proteins or peptides originating from plants or other organisms either alone or in combination (1a). The addition of new antimicrobial secondary metabolites to a species, which can be achieved by adding genes encoding the appropriate biosynthetic enzymes, lies under this strategy. Other strategies involving detoxification, quenching pathogen signals etc also belong in this category. A variant concerns the use of pathogen-inducible promoters to regulate these antimicrobial factors (1b). Plants have their own effective defences – induced resistance, basal resistance and race-specific resistance, which activate the plant's own antimicrobial defences. Therefore, a second strategy concerns the manipulation of the regulation of these processes (2) and can exploit the recognition processes as well as the regulatory signal transduction pathways. A third strategy (3) is pathogen mimicry by which we mean the manipulation of the plant to prime the plant to recognise a specific pathogen. Mechanistically, this is also termed pathogen derived resistance or genetic vaccination. The unique nature of viruses

has made it possible to combat them effectively through gene silencing, which can be considered part of strategy 3.

In this review, we do not attempt to generate a comprehensive review of the many studies which have demonstrated improved disease resistance by transgenic approaches, but illustrate the strategies used with pertinent examples (see Table 1). Indeed, one of the major challenges in writing this review lies in the discrepancy in the quality of documentation at different stages in the process. There are many examples where enhanced resistance has been documented in refereed journals for transgenic plants in the laboratory, but few where documentation extends to field conditions or adoption by practical agriculture.

Herbicide tolerance

Weed control using GM crops has been possible because of the biology of herbicide tolerance. Synthetic herbicides have been developed to be effective

killers of many plants and target different vital processes common to vascular plants. Plants have not been exposed to these substances during evolutionary timescales and natural resistance is therefore not present in target plants. Transgenic herbicide-resistant plants are in commercial use with tolerance against Glyphosate (*N*-(phosphonomethyl) glycine) and Glufosinate, best known under the trade names Roundup® and Basta® (among others), respectively. High durability of the transgenic herbicide tolerance approach has been anticipated, since the main herbicide involved, glyphosate, has proven efficient for more than two decades before being incorporated into the GMO strategy (see Cerdeira and Duke 2006; Senior and Bavage 2003; Senior and Dale 2002 for further discussion of herbicide tolerance).

Insect resistance

In our view, the success with GM insect resistance is attributable both to good fortune, in that appropriate genes are known, and fundamental biological differences between biting insects and pathogens. In contrast to the synthetic herbicides, Bt toxins are natural products of the common soil bacteria *Bacillus thuringiensis* which were originally isolated from moribund insects. This topic has been the subject of several recent reviews (Babu et al. 2003; Christou et al. 2006; Ferry et al. 2004, 2006). As for herbicide tolerance, robust durability of the transgenic approach has been anticipated since Bt toxins have been used for decades as durable insecticides, though some resistance has been observed (e.g., Perez and Shelton 1997). Furthermore, Bt resistance has been demonstrated to be associated with a fitness cost for the insects (Bird and Akhurst 2004; Carriére et al. 2001). There are, nevertheless, several documented examples of Bt resistant insect pests, e.g., (Gahan et al. 2001; Huang et al. 1999) and, if this strategy is to continue to be successful, the use of Bt transgenes must be managed carefully.

Plants do not suffer from producing these proteins, which are toxic to insects. This is an obvious advantage of the approach. Results suggesting toxic effects to mammals are controversial and inconclusive (Séralini et al. 2007), and it should be borne in mind that this technology has been applied in crops used for both human and stock consumption for over a decade without any prior indications of problems.

Many different strains of *B. thuringiensis* have been described which are toxic to different families of insects, where the proteinaceous toxins act in the midgut of the digestive system (Babu et al. 2003). In particular, the *Cry*1a toxin, affecting Lepidoptera, has proven an effective means of controlling several stem-boring insect larvae, especially in maize (*Zea mays*), where the Corn Borers *Sesamia nonagrioides* and *Ostrinia nubilalis* are among the most serious. The *Cry*3a gene, from another *B. thuringiensis* strain, encodes a Bt toxin effective against Coleoptera, and is used extensively against the Cotton Bollworm *Helicoverpa armigera* in cotton (*Gossypium hirsutum*).

It should be noted that, while the known Bt toxins used in plants work against biting insect pests, there are large groups of insect pests, including sucking insects (Homoptera, such as leaf hoppers and aphids) against which no natural Bt toxins are known. Synthetic, chimeric toxins have been developed which extend the range of these toxins to other insect groups (Mehlo et al. 2005). Furthermore, as for fungi, there are examples of transgenic strategies using genes encoding other insecticidal proteins, for example lectins, which have led to resistance against these types of insects in the laboratory (Saha et al. 2006; Yao et al. 2003).

What issues affect plant disease resistance?

The differing biology of the various types of plant pathogens presents substantial problems in developing GM resistant plants. Firstly, the kinds of organisms causing disease are taxonomically highly diverse; the major groups include cellular pathogens (e.g., bacteria, fungi and the algal Oomycetes) and molecular pathogens (i.e. viruses). These are physiologically very different from each other, and therefore no single gene product can be expected to have a direct toxic effect on all types of pathogens. Secondly, pathogens use two major life strategies, namely biotrophy and necrotrophy. Biotrophic pathogens essentially act as a sink for the host's anabolic assimilates, and therefore keep it alive. Meanwhile, necrotrophic pathogens consume the host tissues as invaded. Hemibiotrophs combine both strategies in their life cycle. Consequently, plants have developed quite different ways for dealing with these two strategies (see below).

Table 1 Examples of transgenic strategies resulting in enhanced disease resistance, named in the text

Strategy[a]	Gene	Recipient	Donor	Pathogen	Reference
Constitutive (a) and inducible (b) defences					
1a	AiiA	Potato (Solanum tuberosum)	Bacillus	Pectobacterium (Erwinia) carotovora	(Dong et al. 2001)
1a	Chi11 (chitinase) Tlp (PR-4)	Rice (Oryza sativa)	Rice (Oryza sativa)	Rhizoctonia solani	(Kalpana et al. 2006)
1a	Cry1Ab (Bt toxin)	Maize (Zea mays)	Bacillus thuringiensis	Fusarium spp	(Clements et al. 2003; Hammond et al. 2004)
1a	gf-2.8 (oxalate oxidase)	Soybean (Glycine max)	Wheat (Triticum aestivum)	Sclerotinia sclerotiorum	(Cober et al. 2003; Donaldson et al. 2001)
1a	gf-2.8 (oxalate oxidase)	Sunflower (Helianthus annuus)	Wheat (Triticum aestivum)	Sclerotinia sclerotiorum	(Hu et al. 2003)
1a	gf-2.8 (oxalate oxidase)	Poplar (Populus × euramericana)	Wheat (Triticum aestivum)	Septoria musiva	(Liang et al. 2004)
1a	Gfzhd101	Maize (Zea mays)	Clonostachys rosea	Fusarium graminearum	(Igawa et al. 2007)
1a	KP4	Wheat (Triticum aestivum)	Virus infecting Ustilago maydis	Tilletia caries	(Clausen et al. 2000; Schlaich et al. 2006; Schlaich et al. 2007)
1a	Suppressor 2b (CMV)	Tobacco (Nicotiana tabacum)	Cucumber mosaic virus (CMV)	CMV	(Qu et al. 2007)
1a	Synthetic sequence generating miRNA based on P69 (TYMV) & HC-Pro (TuMV)	Arabidopsis thaliana	Turnip yellow mosaic virus (TYMV) & turnip mosaic virus (TuMV)	TYMV & TuMV	(Niu et al. 2006)
1a	RCH10 (PR-3) & AGLU1 (PR-2)	Tobacco (Nicotiana tabacum)	Rice (Oryza sativa) & alfalfa (Medicago sativa)	Cercospora nicotianae	(Zhu et al. 1994)
1a	Synthetic D4E1	Tobacco (Nicotiana tabacum)	Cecropia (insect)	Colletotrichum destructivum	(Cary et al. 2000)
1a	Synthetic D4E1	Poplar (Populus tremula Populus alba)	Cecropia (insect)	Agrobacterium tumefaciens, Xanthomonas populi pv. populi & Hypoxylon mammatum	(Mentag et al. 2003)
1a	Synthetic D4E1	Cotton (Gossypium hirsutum)	Cecropia (insect)	Thielaviopsis basicola	(Rajasekaran et al. 2007)
1a & 1b	Vst1 (Stilbene (resveratrol) synthase)	Tobacco (Nicotiana tabacum)	Grapevine (Vitis vinifera)	Botrytis cinerea	(Hain et al. 1993)
1a & 1b	Vst1 (Stilbene (resveratrol) synthase) pss (pinosylvin synthase)	Barley (Hordeum vulgare) & Wheat (Triticum aestivum)	Grapevine (Vitis vinifera); Scots Pine (Pinus sylvestris)	Botrytis cinerea; Puccinia recondita f.sp. tritici & Stagonospora (Septoria) nodorum	(Leckband and Lorz 1998; Serebriakova et al. 2005)
1b	9f-2.8 (oxalate oxidase) & TaPERO (peroxidase)	Wheat (Triticum aestivum)	Wheat (Triticum aestivum)	Blumeria graminis f.sp. tritici	(Altpeter et al. 2005)

	Gene	Source	Host	Pathogen	Reference
Detection of pathogen attack					
2a	Rpi-blb2 (NB-LRR)	Solanum bulbocastanum	Potato (Solanum tuberosum)	Phytophthora infestans	(van der Vossen et al. 2003, 2005)
2a	Rxo1 (NBS-LRR)	Maize (Zea mays)	Rice (Oryza sativa)	Xanthomonas oryzae pv. oryzicola	(Zhao et al. 2005)
2a	Vf (Cf)	Malus floribunda	Malus domestica (apple)	Venturia inaequalis	(Belfanti et al. 2004)
2a	Xa21 (NBS-LRR)	Rice (Oryza sativa)	Rice (Oryza sativa)	Xanthomonas oryzae pv. oryzae	(Wang et al. 2007; Zhai et al. 2002)
2a/2b	Pto	Lycopersicon pimpinellifolium	Tomato (Lycopersicon esculentum)	Pseudomonas syringae pv. tomato	(Tang et al. 1999)
Regulation of inducible defences					
2b	NH1 (NPR1)	Rice (Oryza sativa)	Rice (Oryza sativa)	Xanthomonas oryzae pv. oryzae	(Chern et al. 2005)
2b	NPR1	Arabidopsis thaliana	Wheat (Triticum aestivum)	Fusarium graminearum	(Makandar et al. 2006)
2b	NPR1	Arabidopsis thaliana	Arabidopsis thaliana	Pseudomonas syringae and Peronospora parasitica	(Cao et al. 1998)
Pathogen mimicry					
3	C1 (TYLCV)	Tomato Yellow Leaf Curl Virus (TYLCV)	Tomato (Lycopersicon esculentum)	TYLCV	(Fuentes et al. 2006)
3	ORF2b (PLRV)	Potato leaf-roll virus (PLRV)	Potato (Solanum tuberosum)	PLRV	(Vazqez Rovere et al. 2007)
3	PSRV coat protein gene	Papaya ringspot virus (PRSV)	Papaya (Carica papaya)	PRSV	(Fuchs and Gonsalves 2007)
3	REP (TYLCV)	Tomato yellow leaf curl virus (TYLCV)	Tomato (Lycopersicon esculentum)	TYLCV	(Yang et al. 2004)

[a] See Fig. 1 for explanation.

R-genes

The so-called resistance (R) genes have been widely employed for many years, through conventional breeding programmes, with great success. These genes control many plant diseases caused by bio-trophic pathogens such as rusts and powdery mildew fungi. These genes have the advantage of conferring complete resistance against specific races of the pathogen. Many specific resistance genes are available in the major crops (Hovmøller et al. 1997; Hovmøller 2007; McDonald and Linde 2003). Typically, 40–70 specific R-genes and alleles have been described for the rust and powdery mildew diseases of major crops. Most R-genes encode the so-called nucleotide binding-site, leucine-rich repeat (NB-LRR) proteins, which activate down-stream defence to combat the disease, when the pathogen has a specific avirulence-gene (*Avr*) that corresponds to the specific R-gene (see the recent review by (McHale et al. 2006). However, resistance obtained by intro-gression of these types of gene generally has the drawback that pathogen populations eventually adapt to their presence and overcome them (Hovmøller et al. 1997; McDonald and Linde 2003). In other words, when the *Avr*-gene in the pathogen is inactivated by a mutation, the resistance is no longer functional. As the *Avr*-genes often encode effector proteins which have evolved to function in pathogenicity, there is strong balancing selection in natural plant and pathogen populations for polymorphism at the genetic loci in host and pathogen. This means that many *Avr*-alleles are present in natural pathogen populations. However, genotypes carrying a virulent allele of any *Avr*-gene locus will eventually migrate to and invade the resistant plant population, leading to reduced efficiency of the specific resistance gene. These types of resistance genes operate at the recognition stage of an interaction and generally against biotrophic patho-gens, where the expression of resistance is often associated with a form of programmed cell death (PCD), known as the hypersensitive response (HR).

In some cases, R-genes can provide effective protection against pathogens when transformed into new species and even into new genera, and this protection can be broad spectrum, i.e., independent of pathogen race and even species (Oldroyd and Staskawicz 1998; Rommens et al. 1995; Tai et al. 1999). This represents strategy 2a in Fig. 1. Recently, *Rxo1*, an R-gene derived from maize (*Zea mays*), a non-host of the rice bacterial pathogen, *Xanthomonas oryzae* pv. *oryzicola* was successfully transformed into rice (*Oryza sativa*) and shown to confer resistance against *X.o. oryzicola* (Zhao et al. 2005). Thus, the potential of using R-genes as transgenes across natural breed-ing barriers exists. However, inter-species differences may radically influence R-gene function (Ayliffe et al. 2004) and therefore it is preferable to use R-genes from closely related species. The transgenic approach circumvents tedious backcrossing and has successful-ly been accomplished in rice for the R-gene, *Xa21*, conferring broad, but nevertheless race-specific resis-tance to the bacterial leaf blight disease (Wang et al. 2007). *Xa21* has subsequently been transformed into a restorer line for hybrid rice and shown to provide resistance without compromising elite traits (Zhai et al. 2002). Field tests of *Xa21* transgenic rice in The Philippines, China and India have shown satisfactory results (Datta 2004). However, deregulation of trans-genic *Xa21* rice for large scale cultivation is still pending. It should be noted that conventional breed-ing assisted by the use of molecular marker techni-ques has already provided hybrids containing *Xa21*, pyramided with other resistance genes (Joseph et al. 2004; Zhang et al. 2006), thereby creating a compet-itive alternative to the transgenic approach.

An NB-LRR R-gene, Rpi-blb2, derived from the wild potato relative, *Solanum bulbocastanum*, confers broad-spectrum race-nonspecific resistance in potato (*Solanum tuberosum*) against the Oomycete pathogen *Phytophthora infestans* (van der Vossen et al. 2003, 2005) and patent EP20020075565. A representative of the Cf family of R-genes, *Vf* was cloned from the wild apple species *Malus floribunda*, and transferred to cultivated apples (*Malus domestica*) where resis-tance to a presumably mixed population of *Venturia inaequalis* isolates was demonstrated (Belfanti et al. 2004). Compared to conventional breeding, the trans-gene approach will facilitate introduction of more R-genes into a crop at the same time (pyramiding). This will extend the durability of the resistance concerned. The transgenic strategy using R-genes can, however, have negative side effects. For example, over-expres-sion of the *Pto* gene from tomato (*Lycopersicon esculentum*) resulted in a lesion mimic phenotype in the mesophyll tissue (Tang et al. 1999).

Some necrotrophic pathogens are adapted to deliberately use the R-gene-type of recognition in

order to activate PCD through the use of specific toxins that provoke the R-gene signalling (reviewed by Glazebrook 2005; Mayer et al. 2001). R-genes effective against these types of necrotrophic pathogens are simply unknown. This statement needs to read in the light of the fact that many apparently necrotrophic pathogens are really hemibiotrophic and exhibit an albeit brief biotrophic or endophytic phase during early stages of infection, where R-genes can be effective, e.g., *Bipolaris sorokiniana*, *Magnaporthe oryzae*, *Rhynchosporium secalis*, *Phytophthora infestans* (see Parlevliet 2003).

Induced resistance

Plants have effective defences against pathogens. These defences are invariably activated following pathogen attack, though they are not always sufficiently effective to lead to resistance (see the articles by Collinge et al. 2007; Conrath, 2007; Shetty et al., 2007). Briefly, it is often observed that pathogen attack or treatment with various inducer molecules can result in effective local induction of resistance, or less effective systemic resistance which, nevertheless, has a broad effect on many pathogens (Durrant and Dong 2005). Such studies of induced resistance have led to two different strategies aiming for the development of transgenic disease resistant plants. One of these can be considered a first generation strategy, analogous to the strategies used in GM crops to control insect pests, which concerns the use of single gene products that have a direct inhibitory effect on the pathogen (strategy 1 in Fig. 1). Second generation strategies (i.e. more recent studies) are based on an understanding of the mechanisms regulating disease resistance in plants, for example the R-genes as described above (strategy 2). Neither of these strategies has yet led to GM disease-resistant crops in production, though the latter strategy is promising.

First generation strategies for transgenic disease-resistant plants

The study of plant defence mechanisms in the 1970s and 1980s led rapidly to the discovery that various defence proteins (i.e. the PR or pathogenesis-related proteins), certain small peptides and a wealth of secondary metabolites possess direct antimicrobial activities (Broekaert et al. 2000; Castro and Fontes 2007; Field et al. 2006; Hammerschmidt 1999; van Loon et al. 2006). In contrast to the case with Bt toxins and insects, however, no single protein or metabolite has been identified with a major effect on a range of pathogens. For example, early studies based on *in vitro* data suggested that the plant defence enzyme, chitinase (Collinge et al. 1993), was a promising candidate to provide resistance against any fungal pathogens since the substrate, namely chitin, is a major constituent of fungal cell walls. Many studies used single genes encoding antimicrobial proteins, such as chitinase to make transgenic plants (see Broekaert et al. 2000; van Loon et al. 2006). There are a number of examples where it was demonstrated that constitutive expression of single genes gave a significantly improved disease resistance (see Broekaert et al. 2000 for an early comprehensive review), but in no case was the effect more than partial, even when several genes encoding antimicrobial proteins were combined in the same plant (e.g., Kalpana et al. 2006; Zhu et al. 1994). Some examples are discussed below and listed as 1a/1b in Table 1. Antimicrobial proteins can act through a wealth of physiological mechanisms, few of which are really understood. Some act directly to interfere with pathogen physiology or indirectly by interfering with pathogenicity processes necessary for infection.

A concern associated with the production of antimicrobial proteins is that some might be allergenic or toxic to vertebrates, and there are well-established analyses for allergenic or toxicological risks [see e.g., Schlaich et al. (2007) for examples applied in a relevant case]. Another concern is related to the risk of selecting new microflora resistant to future antibiotics of relevance to humans. From a plant science perspective, it would be interesting to use plant antimicrobial proteins, such as defensins (Broekaert et al. 2000), as alternative medical antibiotics, similar to plectasin (Mygind et al. 2005). A means for reducing toxicological and allergenic risks, whilst simultaneously reducing the risk that pathogens will develop tolerance or resistance to specific proteins, is to use promoters which confer tissue-specific or defence-response-specific expression in the transgene. An example of this strategy is the use of epidermis-specific expression of defence genes in wheat (*Triticum aestivum*) where constitutive expression of a wheat peroxidase gene specifically in the epidermis

provided some protection against the powdery mildew fungus (Altpeter et al. 2005).

Analogous to Bt insect resistance, antimicrobial proteins are found in many organisms other than plants, and have been exploited in transgenic strategies. It can be predicted that increased knowledge of the biology of plant–microbe, and microbe–microbe interactions will provide further examples with potential for GM-strategies. Thus transgenic wheat has been prepared with a gene encoding the protein KP4 from a virus, which infects the smut fungus *Ustilago maydis*. These plants exhibited variously 10–30% protection against the smut (*Tilletia caries*) in field and greenhouse tests (Clausen et al. 2000; Schlaich et al. 2006, 2007). Antimicrobial peptides do not need to be natural. For example, transgenic cotton, prepared using a synthetic peptide, D4E1 (derived from an insect antimicrobial peptide), exhibited enhanced resistance against the fungus *Thielaviopsis basicola* (Rajasekaran et al. 2007). Interestingly, the same peptide provided resistance to bacterial pathogens in transgenic poplar (Mentag et al. 2003; Montesinos 2007).

A new approach to protect plants against bacterial diseases is based on interference with the communication system, quorum-sensing, used by several phytopathogenic bacteria to regulate expression of virulence genes according to population density (reviewed by Cui and Harling 2005). The enzyme, AiiA, isolated from bacterial strain, *Bacillus* sp.240B1, was found to degrade the quorum-sensing signalling molecule of the soft rot pathogen, *Erwinia carotovora*, and thereby rendering the bacteria incapable of infecting the host (Dong et al. 2000). Transgenic expression of AiiA *in planta* was subsequently demonstrated to provide significant enhancement of resistance against soft rot in potato (Dong et al. 2001; US patent 7205452). The strategy looks technically very promising since the microbial target is likely to be strongly conserved. However, since similar quorum-sensing is also known for bacterial pathogens of humans (for example, *Pseudomonas aeruginosa*), this strategy also raises the concern that there is a risk that control of bacterial infection in humans will be impaired.

A plethora of different antimicrobial secondary metabolites (known as phytoalexins or phytoanticipins (VanEtten et al. 1994)) are produced in plants. These metabolites can have roles in disease resistance, and in some cases it has been demonstrated that these can indeed limit the host range of specific pathogens (Field et al. 2006; Osbourn 1996). Specific metabolites are often restricted to closely related plant species, and pathogens adapted to a particular plant species need to be able to withstand these antimicrobial metabolites, for example, by detoxifying them. This makes them attractive subjects for exploitation in transgenic strategies. It can be predicted that the pathogens adapted to parasitise one species are not adapted to the phytoalexins of a distantly related species and are therefore incapable of detoxifying them. However, one problem in exploiting secondary antimicrobial metabolites in transgenic disease resistance strategies is that they are usually the products of multi-step biosynthetic pathways, requiring multiple enzymes, each comprised of one or more proteins, which are individually the products of separate genes. Unfortunately, this calls for simultaneous or sequential transformation of many genes into a single plant line. In many cases, the complexities of the biosynthetic pathways remain to be clarified and the necessary genes cloned. The best exploited exception concerns the stilbenes, especially reservatrol. In this case, it has proven possible to make a new phytoalexin following transfer of a single gene, with resulting improved resistance (Hain et al. 1993; Leckband and Lorz 1998; Serebriakova et al. 2005). However, in no case has the desired complete resistance been obtained.

The regulation of disease resistance – the second generation

Whereas cucumber and tobacco, in particular, provided the physiological understanding of induced resistance, mutational studies using *Arabidopsis thaliana* have provided a profound understanding of the nature of regulation of defence mechanisms (Glazebrook 2005). Arabidopsis genetics has, for instance, been instrumental in the analysis of the different mechanisms of resistance operating in biotrophs and necrotrophs. This has led to strategies for utilising regulatory genes in developing GM disease-resistant plants (Campbell et al. 2002). The use of R-genes (included in strategy 2a described above) can be considered to fall under the concept of this approach. A general strategy is to activate defence signalling

pathways and thereby simultaneously stimulate a wider collection of the down-stream response genes, which manifest the resistance (strategy 2b). For this purpose, the knowledge obtained from work on mutant plants that constitutively express defence responses can be explored. Such plants are generally resistant to a number of different pathogens, but they often suffer from being lesion mimics and dwarves (see Lorrain et al. 2003). Examples of such mutants are *lsd1* (Torres et al. 2005), *acd2* (Mach et al. 2001), *acd11* (Brodersen et al. 2005), *cpr1*, *cpr5*, *cpr6* (Clarke et al. 2000) and *syp121 syp122* (Zhang et al. 2007). Some mutations causing lesion mimic phenotypes have occurred in NB-LRR-type R-genes. Here, mutations in specific motifs of the NB domain permanently stimulate resistance as they mimic avirulence-activation of the R-protein (Howles et al. 2005; Takken et al. 2006).

An interesting example concerns the *NPR1* (or *NIM1*) gene, a key defence regulator first identified in *Arabidopsis* (Durrant and Dong 2005). Over-expression of this gene confers broad-spectrum resistance against various pathogens (Cao et al. 1998). The effect is not restricted to *Arabidopsis*, thus over-expression of *Arabidopsis* NPR1 in wheat led to resistance against *Fusarium graminearum* (Makandar et al. 2006). Transgenic rice plants over-expressing the rice NPR1 orthologue (NH1) acquire high levels of resistance to *Xanthomonas oryzae* pv. *oryzae* (Chern et al. 2005).

Because of lesion development and dwarfism, the resistance caused by this kind of mutation cannot be used directly. However, if the regulatory gene can be up or down-regulated according to the function of the protein, so that resistance is activated only when a pathogen attacks, then this would provide a useful strategy for developing disease resistance. Here pathogen inducible gene promoters can become useful, although the choice of promoter is not trivial. Such a promoter must not itself be stimulated by the defence response to be regulated; otherwise a runaway lesion response will occur following the first pathogen attack of the plant.

Specific problems – toxins

Necrotrophic pathogens, in contrast to biotrophs, use pathogenicity factors such as toxins and hydrolytic enzymes to effect successful infection. Indeed, without effective production of the toxin, the pathogen is often incapable of causing infection. Some toxins have the unfortunate side effect of being toxic to mammals and not just the target plant tissue, in which case they fall into the category of mycotoxins. Toxins often accumulate in biologically active concentrations in tissues remote from the site of infection. In some cases, the toxins are therefore a significant factor in crop spoilage disproportionate to actual loss of yield, especially where they are distasteful or poisonous to the consumer of the crop. One strategy for GM disease resistance (falls under strategy 1) is to target the toxin, i.e. cause its degradation, and thereby reduce infection and loss, and simultaneously reduce spoilage where mycotoxins are concerned. Transgenic maize, where the levels of the *Fusarium*-toxin zearalerone were reduced to 10% of the wild-type levels, proves that the approach is feasible (Igawa et al. 2007). A concern for this approach is that it may lead to the accumulation of breakdown products of which little is currently known. This is not likely to be a major problem where removal of the toxin from the system arrests pathogen development.

Oxalic acid has an important role as a toxic pathogenicity factor in several species of necrotroph, of which *Sclerotinia sclerotiorum* is a particular problem in many dicotyledonous species, for example, oil seed rape (*Brassica napus*) and sunflower (*Helianthus annuus*). Several studies have therefore taken the approach of constitutively expressing a heterologous (usually wheat) oxalate oxidase gene in a target crop in order to neutralise the oxalic acid produced by the pathogen. The products of the enzyme include the reactive oxygen species hydrogen peroxide, which itself has an important role in disease resistance (see Shetty et al. 2007 details). Examples where partial resistance has been obtained in the laboratory include sunflower and soybean (*Glycine max*) against *Sclerotinia sclerotiorum* (Cober et al. 2003; Donaldson et al. 2001; Hu et al. 2003) as well as poplar (*Populus × euramericana*) against *Septoria musiva* (Liang et al. 2004).

Bt maize and Fusarium toxins – serendipity

Transgenic maize with Bt toxin genes (specifically the Cry1Ab protein) from *Bacillus thuringiensis* are not

Eur J Plant Pathol (2008) 121:217–231

just insect-resistant but also consistently less attacked by *Fusarium* spp. and contain consistently reduced levels of toxin (Clements et al. 2003; Hammond et al. 2004). The reduced infection is likely to be a consequence of reduced opportunity for fortuitous fungal infection in tissues less wounded by insects (Duvick 2001).

Virus resistance

Numerous reports concern transgenic resistance to plant viruses (reviewed by Sudarshana et al. 2007; Fuchs and Gonsalves 2007) in which RNA-mediated gene silencing especially is a predominant strategy (viral RNA is degraded and viral DNA is inactivated by methylation). Most of these approaches can be categorised as pathogen mimicry (strategy 3, in Fig. 1). RNA-mediated resistance to both DNA and RNA viruses can be obtained without transgenic expression of a protein and this strategy thereby minimises toxicological and allergenic risks. Transgenes constitutively expressed to provide RNA-mediated virus resistance fall into three major types: (A) Sense or antisense viral sequences, (B) Inverted repeats/hairpin RNA of viral sequences, (C) Sequences of engineered microRNAs targeted against viruses. For examples of the three types: (Fuentes et al. 2006; Niu et al. 2006; Smith et al. 2000; Vazqez Rovere et al. 2007; Yang et al. 2004). The three types of transgenes have been compared for their ability to provide disease protection in short term experiments, and a relative order of efficiency (A < B) and (B < C) has been reported for RNA viruses (Smith et al. 2000; Qu et al. 2007). However, a more extensive comparison is needed before general conclusions can be drawn.

The best documented examples of transgenic virus resistance applied in farmers' fields have involved transgenes of type (A) above, although the mechanism of RNA-mediated resistance was initially not known. In the 1990s, the Papaya industry on Hawaii suffered a 50% decline in production due to an outbreak of the potyvirus Papaya ringspot virus, PRSV (Gonsalves 1998). Virus resistance was obtained in a high-yielding papaya hybrid using the viral coat protein sequence as the transgene (type A above). Following distribution of transgenic seeds to farmers, a 50% rebound of total papaya production on Hawaii was achieved within 4 years (Gonsalves

2004). A similar approach has been successfully applied in US cucurbit production, although the situation has been more complicated due to the presence of several different viruses (Fuchs et al. 1997; Fuchs and Gonsalves 2007).

For DNA viruses, the geminiviruses constitute a focal area of intense research in a range of crops (tomato, cassava, maize, legumes). For these viruses, strategies for transgenic resistance involves both RNA-mediated resistance and several approaches using mutated viral proteins exerting transdominant negative effects on viral replication (reviewed by Vanderschuren et al. (2007)). Several of the DNA viruses are whitefly-transmitted, and an inherent problem in crops like tomato is that, even though a transgene may protect against the virus, it will not protect the crop against the substantial damage caused by the whiteflies. An approach targeting both viruses and insects might be valuable, and some inspiration for future research in that direction can be obtained from a recent study in rice: under controlled conditions, it was demonstrated that inhibition of phloem-feeding insects, by transgenic expression of a garlic lectin, subsequently reduced the associated viral disease, rice tungro, vectored by the insects (Saha et al. 2006).

Discussion

Given the effort put into biotechnological approaches for introducing disease resistance into crops over the last two decades and the lack of concrete results in terms of transgenic crops in use, it is pertinent to pose the question as to under which circumstances should one attempt to make disease resistant plants by genetic engineering. Would it not be better to use the resources required, especially public funding, to support classic plant breeding initiatives? The answer probably lies in a balance between the two approaches. Most plant breeding in the developed economies is run effectively by private enterprise. The effectiveness of plant breeding has improved dramatically in recent years through the development of molecular marker technologies, which are particularly beneficial for disease resistance breeding where costly (and potentially harmful) phenotypic screening can be minimised. The investment required for making transgenic plants is enormous and the markets

apparently uncertain due to barriers caused by legislation which in themselves represent a political reaction to public opposition to technologies carrying perceived risk. Indeed, it can be argued that the public reaction, especially in Europe, to the success of transgenic herbicide-tolerant crops has set back the opportunities for plant biotechnology by at least a decade.

There is the issue of ineffectiveness. Most strategies tried to date have resulted in at best partial resistance. Partial resistance provides, of course, a clear advantage over susceptibility; the development of a pathogen on a partially resistant plant is slower, which means that the spread through a population (crop) will be slower. This is widely exploited by breeders. However, given the enormous costs associated with developing GM crops, in most cases partial resistance by GM is not considered attractive for commercial development. A related issue is whether a gene product can be expected to confer protection against many different, or a single pathogen species. Some genes offer prospects for general antimicrobial activities, i.e. strategies effective against different taxa; others will prove very narrow in their mode of action. This will not in itself be a disadvantage where a high value crop is threatened by a specific problem with major economic impact (e.g., potato-late blight, wheat-stem rust, coffee-rust, banana-black sikatoka or Panama disease).

Where strategies are based on the introduction of single genes, there is a risk of rapid breakdown, a problem well known from the introduction of race-specific resistance by conventional breeding (Hovmøller et al. 1997; McDonald and Linde 2003; Parlevliet 2003). The potential and need for pyramiding genes must be evaluated carefully in order to avoid the risk of breakdown and prolong the lifespan of the transgenic crop. In addition, it can be an advantage to ensure that the gene products are produced only when needed by using tissue-specific, pathogen-inducible promoters (Altpeter et al. 2005). Another important issue is to ensure that effective resistance is introduced in a sufficiently broad genetic background to avoid exposure to new risks from pests and pathogens associated with monoculture. Finally, the potential health risks – toxicity and allergenicity, have to be borne in mind.

Much has been written about the ethics of making transgenic plants, especially where synthetic genes are used or where genes are transferred between different Kingdoms or Phyla (from bacteria to plant, from insect to plant). We will not expand on this debate other than to refer to a recent movement – cis-genetics, or "all native" – to emphasize solutions based on gene silencing within species or the alteration of regulation of existing genes (Rommens 2004).

At present, there are no signs that transgenic fungal or bacterial resistance will be introduced in commercial crops in the near future. In contrast, the clear results obtained repeatedly in laboratory and field studies demonstrate that transgenic strategies for virus resistance work effectively. Despite this, virus resistant GM crops have been commercially introduced in only very few cases. Three factors need to be present: the technical solution to a problem which has no other obvious alternative, the economic incentive for implementing the solution, and therefore market and public acceptance. The combination of these factors was present for the Papaya Ringspot Virus in Hawaii. Apparently, continued research into transgenic virus resistance and improved understanding of the mechanisms involved has not led to any significant new introductions of virus resistant GM crops since the late 1990s. Perhaps the expiry of the EU moratorium for the introduction of new transgenic crops in Europe will facilitate this process.

References

Altpeter, F., Varshney, A., Abderhalden, O., Douchkov, D., Sautter, C., Kumlehn, J., et al. (2005). Stable expression of a defense-related gene in wheat epidermis under transcriptional control of a novel promoter confers pathogen resistance. *Plant Molecular Biology, 57*, 271–283.

Ayliffe, M. A., Steinau, M., Park, R. F., Rooke, L., Pacheco, M. G., Hulbert, S. H., et al. (2004). Aberrant mRNA processing of the maize Rp1-D rust resistance gene in wheat and barley. *Molecular Plant-Microbe Interactions, 17*, 853–864.

Babu, R. M., Sajeena, A., Seetharaman, K., Reddy, M. S. (2003). Advances in genetically engineered (transgenic) plants in pest management – An over view. *Crop Protection, 22*, 1071–1086.

Belfanti, E., Silfverberg-Dilworth, E., Tartarini, S., Patocchi, A., Barbieri, M., Zhu, J., et al. (2004). The HcrVf2 gene from a wild apple confers scab resistance to a transgenic cultivated variety. *Proceedings of the National Academy of Sciences, 101*, 886–890.

Bird, L. J., Akhurst, R. J. (2004). Relative fitness of Cry1A-resistant and -susceptible *Helicoverpa armigera* (Lepidoptera: Noctuidae) on conventional and transgenic cotton. *Journal of Economic Entomology, 95*, 1699–1709.

Brodersen, P., Malinovsky, F. G., Hematy, K., Newman, M. A., Mundy, J. (2005). The role of salicylic acid in the induction of cell death in Arabidopsis acd11. *Plant Physiology, 138*, 1037–1045.

Broekaert, W. F., Terras, F. R. G., Cammue, B. P. A. (2000). Induced and preformed antimicrobial proteins. In A. J. Slusarenko, R. S. S. Fraser, L. C. van Loon (Eds.) *Mechanisms of resistance to plant diseases* (pp. 371–477). Dordrecht: Kluwer.

Campbell, M. A., Fitzgerald, H. A., Ronald, P. C. (2002). Engineering pathogen resistance in crop plants. *Transgenic Research, 11*, 599–613.

Cao, H., Li, X., Dong, X. N. (1998). Generation of broad-spectrum disease resistance by overexpression of an essential regulatory gene in systemic acquired resistance. *Proceedings of the National Academy of Sciences of the United States of America, 95*, 6531–6536.

Carriére, Y., Ellers-Kirk, C., Liu, Y-B., Sims, M. A., Patin, A. L., Dennehy, T. J., et al. (2001). Fitness costs and maternal effects associated with resistance to transgenic cotton in the pink bollworm (Lepidoptera: Gelechiidae). *Journal of Economic Entomology, 94*, 1571–1576.

Cary, J. W., Rajasekaran, K., Jaynes, J. M., Cleveland, T. E. (2000). Transgenic expression of a gene encoding a synthetic antimicrobial peptide results in inhibition of fungal growth in vitro and in planta. *Plant Science, 154*, 171–181.

Castro, M. S., Fontes, W. (2007). Plant defense and antimicrobial peptides. *Protein and Peptide Letters, 12*, 11–16.

Cerdeira, A. L., Duke, S. O. (2006). The current status and environmental impacts of glyphosate-resistant crops: A review. *Journal of Environmental Quality, 35*, 1633–1658.

Chern, M. S., Fitzgerald, H. A., Canlas, P. E., Navarre, D. A., Ronald, P. C. (2005). Overexpression of a rice NPR1 homolog leads to constitutive activation of defense response and hypersensitivity to light. *Molecular Plant-Microbe Interactions, 18*, 511–520.

Christou, P., Capell, T., Kohli, A., Gatehouse, J. A., Gatehouse, A. M. R. (2006). Recent developments and future prospects in insect pest control in transgenic crops. *Trends in Plant Science, 11*, 302–308.

Clarke, J. D., Volko, S. M., Ledford, H., Ausubel, F. M., Dong, X. (2000). Roles of salicylic acid, jamonic acid, and ethylene in cpr-induced resistance in Arabidopsis. *The Plant Cell, 12*, 2175–2190.

Clausen, M., Krauter, R., Schachermayr, G., Potrykus, I., Sautter, C. (2000). Antifungal activity of a virally encoded gene in transgenic wheat. *Nature Biotechnology, 18*, 446–449.

Clements, M. J., Campbell, K. W., Maragos, C. M., Pilcher, C., Headrick, J. M., Pataky, J. K., et al. (2003). Influence of Cry1Ab protein and hybrid genotype on fumonisin contamination and fusarium ear rot of corn. *Crop Science, 43*, 1283–1293.

Cober, E. R., Rioux, S., Rajcan, I., Donaldson, P. A., Simmonds, D. H. (2003). Partial resistance to white mold in a transgenic soybean line. *Crop Science, 43*, 92–95.

Collinge, D. B., Jensen, M. K., Lyngkjær, M. F., Rung, J. H. (2007). How can we exploit functional genomics to understand the nature of plant defences? Barley as a case study. *European Journal of Plant Pathology* (this issue).

Collinge, D. B., Kragh, K. M., Mikkelsen, J. D., Nielsen, K. K., Rasmussen, U., Vad, K. (1993). Plant chitinases. *The Plant Journal, 3*, 31–40.

Conrath, U. (2007). Priming: It's all the world to induced disease resistance. *European Journal of Plant Pathology* (this issue).

Cui, X., Harling, R. (2005). N-acyl-homoserine lactone-mediated quorum sensing blockage, a novel strategy for attenuating pathogenicity of Gram-negative bacterial plant pathogens. *European Journal of Plant Pathology, 111*, 327–339.

Datta, S. K. (2004). Rice biotechnology: A need for developing countries. *AgBioForum, 7*, 31–35.

Donaldson, P. A., Anderson, T., Lane, B. G., Davidson, A. L., Simmonds, D. H. (2001). Soybean plants expressing an active oligomeric oxalate oxidase from the wheat gf-2.8 (germin) gene are resistant to the oxalate-secreting pathogen *Sclerotina sclerotiorum. Physiological and Molecular Plant Pathology, 59*, 1096–1178.

Dong, Y-H., Wang, L., Xu, J-L., Zhang, H-B., Zhang, X. F., Zhang, L. H. (2001). Quenching quorum-sensing-dependent bacterial infection by an N-acyl homoserine lactonase. *Nature, 411*, 813–817.

Dong, Y. H., Xu, J. L., Li, X. Z., Zhang, L. H. (2000). AiiA, an enzyme that inactivates the acylhomoserine lactone quorum-sensing signal and attenuates the virulence of *Erwinia carotovora. Proceedings of the National Academy of Sciences, 97*, 3526–3531.

Durrant, W. E., Dong, X. N. (2005). Systemic acquired resistance. *Annual Review of Phytopathology, 42*, 185–209.

Duvick, J. (2001). Prospects for reducing fumonisin contamination of maize through genetic modification. *Environmental Health Perspectives, 109*, 337–342.

Ferry, N., Edwards, M., Gatehouse, J., Capell, T., Christou, P., Gatehouse, A. (2006). Transgenic plants for insect pest control: A forward looking scientific perspective. *Transgenic Research, 15*, 13–19.

Ferry, N., Edwards, M. G., Gatehouse, J. A., Gatehouse, A. M. R. (2004). Plant-insect interactions: Molecular approaches to insect resistance. *Current Opinion in Biotechnology, 15*, 155–161.

Field, B., Jordan, F., Osbourn, A. (2006). First encounters – Deployment of defence-related natural products by plants. *New Phytologist, 172*, 193–207.

Fuchs, M., Ferreira, S., Gonsalves, D. (1997). Management of virus diseases by classical and engineered protection. Molecular Plant Pathology On-Line http://www.bspp.org.uk/mppol/] 1997/0116fuchs.

Fuchs, M., Gonsalves, D. (2007). Safety of virus-resistant transgenic plants two decades after their introduction: Lessons from realistic field risk assessment studies. *Annual Review of Phytopathology, 45*, 173–202.

Fuentes, A., Ramos, P. L., Fiallo, E., Callard, D., Sanchez, Y., Peral, R., et al. (2006). Intron-hairpin RNA derived from replication associated protein C1 gene confers immunity to tomato yellow leaf curl virus infection in transgenic tomato plants. *Transgenic Research, 15*, 291–304.

Gahan, L. J., Gould, F., Heckel, D. G. (2001). Identification of a gene associated with Bt resistance in *Heliothis virescens. Science, 293*, 857–860.

Glazebrook, J. (2005). Contrasting mechanisms of defense against biotrophic and necrotrophic pathogens. *Annual Review of Phytopathology, 43*, 205–227.

Gonsalves, D. (1998). Control of papaya ringspot virus in papaya: A case study. *Annual Review of Phytopathology, 36*, 415–437.

Gonsalves, D. (2004). Transgenic papaya in Hawaii and beyond. *AgBioForum, 7*, 36–40.

Hain, R., Reif, H. J., Krause, E., Langebartels, R., Kindl, H., Vornam, B., et al. (1993). Disease resistance results from foreign phytoalexin expression in a novel plant. *Nature, 361*, 153–156.

Hammerschmidt, R. (1999). Phytoalexins: What have we learned after 60 years? *Annual Review of Phytopathology, 37*, 285–306.

Hammond, B. G., Campbell, K. W., Pilcher, C. D., Degooyer, T. A., Robinson, A. E., McMillen, B. L., et al. (2004). Lower fumonisin mycotoxin levels in the grain of Bt corn grown in the United States in 2000–2002. *Journal of Agricultural and Food Chemistry, 52*, 1390–1397.

Hovmøller, M. S. (2007). Source of seedling and adult plant resistance to *Puccinia strüfomis* f. sp. *tritici* in European wheats. *Plant Breeding, 126*, 225–233.

Hovmøller, M. S., Østergård, H., Munk, L. (1997). Modelling virulence dynamics of airborne plant pathogens in relation to selection by host resistance. In I. R. Crute, E. Holub, J. J. Burdon (Eds.) *The gene-for-gene relationship in plant–parasite interactions. The gene for gene relationship in plant parasite interactions* (pp. 173–190). Wallingford, UK: CABI International.

Howles, P., Lawrence, G., Finnegan, J., McFadden, H., Ayliffe, M., Dodds, P., et al. (2005). Autoactive alleles of the Flax L6 rust resistance gene induce non-race-specific rust resistance associated with the hypersensitive response. *Molecular Plant-Microbe Interactions, 18*, 570–582.

Hu, X., Bidney, D. L., Yalpani, N., Duvick, J. P., Crasta, O., Folkerts, O., et al. (2003). Overexpression of a gene encoding hydrogen peroxide-generating oxalate oxidase evokes defense responses in sunflower. *Plant Physiology, 133*, 170–181.

Huang, F., Buschman, L. L., Higgins, R. A., McGaughey, W. H. (1999). Inheritance of resistance to *Bacillus thuringiensis* toxin (Dipel ES) in the European corn borer. *Science, 284*, 965–967.

Igawa, T., Takahashi-Ando, N., Ochiai, N., Ohsato, S., Shimizu, T., Kudo, T., et al. (2007). Reduced contamination by the Fusarium mycotoxin Zearalenone in maize kernels through genetic modification with a detoxification gene. *Applied and Environmental Microbiology, 73*, 1622–1629.

James, C. (2006) Global status of commercialized biotech/GM crops: 2006. ISAAA Brief 35: http://www.isaaa.org/Resources/publications/briefs/35/highlights/default.html.

Joseph, M., Gopalakrishnan, S., Sharma, R. K., Singh, V. P., Singh, A. K., Singh, N. K., et al. (2004). Combining bacterial blight resistance and Basmati quality characteristics by phenotypic and molecular marker-assisted selection in rice. *Molecular Breeding, 13*, 377–387.

Kalpana, K., Maruthasalam, S., Rajesh, T., Poovannan, K., Kumar, K. K., Kokiladevi, E., et al. (2006). Engineering sheath blight resistance in elite indica rice cultivars using genes encoding defense proteins. *Plant Science, 170*, 203–215.

Leckband, G., Lorz, H. (1998). Transformation and expression of a stilbene synthase gene of *Vitis vinifera* L. in barley and wheat for increased fungal resistance. *Theoretical and Applied Genetics, 96*, 1004–1012.

Liang, H., Maynard, C. A., Allen, R. D., Powell, W. A. (2004). Increased *Septoria musiva* resistance in transgenic hybrid poplar leaves expressing a wheat oxalate oxidase gene. *Plant Molecular Biology, 45*, 619–629.

Lorrain, S., Vailleau, F., Balagué, C., Roby, D. (2003). Lesion mimic mutants: Keys for deciphering cell death and defense pathways in plants? *Trends in Plant Science, 8*, 263–271.

Mach, J. M., Castillo, A. R., Hoogstraten, R., Greenberg, J. T. (2001). The Arabidopsis-accelerated death gene ACD2 encodes red chlorophyll catabolite reductase and suppresses the spread of disease symptoms. *Proceedings of the National Academy of Sciences of the United States of America, 98*, 771–776.

Makandar, R., Essig, J. S., Schapaugh, M. A., Trick, H. N., Shah, J. (2006). Genetically engineered resistance to Fusarium head blight in wheat by expression of Arabidopsis NPR1. *Molecular Plant-Microbe Interactions, 19*, 123–129.

Mayer, A. M., Staples, R. C., Gil-ad, N. L. (2001). Mechanisms of survival of necrotrophic fungal plant pathogens in hosts expressing the hypersensitive response. *Phytochemistry, 58*, 33–41.

McDonald, B. A., Linde, C. (2003). The population genetics of plant pathogens and breeding strategies for durable resistance. *Euphytica, 124*, 163–180.

McHale, L., Tan, X. P., Koehl, P., Michelmore, R. W. (2006). Plant NBS-LRR proteins: adaptable guards. *Genome Biology 7:* http://genomebiology.com/2006-7/4/212/abstract.

Mehlo, L., Gahakwa, D., Nghia, P. T., Loc, N. T., Capell, T., Gatehouse, J. A., et al. (2005). An alternative strategy for sustainable pest resistance in genetically enhanced crops. *Proceedings of the National Academy of Sciences of the United States of America, 102*, 7812–7816.

Mentag, R., Luckevich, M., Morency, M. J., Seguin, A. (2003). Bacterial disease resistance of transgenic hybrid poplar expressing the synthetic antimicrobial peptide D4E1. *Tree Physiology, 23*, 405–411.

Montesinos, E. (2007). Antimicrobial peptides and plant disease control. *FEMS Microbiology Letters, 270*, 1–11.

Mygind, P. H., Fischer, R. L., Schnorr, K. M., Hansen, M. T., Sonksen, C. P., Ludvigsen, S., et al. (2005). Plectasin is a peptide antibiotic with therapeutic potential from a saprophytic fungus. *Nature, 437*, 975–980.

Niu, Q. W., Lin, S. S., Reyes, J. L., Chen, K. C., Hu, H. W., Yeh, S. D., et al. (2006). Expression of artificial micro-RNA in transgenic Arabidopsis thaliana confers virus resistance. *Nature Biotechnology, 24*, 1420–1428.

Oldroyd, G. E. D., Staskawicz, B. J. (1998). Genetically engineered broad-spectrum disease resistance in tomato. *Proceedings of the National Academy of Sciences of the United States of America, 95*, 10300–10305.

Osbourn, A. (1996). Saponins and plant defence – A soap story. *Trends in Plant Science, 1*, 4–9.

Parlevliet, J. E. (2003). Durability of resistance against fungal, bacterial and viral pathogens; present situation. *Euphytica, 124*, 147–156.

Perez, C. J., Shelton, A. M. (1997). Resistance of *Plutella xylostella* (Lepidoptera: Plutellidae) to *Bacillus thuringiensis* Berliner in Central America. *Journal of Economic Entomology, 90*, 87–93.

Qu, J., Ye, J., Fang, R. (2007). Artificial miRNA-mediated virus resistance in plants. Journal of Virology doi: 10.1128/JVI.02457-06

Rajasekaran, K., Cary, J. W., Jaynes, J. M., Cleveland, T. E. (2007). Disease resistance conferred by the expression of a gene encoding a synthetic peptide in transgenic cotton (*Gossypium hirsutum* L.) plants. *Plant Biotechnology Journal, 3*, 545–554.

Rommens, C. M. (2004). All-native DNA transformation: A new approach to plant genetic engineering. *Trends in Plant Science, 9*, 457–464.

Rommens, C. M. T., Salmeron, J. M., Oldroyd, G. E. D., Staskawicz, B. J. (1995). Intergeneric transfer and functional expression of the tomato disease resistance gene Pto. *The Plant Cell, 7*, 1537–1544.

Saha, P., Dasgupta, I., Das, S. (2006). A novel approach for developing resistance in rice against phloem limited viruses by antagonizing the phloem feeding hemipteran vectors. *Plant Molecular Biology, 62*, 735–752.

Schlaich, T., Urbaniak, B. M., Malgras, N., Ehler, E., Birrer, C., Meier, L., et al. (2006). Increased field resistance to Tilletia caries provided by a specific antifungal virus gene in genetically engineered wheat. *Plant Biotechnology Journal, 4*, 63–75.

Schlaich, T., Urbabiak, B., Plissonnier, M-L., Malgras, N., Sautter, C. (2007). Exploration and Swiss field testing of a viral gene for specific quantitative resistance against smuts and bunts in wheat. *Advances in Biochemical Engineering and Biotechnology, 107*, 97–112.

Senior, I. J., Bavage, A. D. (2003). Comparison of genetically modified and conventionally derived herbicide tolerance in oilseed rape: A case study. *Euphytica, 132*, 217–226.

Senior, I. J., Dale, P. J. (2002). Herbicide-tolerant crops in agriculture: Oilseed rape as a case study. *Plant Breeding, 121*, 97–107.

Séralini, G. E., Cellier, D., de Vendomois, J. S. (2007). New analysis of a rat feeding study with a genetically modified maize reveals signs of hepatorenal toxicity. *Archives of Environmental Contamination and Toxicology, 52*, 596–602.

Serebriakova, L., Oldach, K. H., Lorz, H. (2005). Expression of transgenic stilbene synthases in wheat causes the accumulation of unknown stilbene derivatives with antifungal activity. *Journal of Plant Physiology, 162*, 985–1002.

Shetty, N. P., Jørgensen, H. J. L., Sharathchandra, R. G., Collinge, D. B., Shetty, H. S. (2007). Roles of reactive oxygen species in interactions between plants and eucaryotic pathogens. European Journal of Plant Pathology (this issue).

Smith, N. A., Singh, S. P., Wang, M. B., Stoutjesdijk, P. A., Green, A. G., Waterhouse, P. M. (2000). Total silencing by intron-spliced hairpin RNAs. *Nature, 407*, 319–321.

Sudarshana, M. R., Roy, G., Falk, B. W. (2007). Methods for engineering resistance to plant viruses. *Methods Molecular Biology, 354*, 183–195.

Tai, T. H., Dahlbeck, D., Clark, E. T., Gajiwala, P., Pasion, R., Whalen, M. C., et al. (1999). Expression of the Bs2 pepper gene confers resistance to bacterial spot disease in tomato. *Proceedings of the National Academy of Sciences, 96*, 14153–14158.

Takken, F. L. W., Albrecht, M., Tameling, W. I. L. (2006). Resistance proteins: Molecular switches of plant defence. *Current Opinion in Plant Biology, 9*, 383–390.

Tang, X. Y., Xie, M. T., Kim, Y. J., Zhou, J. M., Klessig, D. F., Martin, G. B. (1999). Overexpression of Pto activates defense responses and confers broad resistance. *The Plant Cell, 11*, 15–29.

Torres, M. A., Jones, J. D. G., Dangl, J. L. (2005). Pathogen-induced, NADPH oxidase-derived reactive oxygen intermediates suppress spread of cell death in *Arabidopsis thaliana*. *Nature Genetics, 37*, 1130–1134.

van der Vossen, E. A. G., Gros, J., Sikkema, A., Muskens, M., Wouters, D., Wolters, P., et al. (2005). The Rpi-blb2 gene from *Solanum bulbocastanum* is an Mi-1 gene homolog conferring broad-spectrum late blight resistance in potato. *The Plant Journal, 44*, 208–222.

van der Vossen, E., Sikkema, A., Hekkert, B. T. L., Gros, J., Stevens, P., Muskens, M., et al. (2003). An ancient R gene from the wild potato species *Solanum bulbocastanum* confers broad-spectrum resistance to *Phytophthora infestans* in cultivated potato and tomato. *Plant Journal, 36*, 867–882.

van Loon, L. C., Rep, M., Pieterse, C. M. J. (2006). Significance of inducible defense-related proteins in infected plants. *Annual Review of Phytopathology, 44*, 135–162.

Vanderschuren, H., Stupak, M., Futterer, J., Gruissem, W., Zhang, P. (2007). Engineering resistance to geminiviruses – Review and perspectives. *Plant Biotechnology Journal, 5*, 207–220.

VanEtten, H. D., Mansfield, J. W., Bailey, J. A., Farmer, E. E. (1994). Two classes of plant antibiotics – Phytoalexins versus phytoanticipins. *The Plant Cell, 6*, 1191–1192.

Vazqez Rovere, C., Asurmendi, S., Hopp, H. E. (2007). Transgenic resistance in potato plants expressing potato leaf roll virus (PLRV) replicase gene sequences is RNA mediated and suggests the involvement of post-transcriptional gene silencing. *Archives of Virology, 146*, 1337–1353.

Wang, G-L., Song, W. Y., Ruan, D. L., Sideris, S., Ronald, P. C. (2007). The cloned gene, Xa21, confers resistance to multiple *Xanthomonas oryzae* pv. *oryzae* isolates in transgenic plants. *Molecular Plant-Microbe Interactions, 9*, 855.

Yang, Y., Sherwood, T. A., Patte, C. P., Hiebert, E., Polston, J. E. (2004). Use of tomato yellow leaf curl virus (TYLCV) rep gene to engineer TYLCV resistance in tomato. *Phytopathology, 94*, 490–496.

Yao, J. H., Pang, Y. Z., Qi, H. X., Wan, B. L., Zhao, X. Y., Kong, W. W., et al. (2003). Transgenic tobacco expressing *Pinellia ternata* agglutinin confers enhanced resistance to aphids. *Transgenic Research, 12*, 715–722.

Zhai, W. X., Wang, W. M., Zhou, Y. L., Li, X. B., Zheng, X. W., Zhang, Q., et al. (2002). Breeding bacterial blight-resistant hybrid rice with the cloned bacterial blight resistance gene Xa21. *Molecular Breeding, 8*, 285–293.

Zhang, Z. G., Feechan, A., Pedersen, C., Newman, M. A., Qiu, J. L., Olesen, K. L., et al. (2007). A SNARE-protein has opposing functions in penetration resistance and defence signalling pathways. *The Plant Journal, 49*, 302–312.

Zhang, J., Li, X., Jiang, G., Xu, Y., He, Y. (2006). Pyramiding of Xa7 and Xa21 for the improvement of disease resistance to bacterial blight in hybrid rice. *Plant Breeding, 125*, 600–605.

Zhao, B. Y., Lin, X. H., Poland, J., Trick, H., Leach, J., Hulbert, S. (2005). From the cover: A maize resistance gene functions against bacterial streak disease in rice. *Proceedings of the National Academy of Sciences, 102*, 15383–15388.

Zhu, Q., Maher, E. A., Masoud, S., Dixon, R. A., Lamb, C. J. (1994). Enhanced protection against fungal attack by constitutive coexpression of chitinase and glucanase genes in transgenic tobacco. *Bio-Technology, 12*, 807–812.

Eur J Plant Pathol (2008) 121:233–242
DOI 10.1007/s10658-007-9251-4

Priming: it's all the world to induced disease resistance

Katharina Goellner · Uwe Conrath

Received: 23 May 2007 / Accepted: 5 November 2007

Abstract After infection by a necrotising pathogen, colonisation of the roots with certain beneficial microbes, or after treatment with various chemicals, many plants establish a unique physiological situation that is called the 'primed' state of the plant. In the primed condition, plants are able to 'recall' the previous infection, root colonisation or chemical treatment. As a consequence, primed plants respond more rapidly and/or effectively when re-exposed to biotic or abiotic stress, a feature that is frequently associated with enhanced disease resistance. Though priming has been known as a component of induced resistance for a long time, most progress in the understanding of the phenomenon has been made over the past few years. Here we summarize the current knowledge of priming and its relevance for plant protection in the field.

Keywords Benzothiadiazole · 2,6-dichloroisonicotinic acid · Potentiation of defence responses · Salicylic acid · Sensitisation · Stress resistance

K. Goellner · U. Conrath (✉)
Plant Biochemistry & Molecular Biology Group,
Department of Plant Physiology,
RWTH Aachen University,
52056 Aachen, Germany
e-mail: uwe.conrath@bio3.rwth-aachen.de

Abbreviations

BABA	β-aminobutyric acid
IR	induced resistance
ISR	induced systemic resistance
MAMP	microbe-associated molecular pattern
SA	salicylic acid
SAR	systemic acquired resistance

Introduction

When a plant becomes infected by a necrotising pathogen it often develops an enhanced resistance to a broad and distinctive spectrum of pathogens in organs distant from the site of infection. This type of induced resistance (IR) is called systemic acquired resistance (SAR; for reviews see Conrath 2006; Durrant and Dong 2004). The identity of the long-distance signal (s) that moves from the primary infection site to the remote organs to boost the plant's disease resistance is still unknown (Conrath 2006; Durrant and Dong 2004). In the 1990s, however, studies with transgenic tobacco and *Arabidopsis* plants constitutively expressing a bacterial salicylic acid (SA) hydroxylase (Delaney et al. 1994; Gaffney et al. 1993), and more recent work with *Arabidopsis* mutants affected in either SA production or SA signalling (reviewed by Dong 2001), clearly demonstrated an essential role for SA in the establishment of SAR in remote tissue.

Colonisation of plant roots with selected strains of non-pathogenic growth-promoting rhizosphere bacteria can also provoke broad-spectrum disease resistance in plants. This type of IR is called induced systemic resistance (ISR; reviewed by Van Loon et al. 1998). In contrast to SAR, the ISR response does not require SA. It depends rather on responsiveness to jasmonic acid and ethylene (Pieterse et al. 1998). In addition to ISR, other interactions between plants and beneficial micro-organisms can also elicit systemic broad-spectrum disease resistance in plants. For example, during its symbiosis with barley roots, the endophytic basidio-mycete *Piriformospora indica* confers systemic resistance to fungal diseases and salt stress (Waller et al. 2005). Furthermore, enhanced pathogen resistance was reported from plants whose roots have been colonized by mycorrhizal symbionts (Pozo et al. 2005).

Induction of IR by chemicals

Many different organic and inorganic compounds have been shown to activate IR in plants (Kuć 2001). When SA was identified as an essential endogenous signal for the SAR response (see the above text), an intensive search was initiated in order to identify synthetic chemicals able to mimic SA in SAR induction. 2,6-dichloroisonicotinic acid and its methyl ester (both are named INA) were the first synthetic compounds reported to activate the *bona fide* SAR response in plants (Kessmann et al. 1994). Later, benzo(1,2,3) thiadiazole-7-carbothioic acid *S*-methyl ester (BTH) became an attractive synthetic SAR activator (Friedrich et al. 1996; Görlach et al. 1996; Lawton et al. 1996). SA, INA and BTH are assumed to activate SAR via the same signalling pathway (Ryals et al. 1996).

Priming is a part of IR in plants

In addition to the direct activation of some anti-microbial defence reactions, systemic resistance responses in plants are frequently associated with a primed state in which the plants are able to 'recall' previous infection, root colonisation or chemical treatment (Fig. 1). In consequence, primed plants respond more rapidly and/or effectively when re-exposed to biotic or abiotic stress (reviewed by Conrath et al. 2002, 2006). Although priming has been known to represent part of the defence arsenal of plants for decades (reviewed by Kuć 1987), the

phenomenon did not attract much attention until the early 1990s. In a thorough analysis of the priming phenomenon, Conrath and associates employed a parsley cell culture and a microbe-associated molecular pattern (MAMP) from the cell wall of *Phytophthora sojae* to elucidate molecular aspects of priming and the associated amplification of MAMP-induced defence responses (for reviews see Conrath et al. 2002, 2006). By doing so, it was demonstrated that pre-treatment with SA, INA or BTH in a time-dependent manner primed the parsley cells for stronger activation by low MAMP doses of various cellular defence responses. These included the so-called early oxidative burst, rapidly induced ion transport changes at the plasma membrane, the synthesis and secretion of phytoalexins (coumarin derivatives), cell wall phenolics and a lignin-like polymer, as well as the accumulation of transcripts for various defence- associated genes (Katz et al. 1998; Kauss and Jeblick 1995; Kauss et al. 1992; 1993; Thulke and Conrath 1998). In ensuing investigations with the parsley cells, it was demonstrated that the effect of SAR inducers on defence gene activation depended on both the gene whose expression was being assayed and the dose of the inducer to be applied (Katz et al. 1998; Thulke and Conrath 1998). Whereas certain defence genes were found to be directly responsive to moderate concentrations of SA or BTH, other defence genes were not expressed. However, this second set of defence genes exhibited a very intense activation after pre-treatment with moderate concentrations of SA or BTH followed by treatment of the cells with a very low, sub-optimal MAMP dose (Katz et al. 1998; Thulke and Conrath 1998). These results documented a dual role for SAR inducers in the activation of plant defence responses: on the one hand, moderate doses of the SAR activators directly induced certain defence-associated genes, while on the other they primed cells for boosted expression of some other defence genes induced by a challenge treatment.

Mur et al. (1996) provided the first in-depth analysis of the priming phenomenon in intact plants. The authors reported that soil drench pre-application of SA to transgenic tobacco plants expressing chimeric *PR-1::GUS* or *PAL-3::GUS* defence genes did not cause significant gene activation. However, upon infection with *Pseudomonas syringae* (*Ps*) pv. *syringae* or after wounding, activation of the reporter genes was much stronger in the SA pre-treated plants

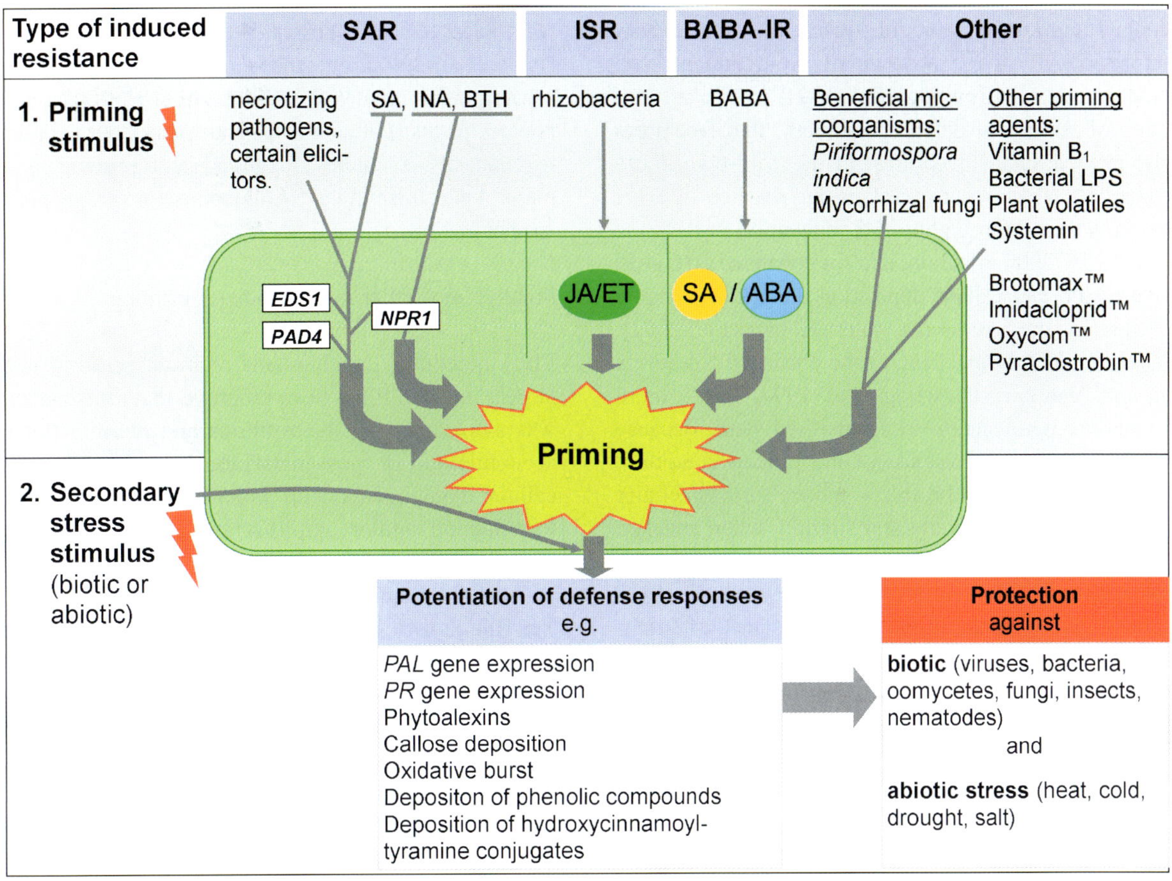

Type of induced resistance	SAR		ISR	BABA-IR	Other	
1. Priming stimulus	necrotizing pathogens, certain elicitors.	SA, INA, BTH	rhizobacteria	BABA	Beneficial microorganisms: *Piriformospora indica* Mycorrhizal fungi	Other priming agents: Vitamin B_1 Bacterial LPS Plant volatiles Systemin Brotomax™ Imidacloprid™ Oxycom™ Pyraclostrobin™

EDS1
PAD4
NPR1

JA/ET   SA / ABA

Priming

2. Secondary stress stimulus (biotic or abiotic)

Potentiation of defense responses e.g.

PAL gene expression
PR gene expression
Phytoalexins
Callose deposition
Oxidative burst
Depositon of phenolic compounds
Deposition of hydroxycinnamoyl-tyramine conjugates

Protection against

biotic (viruses, bacteria, oomycetes, fungi, insects, nematodes)
and
abiotic stress (heat, cold, drought, salt)

Fig. 1 Events associated with induced resistance phenomena in plants

than in the plants that had not been pre-treated with SA (Mur et al. 1996). Similarly, pre-treatment with BTH was shown to prime *Arabidopsis* for expression of *PAL* by *Ps* pv. *tomato* DC3000 that was much stronger than in a control plant not primed with BTH and then exposed to *Ps* pv. *tomato* DC3000 challenge (Kohler et al. 2002). BTH-induced priming also boosted *PAL* gene activation and callose deposition when these responses were induced either by mechanically wounding the leaves with forceps or by infiltration with water (Kohler et al. 2002). The observations made with *Arabidopsis* suggested that priming might be a common component that mediates cross-talk between pathogen defence reactions on one hand, and wound and osmotic stress responses on the other (Kohler et al. 2002).

The *Arabidopsis edr* (*enhanced disease resistance*) *1* mutant has constitutively enhanced resistance to the bacterium *Ps* pv. *tomato* DC3000 and to the fungus *Erisyphe cichoracearum*. Interestingly, *edr1* differs

from other mutants with enhanced disease resistance in that it displays no constitutive activation of the defence genes *PR-1* and *BGL2*, though transcripts for both these genes accumulate after pathogen infection (Frye and Innes 1998). This fact, and the finding that *edr1* displays more intense induction of defence responses after infection than the wild type, strongly suggest an engagement of EDR1 in priming.

In response to infection with avirulent pathogens, the *Arabidopsis non-expresser of PR genes* (*npr*) *1* mutant (also named *nim1* or *sai1*) accumulates SA levels comparable to those of the wild-type. However, *npr1* is unable to express biologically or chemically induced SAR (Cao et al. 1994; Delaney et al. 1995). Intriguingly, the higher activation by BTH-mediated priming of *Ps* pv. *tomato*-induced, wound-elicited and water infiltration-activated defence response are absent in *npr1* (Kohler et al. 2002). This result indicates that a functional *NPR1* gene is required for priming. Conversely, the *constitutive expresser of PR genes* (*cpr*) *1* and *cpr5* mutants of *Arabidopsis*, which both express SAR in the absence of pre-treatment with SAR activators (Bowling et al. 1994, 1997), are in a state of 'enhanced defence readiness' that resembles the primed state. In contrast to the wild-type, these plants are already sensitised in the absence of BTH pre-treatment, for higher *PAL* activation by *Ps* pv. *tomato* DC3000 infection, and for enhanced *PAL* gene expression and callose deposition upon wounding or infiltrating the leaves with water (Kohler et al. 2002). Thus, it is likely that, due to the enhanced SA levels in *cpr1* and *cpr5* (Bowling et al. 1994, 1997), these plants are permanently in a sensitised state. Because of the permanent presence of the primed condition, *cpr1* and *cpr5* might be able to induce various cellular defence reactions rapidly and strongly when attacked by pathogens, wounded or infiltrated on the leaves with water. In this context, it is noteworthy that the constitutively enhanced pathogen resistance of another *Arabidopsis* mutant referred to as *cpr5-2* has been ascribed to the boosted expression of the *PR-1* defence gene (Boch et al. 1998).

The close correlation between the SAR state and presence of the primed condition supported the assumption that priming is a key mechanism in the *bona fide* SAR response in plants. The assumption was further supported by the close correlation between the capability of various chemicals to activate SAR against *Tobacco mosaic virus* in tobacco (Conrath et al. 1995) and their capacity to prime for enhanced *PAL* expression induced by MAMP treatment in parsley cells (Katz et al. 1998; Thulke and Conrath 1998) or *Ps* pv. *tomato* DC3000 infection, wounding or water infiltration in intact *Arabidopsis* plants (Kohler et al. 2002). Moreover, an attenuation of priming and the associated loss of the enhanced activation of the oxidative burst have been associated with a lack of resistance to avirulent bacterial pathogens in tobacco (Mur et al. 2000). Finally, over-expression of the disease resistance-associated gene *PTI5* in tomato boosts pathogen-induced defence gene expression and enhances the resistance to *Ps* pv. *tomato* (He et al. 2001).

Observations similar to those made with parsley cell cultures, and tobacco, tomato, and *Arabidopsis* plants have been reported from SA-treated soybean cell cultures infected with *Ps* pv. *glycinea* (Shirasu et al. 1997), from BTH-primed and elicited *Agastache rugosa* suspension cells (Kim et al. 2001) and from

SAR-induced and subsequently infected sunflower (Prats et al. 2002), cucumber (Cools and Ishii 2002), asparagus (He and Wolyn 2005; He et al. 2002), and cowpea plants (Latunde-Dada and Lucas 2001). Taken together, these studies suggested that priming is a major mechanism in the SAR response in various plant species.

Putative molecular mechanisms of priming

The molecular mechanisms behind priming are largely unclear. It has been proposed that sensitisation was associated with biosynthesis and pre-infectional accumulation or post-translational modification of cellular components with important roles in signal transduction and/or amplification. Accumulation or modification of the components per se would not activate the majority of the plant's defence responses. Yet, due to their enhanced level in primed cells, there is increased activation of the signal transmission components and thus, potentiated activation of down-stream defence responses, but only after subsequent exposure to biotic or abiotic stress (Conrath et al. 2006). Recently, mitogen-activated protein kinases have been identified that accumulate after priming in *Arabidopsis* plants without displaying enzyme activity (G. J. M. Beckers, Y. Liu, S. Zhang and U. Conrath, unpubl.). Probably due to the enhanced level of the inactive mitogen-activated protein kinases in primed plants, there is increased MAP kinase activity after stimulation by biotic or abiotic stress associated with boosted induction of defence responses (G. J. M. Beckers, Y. Liu, S. Zhang and U. Conrath, unpubl.). Thus, certain mitogen-activated protein kinases are likely potential candidates for cellular components that mediate priming. Future genetic analysis will probably yield more molecular markers of priming for enhanced stress resistance.

Priming in IR to herbivores and between plant species in nature

In response to wounding or herbivore attack, plants often release extrafloral nectar or volatile organic compounds (VOCs). Whereas some of these traits serve to attract parasitic or predatory natural enemies of the herbivores (Paré and Tumlinson 1999), others have a role in enhancing the disease resistance in the wounded or herbivore-attacked plants themselves (Heil and Silva

Bueno 2007) or neighbouring, unharmed plants (Baldwin and Schultz 1983; Heil and Kost 2006). During past years, there is increasing evidence to suggest that the VOC-induced disease resistance is mediated by priming. In a pioneering paper, Engelberth et al. (2004) showed that maize seedlings, when exposed to certain volatiles from neighbouring plants and subsequently challenged by a combination of mechanical damage and exposure to caterpillar regurgitant, show a higher production of volatile sesquiterpenes and jasmonic acid than triggered plants not exposed to the volatiles previously. In a subsequent study, it was shown that VOC-induced priming for boosted activation of defence genes and potentiated emission of aromatic and terpenoid volatiles in maize correlates with reduced caterpillar feeding and improved attraction of the parasitoid *Cotesia marginiventris*, respectively (Ton et al. 2006). In field studies, Heil and Kost (2006) demonstrated that lima bean plants respond to leaf damage with the secretion of extrafloral nectar. This response was much higher in plants that had been previously primed by exposure to VOCs. Intriguingly, Kessler et al. (2006) also provided evidence that priming can even be the result of plant–plant communication in nature and that VOCs can even serve as priming-inducing signals between different plant species. The authors reported that VOCs from clipped sagebrush (*Artemesia tridentata*) prime nearby tobacco plants for accelerated production of trypsin proteinase inhibitors concomitant with lower total herbivore damage and a higher mortality rate of young *Manduca sexta* caterpillars (Kessler et al. 2006). Thus, plants can use chemical signals in their environment to assess the risk of herbivory and use this information to adjust their overall defence strategy.

Costs and benefits of priming

The activation of direct defence reactions by external application of high doses of SA, jasmonic acid or by the action of resistance (R)-genes was shown to reduce plant fitness traits such as growth and fruit or seed set under pathogen-free conditions (Agrawal et al. 1999; Baldwin 1998; Cipollini 2002; Heidel et al. 2004; Heil et al. 2000; Korves and Bergelson 2004; Tian et al. 2003; Van Dam and Baldwin 2001). Also, plants transformed with genes encoding SA biosynthesis enzymes (Mauch et al. 2001) or gain-of-resistance mutations in *Arabidopsis* such as *cpr1, cpr5* and *cpr6*,

which all contain constitutively high levels of SA, permanent expression of defence-related *PR* genes and a dwarf phenotype, have been associated with reduced fitness (Bowling et al. 1994, 1997). These observations were also made in the field when Heidel et al. (2004) demonstrated that *Arabidopsis* mutants blocked in SA-inducible defence, as well as mutants showing constitutive expression of these defences, were affected in growth and seed set. The authors reasoned that optimal plant fitness is reached at a certain level of resistance that balances fitness and defence (Heidel et al. 2004). Similar conclusions were drawn from studies on the costs of jasmonic acid-inducible defences, which seem to be affordable only when the plant is actually exposed to herbivore attack (Agrawal et al. 1999; Baldwin 1998).

The trade-off dilemma between disease resistance and costs of defence activation can probably be overcome by priming. During a comparative study by Van Hulten et al. (2006), the costs and benefits of priming in *Arabidopsis* were determined and compared to those of the direct induction of defence: application of low doses of the non-protein amino acid β-aminobutyric acid (BABA) induced a primed state, caused only minor reductions in growth and had no obvious effect on seed production. In contrast, direct induction of defence responses by high doses of either BABA or BTH strongly reduced both these fitness traits. The effects in primed plants, though being minor when compared to those of the direct defence activation, could be caused by an enhanced expression of genes encoding signalling compounds (see the above text) (Maleck et al. 2000; Van Hulten et al. 2006). As a consequence, it has been suggested that priming has a smaller effect on fitness than directly induced defence (Van Hulten et al. 2006). Intriguingly, when under attack by pathogens, primed plants displayed even higher fitness than non-primed ones. Thus, in environments of pathogen challenge, the costs of fitness of the primed state seem to be outweighed by its benefits (Van Hulten et al. 2006).

Employing priming in the greenhouse and field

Many natural and synthetic compounds are able to prime plants (summarized by Conrath et al. 2006). They include Brotomax®, a commercial product derived from urea, copper lignosulphonate, manganese lignosulphonate, and zinc lignosulphonate (Fuster et al.

1995; Ortuño et al. 1997), bacterial lipopolysaccharides (Newmann et al. 2007), vitamin B_1 (Ahn et al. 2005, 2007), systemin (Stennis et al. 1998), and BABA (Zimmerli et al. 2000). In laboratory trials, BABA-induced priming for enhanced stress responses was associated with augmented resistance not only to biotic but also to abiotic challenges such as drought and salt stress (Jakab et al. 2005). Moreover, BABA and some other priming-inducing compounds were also shown to be potent inducers of stress tolerance in the greenhouse and field (Cohen 2002). Foliar application of BABA protected field-grown grape against *Plasmopara viticola*. BABA also suppressed disease symptoms caused by *Phytophthora infestans* on potato and tomato plants in the field. BABA also shielded melon from the induction of sudden wilt disease by *Monosporascus cannonballus* (Cohen 2002). In addition, BABA decreased disease symptoms induced by *Cercosporidium personatum* on peanut in both greenhouse and field trials (Cohen 2002). As the compound also exhibited synergistic interaction with certain fungicides (Cohen 2002), BABA may successfully be integrated into plant disease management in the field.

INA was the first synthetic compound shown to induce both priming for improved activation of defence responses (Kauss et al. 1992) and resistance to fungal and bacterial pathogens on various crops, in the greenhouse as well as in the field (Kessmann et al. 1994). However both INA and SA were insufficiently tolerated by some crops to allow practical use as plant protection compounds (Ryals et al. 1996). BTH (synonym: acibenzolar-S-methyl) is another synthetic chemical and was synthesized some years after INA. BTH primed plant cell cultures and intact plants for better induction of defence responses (Katz et al. 1998; Kohler et al. 2002) and also protected various crops against many diseases in the field (Ryals et al. 1996). As BTH was sufficiently tolerated by most crop plants, the compound became attractive for practical agronomic use. In 1996, BTH was introduced as a 'plant activator' (Ruess et al. 1996) with the trade names Bion®, Actigard® or Boost®. However, the economic success of BTH was limited. BTH exerts protective rather than curative activity. Thus, to serve as a protectant the compound must be applied some time before a potential pathogen attack. Because of this strictly prophylactic activity, BTH was not sufficiently accepted by farmers who favoured the application of curative standard fungicides.

Due to the general lack of consumer acceptance of BTH, it became opportune to identify plant-protecting compounds teaming both direct action on the pathogen and priming-inducing activity in the plant. Some strobilurin fungicides seem to combine both these activities (for a review on strobilurins, see Sauter 2007). In the laboratory, for example, the strobilurin fungicide Pyraclostrobin (trade names: Cabrio®, Headline®) primed the tobacco cv. Xanthi nc for more rapid accumulation of antimicrobial PR-1 defence proteins after infection with *Tobacco mosaic virus* and the wildfire pathogen *Ps* pv. *tabaci*. The Pyraclostrobin-induced priming for enhanced PR-1 accumulation in response to pathogen attack was associated with enhanced disease resistance (Herms et al. 2002). The enhanced resistance to pathogenic viruses and bacteria in Pyraclostrobin-treated plants was also seen on various crops and ornamental plants in both the greenhouse and field (Koehle et al. 2003, 2006). It is interesting that in the field, Pyraclostrobin-induced priming was associated with enhanced resistance also to abiotic stresses, including drought (www.agweb.com/aims/files/HeadlineAdvantage.pdf). In addition, treatment with Pyraclostrobin increased crop yield in the field (http://www.ag.iastate.edu/farms/05reports/n/SoybeanYield.pdf). Also, in various crops Pyraclostrobin and other strobilurin fungicides induce a 'greening effect.' The term refers to the phenomenon of delayed leaf senescence and an increased grain-filling period resulting in enhanced biomass and yield (Bartlett et al. 2002). Together, the findings made with Pyraclostrobin suggest that this chemistry, in addition to exerting direct antifungal activity, may also protect plants by priming them for a boosted activation of subsequently stress-induced defence responses. This conclusion is consistent with an earlier report demonstrating that another commercial fungicide, Oryzemate®, enhanced the resistance to *Tobacco mosaic virus* in tobacco (Koganezawa et al. 1998) and to a bacterial and an oomycete pathogen in *Arabidopsis* (Yoshioka et al. 2001). Oryzemate® contains Probenazole as the active ingredient which is metabolized to saccharin in treated plants (Koganezawa et al. 1998). The latter compound seems to elicit priming in Oryzemate®-treated plants (Siegrist et al. 1998).

Observations similar to those with Pyraclostrobin have recently been made in laboratory and field trials with the insecticide Imidacloprid (trade names: Admire®, Confidor®, Gaucho®, Merit®, Trimax™,

etc). One of its major degradation products, 6-chloronicotinic acid, has a structure very similar to INA and is suspected of causing the so-called 'stress shield effect' on crops by priming them for boosted expression of defence genes, enhancing their tolerance to biotic and abiotic stresses, and increasing plant growth and yield (Thielert 2006).

Various associations of plants with beneficial microbes in the soil have also been demonstrated to induce the primed state and to enhance the resistance of plants to above-ground pathogens and abiotic stresses in the greenhouse and field (Fig. 1). For example, growth-promoting *Pseudomonas fluorescens* not only induces priming in plants (Verhagen et al. 2004), but also suppresses Fusarium wilt disease and improves yield in radish in the greenhouse (Leeman et al. 1995). Furthermore, when growing various crops and ornamental plants after treatment of their seeds with a mixture of eight endo- and ectomycorrhizal fungi (available as MycoGrow™ Micronized Endo/Ecto Seed Mix), there is enhanced growth and augmented disease resistance in the greenhouse as well as under field conditions (www.fungi.com/index.html). Similarly, the root-endophyte *Piriformospora indica* systemically primes barley against biotic and abiotic challenges (Waller et al. 2005), and increases growth and yield of the medicinal plants *Spilanthes calva* and *Withania somnifera* in the field (Rai et al. 2001).

Conclusions

Over the past decade, it has become increasingly clear that priming is an important part of various induced stress resistance phenomena in plants (Fig. 1). In addition to being interesting for studying signal transduction and stress physiology, priming has the potential to emerge as a successful additional strategic tool for modern plant protection. Priming allows plants to activate defence responses more quickly and/or effectively when exposed to biotic or abiotic stress. Due to its advantageous economic features for the plant, priming also represents an ecologically important adaptation to withstand environmental challenges. The phenomenon can be interesting for the development of new concepts for disease control, since priming provides broad-spectrum disease resistance without significantly affecting growth and fruit or seed set. Together, priming offers a smart, effective and realistic option for effective plant protection, especially when combined with the performance of traditional pesticides. The utilization of the natural, broad-spectrum defence capacity of plants in the field will be facilitated by a better understanding of the molecular, physiological and ecological aspects of priming, which constitute an exciting challenge for future research.

Acknowledgements Research on priming in the Plant Biochemistry & Molecular Biology Group is supported by BASF, BASF Plant Science, Bayer CropScience, the German Science Foundation (DFG) and the Peter and Traudl Engelhorn Foundation.

References

Agrawal, A. A., Strauss, S. Y., & Stout, M. J. (1999). Costs of induced responses and tolerance to herbivory in male and female fitness components of wild radish. *Evolution, 53*, 1093–1104.

Ahn, I.-P., Kim, S., & Lee, Y.-H. (2005). Vitamin B$_1$ functions as an activator of plant disease resistance. *Plant Physiology, 138*, 1505–1515.

Ahn, I.-P., Kim, S., Lee, Y.-H., & Suh, S.-C. (2007). Vitamin B$_1$-induced priming is dependent on hydrogen peroxide and the *NPR1* gene in *Arabidopsis*. *Plant Physiology, 143*, 838–848.

Baldwin, I. T. (1998). Jasmonate-induced responses are costly but benefit plants under attack in native populations. *Proceedings of the National Academy of Sciences of the USA, 95*, 8113–8118.

Baldwin, I. T., & Schultz, J. C. (1983). Rapid changes in tree chemistry induced by damage: evidence for communication between plants. *Science, 221*, 277–279.

Bartlett, D. W., Clough, J. M., Godwin, J. R., Hall, A. A., Hamer, M., & Parr-Dobrzanski, B. (2002). The strobilurin fungicides. *Pest Management Science, 58*, 649–662.

Boch, J., Verbsky, M. L., Robertson, T. L., Larkin, J. C., & Kunkel, B. N. (1998). Analysis of resistance gene-mediated defence responses in *Arabidopsis thaliana* plants carrying a mutation in *CPR5*. *Molecular Plant-Microbe Interactions, 12*, 1196–1206.

Bowling, S. A., Clarke, J. D., Liu, Y., Klessig, D. F., & Dong, X. (1997). The *cpr5* mutant of *Arabidopsis* expresses both NPR1-dependent and NPR1-independent resistance. *Plant Cell, 9*, 1573–1584.

Bowling, S. A., Guo, A., Cao, H., Gordon, A. S., Klessig, D., & Dong, X. (1994). A mutation in *Arabidopsis* that leads to constitutive expression of systemic acquired resistance. *Plant Cell, 6*, 1845–1857.

Cao, H., Bowling, S. A., Gordon, A. S., & Dong, X. (1994). Characterization of an *Arabidopsis* mutant that is nonresponsive to inducers of systemic acquired resistance. *Plant Cell, 8*, 1583–1592.

Cipollini, D. F. (2002). Does competition magnify the fitness costs of induced responses in *Arabidopsis thaliana*? A manipulative approach. *Oecologia, 131*, 514–520.

Cohen, Y. R. (2002). β-Aminobutyric acid-induced resistance against plant pathogens. *Plant Disease, 86,* 448–457.

Conrath, U. (2006). Systemic acquired resistance. *Plant Signaling & Behavior, 1,* 179–184.

Conrath, U., Beckers, G. J. M., Flors, V., García-Agustín, P., Jakab, G., Mauch, F., Prime-A-Plant Group, et al. (2006). Priming: getting ready for battle. *Molecular Plant-Microbe Interactions 2006, 19,* 1062–1071.

Conrath, U., Chen, Z., Ricigliano, J. R., & Klessig, D. F. (1995). Two inducers of plant defence responses, 2,6-dichloroisonicotinic acid and salicylic acid, inhibit catalase activity in tobacco. *Proceedings of the National Academy of Sciences of the USA, 92,* 7143–7147.

Conrath, U., Pieterse, C. M. J., & Mauch-Mani, B. (2002). Priming in plant–pathogen interactions. *Trends in Plant Sciences, 7,* 210–216.

Cools, H. J., & Ishii, H. (2002). Pretreatment of cucumber plants with acibenzolar-S-methyl systemically primes a phenylalanine ammonia-lyase (*PAL1*) for enhanced expression upon attack with a pathogenic fungus. *Physiological and Molecular Plant Pathology, 61,* 273–280.

Delaney, T. P., Friedrich, L., & Ryals, J. A. (1995). *Arabidopsis* signal transduction mutant defective in chemically and biologically induced disease resistance. *Proceedings of the National Academy of Sciences of the USA, 92,* 6602–6606.

Delaney, T. P., Uknes, S., Vernooij, B., Friedrich, L., Weymann, K., Negrotto, D., et al. (1994). A central role of salicylic acid in plant disease resistance. *Science, 266,* 1247–1249.

Dong, X. (2001). Genetic dissection of systemic acquired resistance. *Current Opinion in Plant Biology, 4,* 309–314.

Durrant, W. E., & Dong, X. (2004). Systemic acquired resistance. *Annual Review of Phytopathology, 42,* 185–209.

Engelberth, J., Alborn, H. T., Schmelz, E. A., & Tumlinson, J. H. (2004). Airborne signals prime plants against insect herbivore attack. *Proceedings of the National Academy of Sciences of the USA, 101,* 1781–1785.

Friedrich, L., Lawton, K., Ruess, W., Masner, P., Specker, N., Gut-Rella, M., et al. (1996). A benzothiadiazole derivative induces systemic acquired resistance in tobacco. *Plant Journal, 10,* 61–70.

Frye, C. A., & Innes, R. W. (1998). An *Arabidopsis* mutant with enhanced resistance to powdery mildew. *Plant Cell, 10,* 947–956.

Fuster, M. D., García-Puig, D., Ortuño, A., Botía, J. M., Sabater, F., Porras, I., et al. (1995). Selection of Citrus highly productive in secondary metabolites of industrial interest. Modulation of synthesis and/or accumulation processes. In C. García-Viguera, M. Castañer, M. I. Gil, F. Ferreres, & F. A. Tomás-Barberán (Eds.) *Current trends in fruit and vegetable phytochemistry* (pp. 81–85). Madrid: CSIC.

Gaffney, T., Friedrich, L., Vernoij, B., Negrotto, D., Nye, G., Uknes, S., et al. (1993). Requirement of salicylic acid for the induction of systemic acquired resistance. *Science, 261,* 754–756.

Görlach, J., Volrath, S., Knauf-Beiter, G., Hengy, G., Beckhove, U., Kogel, K.-H., et al. (1996). Benzothiadiazole, a novel class of inducers of systemic acquired resistance, activates gene expression and disease resistance in wheat. *Plant Cell, 8,* 629–643.

He, C. Y., Hsiang, T., & Wolyn, D. J. (2002). Induction of systemic disease resistance and pathogen defence responses in *Asparagus officinalis* inoculated with non-pathogenic strains of *Fusarium oxysporum. Plant Pathology, 51,* 225–230.

He, P., Warren, R. F., Zhao, T., Shan, L., Zhu, L., Tang, X., et al. (2001). Overexpression of *PTI5* in tomato potentiates pathogen-induced defence gene expression and enhances disease resistance to *Pseudomonas syringae* pv. *tomato. Molecular Plant-Microbe Interactions, 14,* 1453–1457.

He, C. Y., & Wolyn, D. J. (2005). Potential role for salicylic acid in induced resistance of asparagus roots to *Fusarium oxysporum* f.sp. *asparagi. Plant Pathology, 54,* 227–232.

Heidel, A. J., Clarke, J. D., Antonovics, J., & Dong, X. (2004). Fitness costs of mutations affecting the systemic acquired resistance pathway in *Arabidopsis thaliana. Genetics, 168,* 2197–2206.

Heil, M., Hilpert, A., Kaiser, W., & Linsenmair, K. E. (2000). Reduced growth and seed set following chemical induction of pathogen defence: Does systemic acquired resistance (SAR) incur allocation costs? *Journal of Ecology, 88,* 645–654.

Heil, M., & Kost, C. (2006). Priming of indirect defences. *Ecology Letters, 9,* 813–817.

Heil, M., & Silva Bueno, J. C. (2007). Within-plant signalling by volatiles leads to induction and priming of an indirect plant defence in nature. *Proceedings of the National Academy of Sciences of the USA, 104,* 5467–5472.

Herms, S., Seehaus, K., Koehle, H., & Conrath, U. (2002). A strobilurin fungicide enhances the resistance of tobacco against *Tobacco mosaic virus* and *Pseudomonas syringae* pv. *tabaci. Plant Physiology, 130,* 120–127.

Jakab, G., Ton, J., Flors, V., Zimmerli, L., Métraux, J.-P., & Mauch-Mani, B. (2005). Enhancing *Arabidopsis* salt and drought stress tolerance by chemical priming for its abscisic acid responses. *Plant Physiology, 139,* 267–274.

Katz, V. A., Thulke, O. U., & Conrath, U. (1998). A benzothiadiazole primes parsley cells for augmented elicitation of defence responses. *Plant Physiology, 117,* 1333–1339.

Kauss, H., Franke, R., Krause, K., Conrath, U., Jeblick, W., Grimmig, B., et al. (1993). Conditioning of parsley (*Petroselinum crispum*) suspension cells increases elicitor-induced incorporation of cell wall phenolics. *Plant Physiology, 102,* 459–466.

Kauss, H., & Jeblick, W. (1995). Pretreatment of parsley suspension cultures with salicylic acid enhances spontaneous and elicited production of H_2O_2. *Plant Physiology, 108,* 1171–1178.

Kauss, H., Theisinger-Hinkel, E., Mindermann, R., & Conrath, U. (1992). Dichloroisonicotinic and salicylic acid, inducers of systemic acquired resistance, enhance fungal elicitor responses in parsley cells. *Plant Journal, 2,* 655–660.

Kessler, A., Halitschke, R., Diezel, C., & Baldwin, I. T. (2006). Priming of plant defence responses in nature by airborne signalling between *Artemisia tridentata* and *Nicotiana attenuata. Oecologia, 148,* 280–292.

Kessmann, H., Staub, T., Hofmann, C., Maetzke, T., Herzog, J., Ward, E., et al. (1994). Induction of systemic acquired disease resistance in plants by chemicals. *Annual Review of Phytopathology, 32,* 439–459.

Kim, H. K., Oh, S.-R., Lee, H.-K., & Huh, H. (2001). Benzothiadiazole enhances the elicitation of rosmarinic acid production in a suspension culture of *Agastache rugosa*. *Biotechnology Letters*, *23*, 55–60.

Koehle, H., Conrath, U., Seehaus, K., Niedenbrueck, M., Tavares-Rodrigues, M.-A., Sanchez, W., et al. (2006). Method of inducing virus tolerance of plants. US Patent 20060172887.

Koehle, H., Herms, S., & Conrath, U. (2003). Method for immunizing plants against bacterioses. Patent Application No. WO2003075663.

Koganezawa, H., Sato, T., & Sasaya, T. (1998). Effects of Probenazole and saccharin on symptom appearance of *Tobacco mosaic virus* in tobacco. *Annals of the Phytopathological Society of Japan*, *64*, 80–84.

Kohler, A., Schwindling, S., & Conrath, U. (2002). Benzothiadiazole-induced priming for potentiated responses to pathogen infection, wounding, and infiltration of water into leaves requires the *NPR1/NIM1* gene in *Arabidopsis*. *Plant Physiology*, *128*, 1046–1056.

Korves, T., & Bergelson, J. (2004). A novel cost of R gene resistance in the presence of disease. *The American Naturalist*, *163*, 489–504.

Kuć, J. (1987). Translocated signals for plant immunization. *Annals of the New York Academy of Sciences*, *494*, 221–223.

Kuć, J. (2001). Concepts and direction of induced systemic resistance in plants and its application. *European Journal of Plant Pathology*, *107*, 7–12.

Latunde-Dada, A. O., & Lucas, J. A. (2001). The plant defence activator acibenzolar-S-methyl primes cowpea [*Vignia unguiculata* (L.) Walp.] seedlings for rapid induction of resistance. *Physiological and Molecular Plant Pathology*, *58*, 199–208.

Lawton, K. A., Friedrich, L., Hunt, M., Weymann, K., Delaney, T., Kessmann, H., et al. (1996). Benzothiadiazole induces disease resistance in *Arabidopsis* by activation of the systemic acquired resistance signal transduction pathway. *Plant Journal*, *10*, 71–82.

Leeman, M., van Pelt, J. A., Hendrickx, M. J., Scheffer, R. J., Bakker, P. A. H. M., & Schippers, B. (1995). Biocontrol of Fusarium wilt of radish in commercial greenhouse trials by seed treatment with *Pseudomonas fluorescens* WCS374. *Phytopathology*, *85*, 1301–1305.

Maleck, K., Levine, A., Eulgem, T., Morgan, A., Schmid, J., Lawton, K. A., et al. (2000). The transcriptome of *Arabidopsis thaliana* during systemic acquired resistance. *Nature Genetics*, *26*, 403–410.

Mauch, F., Mauch-Mani, B., Gaille, C., Kull, B., Haas, D., & Reimmann, C. (2001). Manipulation of salicylate content in *Arabidopsis thaliana* by the expression of an engineered bacterial salicylate synthase. *Plant Journal*, *25*, 66–67.

Mur, L. A. J., Brown, I. R., Darby, R. M., Bestwick, C. S., Bi, Y.-M., Mansfield, J. W., et al. (2000). A loss of resistance to avirulent bacterial pathogens in tobacco is associated with the attenuation of a salicylic acid-potentiated oxidative burst. *Plant Journal*, *23*, 609–621.

Mur, L. A. J., Naylor, G., Warner, S. A. J., Sugars, J. M., White, R. F., & Draper, J. (1996). Salicylic acid potentiates defence gene expression in tissue exhibiting acquired resistance to pathogen attack. *Plant Journal*, *9*, 559–571.

Newmann, M.-A., Dow, J. M., Molinaro, A., & Parrilli, M. (2007). Priming, induction and modulation of plants defence responses by bacterial lipopolysaccharides. *Journal of Endotoxin Research*, *13*, 69–84.

Ortuño, A., Botia, J. M., Fuster, M. D., Porras, I., García-Lidón, A., & del Río, J. A. (1997). Effect of scoparone (6–7-dimethoxicoumarin) biosynthesis on the resistance of tangelo Nova, *Citrus paradisi* and *Citrus aurantium* fruits against *Phytophthora parasitica*. *Journal of Agriculture and Food Chemistry*, *45*, 2740–2743.

Paré, P. W., & Tumlinson, J. H. (1999). Plant volatiles as a defence against insect herbivores. *Plant Physiology*, *121*, 325–332.

Pieterse, C. M. J., van Wees, S. C. M., van Pelt, J. A., Knoester, M., Laan, G., Gerrits, H., et al. (1998). A novel signaling pathway controlling induced systemic resistance in *Arabidopsis*. *Plant Cell*, *10*, 1571–1580.

Pozo, M. J., Van Loon, L. C., & Pieterse, C. M. J. (2005). Jasmonates – Signals in plant–microbe interactions. *Journal of Plant Growth Regulation*, *23*, 211–222.

Prats, E., Rubiales, D., & Jorrín, J. (2002). Acibenzolar-methyl-induced resistance to sunflower rust (*Puccinia helianthi*) is associated with enhancement of coumarins on foliar surface. *Physiological and Molecular Plant Pathology*, *60*, 155–162.

Rai, M., Acharya, D., Singh, A., & Varma, A. (2001). Positive growth responses of the medicinal plants *Spilanthes calva* and *Withania somnifera* to inoculation by *Piriformospora indica* in a field trial. *Mycorrhiza*, *11*, 123–128.

Ruess, W., Mueller, K., Knauf-Beiter, G., Kunz, W., & Staub, T. (1996). Plant activator CGA 245704: An innovative approach for disease control in cereals and tobacco. *Proceedings of the Brighton Crop Protect Conference – Pests and Diseases*, 53–60.

Ryals, J. A., Neuenschwander, U. H., Willits, M. G., Molina, A., Steiner, H.-Y., & Hunt, M. D. (1996). Systemic acquired resistance. *Plant Cell*, *8*, 1809–1819.

Sauter, H. (2007). Strobilurins and other complex III inhibitors. In W. Krämer, & U. Schirmer (Eds.) *Modern crop protection compounds* (pp. 341–366). Weinheim: VCH-Wiley.

Shirasu, K., Nakajima, H., Rajasekhar, K., & Dixon, R. A. (1997). Salicylic acid potentiates an agonist-dependent gain control that amplifies pathogen signals in the activation of defence mechanisms. *Plant Cell*, *9*, 261–270.

Siegrist, J., Muehlenbeck, S., & Buchenauer, H. (1998). Cultured parsley cells, a model system for the rapid testing of abiotic and natural substances as inducers of systemic acquired resistance. *Physiological and Molecular Plant Pathology*, *53*, 223–238.

Stennis, M. J., Chandra, S., Ryan, C. A., & Low, P. S. (1998). Systemin potentiates the oxidative burst in cultured tomato cells. *Plant Physiology*, *117*, 1031–1036.

Thielert, W. (2006). A unique product: The story of the Imidacloprid stress shield. *Pflanzenschutz-Nachrichten Bayer*, *59*, 73–86.

Thulke, O. U., & Conrath, U. (1998). Salicylic acid has a dual role in the activation of defence-related genes in parsley. *Plant Journal*, *14*, 35–42.

Tian, D., Traw, M. B., Chen, J. Q., Kreitman, M., & Bergelson, J. (2003). Fitness costs of R-gene-mediated resistance in *Arabidopsis thaliana*. *Nature*, *423*, 74–77.

Ton, J., D'Allessandro, M., Jourdie, V., Jakab, G., Karlen, D., Held, M., et al. (2006). Priming by airborne signals boosts direct and indirect resistance in maize. *Plant Journal*, *49*, 16–26.

Van Dam, N. M., & Baldwin, I. T. (2001). Competition mediates costs of jasmonate-induced defences, nitrogen acquisition and transgenerational plasticity in *Nicotiana attenuata*. *Functional Ecology, 15*, 406–415.

Van Hulten, M., Pelser, M., van Loon, L. C., Pieterse, C. M. J., & Ton, J. (2006). Costs and benefits of priming for defense in *Arabidopsis*. *Proceedings of the National Academy of Sciences of the USA, 103*, 5602–5607.

Van Loon, L. C., Bakker, P. A. H. M., & Pieterse, C. M. J. (1998). Systemic resistance induced by rhizosphere bacteria. *Annual Review of Phytopathology, 36*, 453–483.

Verhagen, B. W. M., Glazebrook, J., Zhu, T., Chang, H.-S., van Loon, L. C., & Pieterse, C. M. J. (2004). The transcriptome of rhizobacteria-induced systemic resistance in *Arabidopsis*. *Molecular Plant-Microbe Interactions, 17*, 895–908.

Waller, F., Ahatz, B., Baltruschat, H., Fodor, J., Becker, K., Fischer, M., et al. (2005). The endophytic fungus *Piriformospora indica* reprograms barley to salt-stress tolerance, disease resistance, and higher yield. *Proceedings of the National Academy of Sciences of the USA, 102*, 13386–13391.

Yoshioka, K., Nakashita, H., Klessig, D. F., & Yamaguchi, I. (2001). Probenazole induces systemic acquired resistance in *Arabidopsis* with a novel type of action. *Plant Journal, 25*, 149–157.

Zimmerli, L., Jakab, G., Métraux, J.-P., & Mauch-Mani, B. (2000). Potentiation of pathogen-specific defence mechanisms in *Arabidopsis* by β-aminobutyric acid. *Proceedings of the National Academy of Sciences of the USA, 97*, 12920–12925.

Eur J Plant Pathol (2008) 121:243–255
DOI 10.1007/s10658-007-9187-8

REVIEW

Resistance proteins: scouts of the plant innate immune system

Wladimir I. L. Tameling · Frank L. W. Takken

Accepted: 9 July 2007 / Published online: 27 September 2007
© KNPV 2007

Abstract Recognition of non-self in plants is mediated by specialised receptors that upon pathogen perception trigger induction of host defence responses. Primary, or basal, defence is mainly triggered by trans-membrane receptors that recognise conserved molecules released by a variety of (unrelated) microbes. Pathogens can overcome these basal defences by the secretion of specific effectors. Subsequent recognition of these effectors by specialised receptors (called resistance proteins) triggers induction of a second layer of plant defence responses. These responses are qualitatively similar to primary defence responses; however, they are generally faster and stronger. Here we give an overview of the predicted (domain) structures of resistance proteins and their proposed mode of action as molecular switches of plant innate immunity. We also highlight recent advances revealing that some of these proteins act in the plant nucleus as transcriptional co-regulators and that crosstalk can occur between members of different resistance protein families.

Keywords Disease resistance · Effector · NB-LRR proteins · Nucleotide binding · Nucleus · PAMP

Abbreviations

BED	BEAF and DREF proteins zinc-finger DNA-binding domain
CC	Coiled Coil
CNL	CC-NB-LRR
ETI	effector-triggered immunity
LRR	leucine rich repeat
MAMP	microbe-associated molecular pattern
NB-	nucleotide binding domain shared by
ARC	Apaf-1, some R proteins and CED4
PAMP	pathogen-associated molecular pattern
PTI	PAMP-triggered immunity
RLK	receptor-like kinase
RLP	receptor-like protein
SD	solanaceous domain
STAND	signal transduction ATPases with numerous domains
TIR	Toll/interleukin-1 receptor like
TNL	TIR-NB-LRR

W. I. L. Tameling
Laboratory of Phytopathology, Wageningen University,
Binnenhaven 5, 6709 PD Wageningen, The Netherlands
e-mail: Wladimir.Tameling@wur.nl

F. L. W. Takken (✉)
Plant Pathology, Swammerdam Institute for Life Sciences,
University of Amsterdam,
PO Box 94062, 1090 Amsterdam, The Netherlands
e-mail: F.L.W.Takken@uva.nl

Introduction

The ability to distinguish self from non-self is the most fundamental aspect of an immune system. Recognition of invaders in both plants and animals is mediated by extra- and intracellular immune receptors. Unlike vertebrates, which have adaptive molecular receivers in specialised mobile cells, plants rely on a spectrum of predetermined receptors expressed in non-mobile cells. Therefore, in plants pathogen-arrest is orchestrated by the cells encountering the pathogen and the systemic signals that originate from these cells.

It has been hypothesized that early land plants contained trans-membrane receptors at their cell surface capable of recognizing microbe- or pathogen-associated molecular patterns (MAMPS or PAMPS) such as cell wall fragments, chitin or peptide motifs in bacterial flagella (Ausubel 2005; Chisholm et al. 2006; Nürnberger and Kemmerling 2006). Recognition of these common and slowly evolving PAMPs triggers the induction of the primary or basal defence responses, nowadays also referred to as PTI (PAMP-triggered immunity; Jones and Dangl 2006). Evolution of this ancient immune system put a constraint on pathogenic microbial populations as it limited their host range and it forced them to develop counter strategies to overcome PTI. This selection pressure has likely to have resulted in the acquisition of virulence effector proteins that suppress basal plant defence. Many plant pathogens have been shown to produce, and deliver, effector proteins in the host (Birch et al. 2006; Catanzariti et al. 2007; Grant et al. 2006; Jones and Dangl 2006). In the subsequent evolutionary struggle to combat these pathogens

plants evolved means to recognise the secreted effector proteins and to mount a robust amplified defence response. This type of secondary defence is referred to as effector-triggered immunity (ETI) and is mediated by resistance (R) proteins. In broad terms the defence responses associated with both PTI and ETI are qualitatively similar; however, those associated with the latter are generally faster and stronger and are often accompanied by localized cell death around the infection site (Jones and Dangl 2006). Although what actually stops pathogen proliferation is still unclear in most cases, new data has recently become available on the receptors that switch on defence after pathogen recognition. This review aims to provide a current overview of the structure and function of these R proteins and highlights recent advances.

Resistance proteins

A common feature of receptors involved in pathogen perception is the leucine-rich repeat (LRR) domain (Fig. 1). This domain is present both in PAMP receptors, where it is fused to a transmembrane domain and a cytoplasmic kinase domain [receptor-like kinase (RLK)], and in the majority of R proteins (Nürnberger and Kemmerling 2006). Some R proteins structurally resemble the PAMP RLK receptors, such

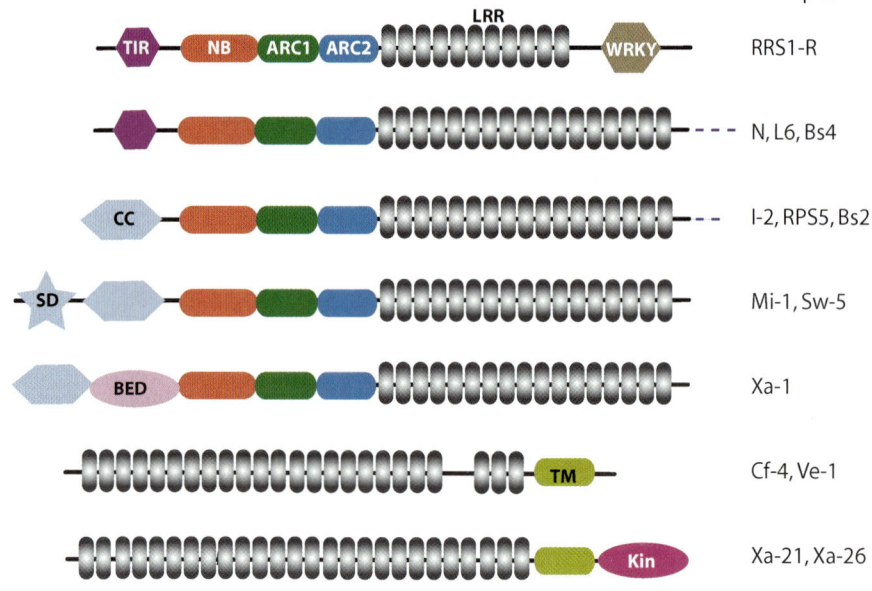

Fig. 1 Schematic representation of domains found in plant LRR R proteins. Domains are not drawn to scale. *TIR* Toll/interleukin-1 receptor, *CC* coiled coil, *NB* nucleotide binding, *ARC1/2* APAF1, R protein and CED4, *LRR* leucine rich repeat, *SD* solanaceous domain, *BED* BEAF/DREAF zinc finger domain, *TM* transmembrane, *Kin* kinase, *WRKY* WRKY transcription factor

as the rice Xa21 and Xa26 proteins (Sun et al. 2004). In others, like the tomato *Cladosporium fulvum* resistance (Cf) proteins (Rivas and Thomas 2005), the extracellularly localized LRR is also fused to a transmembrane domain. However, in these proteins [receptor-like protein (RLP)] no recognisable intracellular signalling domains can be discerned (Rivas and Thomas 2005).

In the majority of currently identified R proteins, however, the LRR resides in the cell and is fused to a nucleotide binding (NB) domain. The core nucleotide binding fold in these proteins is part of a larger entity called the NB-ARC domain due to its presence in Apaf-1 (apoptotic protease-activating factor-1), R proteins and CED-4 (*Caenorhabditis elegans* death-4 protein; van der Biezen and Jones 1998). Database searches have revealed a structurally related domain in animal proteins named NACHT (NAIP, CIITA, HET-E, and TP1) or NOD (for nucleotide-oligomerisation domain) domain (Leipe et al. 2004; Rairdan and Moffett 2007; Ting et al. 2006). Many of these proteins act as receptors sensing intracellular perturbations, such as the presence of microbial compounds (e.g. MAMP recognition by NACHT-LRRs), or cytochrome c leaking from mitochondria (e.g. Apaf-1). Like in R proteins, the NB-ARC/NACHT/NOD domains in these proteins are fused to a repeat structure such as an LRR or WD40 repeat domain (Leipe et al. 2004). Although these proteins share similar mechanistic and structural features, they appear to have evolved independently (Ausubel 2005; Leipe et al. 2004). The ubiquitous use of fused NB-repeat structures throughout the plant and animal kingdom probably reflects the biochemical suitability of such a module for coupled ligand recognition and subsequent activation of downstream signal transduction.

The NB-LRR core of plant R proteins is often equipped with variable amino- and sometimes also carboxy-terminal domains. Figure 1 gives an overview of the various structural domain decorations found in the different subfamilies of NB-LRR R proteins. The two major NB-LRR subfamilies are distinguished by either the presence or absence of an amino-terminal Toll/interleukin-1 receptor-like domain (TIR; Meyers et al. 1999). As many non-TIR NB-LRR proteins contain predicted coiled coil (CC) motifs, this family is collectively referred to as CC-NB-LRRs.

The TIR-NB-LRR (TNL) and CC-NB-LRR (CNL) members do not only differ in their amino-terminal extensions but also in certain motifs in their NB domains, thereby separating them into two evolutionary divergent classes (see below; Meyers et al. 1999; Pan et al. 2000). Members of the CNL group can be further divided based on the presence of additional domains at their amino-terminus. One example is a long extension, which has so far only been found in the Solanaceae and is therefore referred to as solanaceous domain (SD; Mucyn et al. 2006; Rairdan and Moffett 2007). Another example is the BED (named after the BEAF and DREF proteins) zinc-finger DNA- binding domain found in rice Xa1 and in NB-LRRs of poplar (Aravind 2000; Tuskan et al. 2006). At the carboxy-terminus flanking the LRR some R proteins carry extensions without recognizable domains, an exception being the *Arabidopsis* RRS-1-R protein that contains a typical WRKY DNA-binding domain at its carboxy-terminus (Deslandes et al. 2003).

Characteristic features of R protein domain structures

The class of extracellular RLP R proteins, founded by the Cf proteins, has mainly been found in Solanaceous species. These consist mainly of an extracellular LRR domain anchored in the plasma membrane (Rivas and Thomas 2005). Recently, in rice a new type of RLK R protein was identified: Pi-d2. In Pi-d2, which confers resistance to rice blast, the LRR is replaced by a B-lectin domain (lecRLK). Similar to RLPs and RLKs the extracellular domain is proposed to be involved in detection of the pathogen (Chen et al. 2006).

The intracellular NB-LRR R proteins are numerous and are present in large gene families in *Arabidopsis* (~150), rice (~400) and poplar (~400) (Meyers et al. 2003; Monosi et al. 2004; Tuskan et al. 2006). They are among the largest proteins found in plants and range from 860–1900 amino acid residues in size (McHale et al. 2006). As mentioned before, these NB-LRRs often contain four domains connected by linkers; a variable N-terminus, the NB-ARC domain, the LRR domain and a variable C-terminal extension. Unfortunately, so far, no crystal structure has been determined for any plant R protein or parts thereof.

However, 3D modelling templates are available for the LRR and NB-ARC domains to predict their structure (Albrecht and Takken 2006; McHale et al. 2006; Takken et al. 2006). Specific features of each domain will be discussed below.

The LRR domain

The LRR domain represents the major part of RLP R proteins and is composed of a variable number of repeats fitting the 24-amino acid residue consensus motif LxxLxxLxLxxNxLxGxIPxxLGx (L, leucine; x, any amino acid; N, asparagine; G, glycine; I, isoleucine; P, proline; Kajava 1998). In most RLPs this extracellular LRR domain is interrupted by a spacer region not fitting the consensus sequence, thereby dividing the LRR into three subdomains. The largest, amino-terminal, part consists of 21–28 hyper-variable repeats, the middle part represents the spacer and the remainder consists of three or four relatively conserved LRRs (Rivas and Thomas 2005). In RLK R proteins a division of the LRR domain into sub-domains has not been observed (Sun et al. 2006).

Unlike the LRR in RLPs and RLKs, the LRR in NB-LRR R proteins fits a shorter consensus motif that consists of 14 residues [LxxLxxLxLxxC/Nxx (C, cysteine; other symbols as above)] embedded in a repeat with a typical length of 24–28 residues (Kajava 1998). Based on crystal structures of non-plant LRRs the 14-residue core is predicted to form a β-sheet and an attached loop region. The remaining part of the repeat forms a spacer allowing the β-sheets to stack, thereby forming a large right-handed super helical β-sheet (Kobe and Deisenhofer 1994). So far the crystal structures of two plant LRRs, polygalac-turonase-inhibiting protein-2 (PGIP2; containing an extracellular LRR), and the cytoplasmic TIR-1 auxin receptor have been elucidated (Di Matteo et al. 2003; Tan et al. 2007). These structures revealed differences with non-plant LRRs. The PGIP2-LRR has two β-sheets in each repeat, connecting the first with an α-helix in the spacer, resulting in an extended and slightly curved super-structure (Di Matteo et al. 2003). The LRR of TIR-1 forms a horse shoe-like structure in which the β-strands lie at the concave side, whereas the mainly α-helical spacers lie at the convex side. In contrast to other crystallised LRR proteins, the TIR-1 LRR has a cofactor, the inositol-6-phosphate (InsP6), tightly bound in the middle of the

horseshoe that provides a 'floor' for the auxin-binding pocket. Surprisingly, the auxin-binding interface is not formed by residues embedded in the β-sheet, but by three intra-repeat loops that stick out of the plane of the horseshoe (Tan et al. 2007). It will be interesting to determine whether the LRRs of NB-LRR proteins adopt a similar structure that is distinct from that of non-plant LRRs. The LRR domain structure is perfectly suited to mediate protein-protein inter-actions and ligand binding (Kobe and Deisenhofer 1994). Basically two different types of R protein LRR classes can be discerned, those with high genetic diversity and the others showing little variation. This difference has been proposed to reflect the recognition mechanism of the pathogen's effector, direct versus indirect (Ellis et al. 2007). No effector protein, however, has yet been identified that directly binds the LRR. The LRR interactors identified are chaper-ones that might be required for proper folding of the LRR domain (Azevedo et al. 2006; Bieri et al. 2004; Holt et al. 2005; Takahashi et al. 2003). These chaperones include heat-shock proteins such as HSP90 and HSP17 and co-chaperones such as protein phosphatase 5, SGT1 and RAR1 (De la Fuente van Bentem et al. 2005; Hubert et al. 2003; Liu et al. 2004; Liu et al. 2002; Takahashi et al. 2003). Interestingly, the LRR domain appears to interact with the N-terminal part of NB-LRR proteins as exemplified by Rx, Bs2 and N (Leister et al. 2005; Moffett et al. 2002; Ueda et al. 2006). As discussed below, these intramolecular interactions are probably important for the regulation of NB-LRR protein activity and thus for the induction of ETI.

The amino-terminus of plant NB-LRR proteins

In animals, both CC and TIR domains have been implicated in protein-protein interactions, which is also predicted to occur in plant NB-LRR proteins as they are thought to interact with domain-specific downstream signalling components (Feys and Parker 2000). Recent observations indicate that the amino-terminus of at least some NB-LRR proteins also binds host proteins that are subject to attack by pathogen effectors (so-called virulence targets or guardees) in order to guard them and monitor their perturbations. Observations supporting the role of the amino terminus in this guard function are the binding of RIN4 to the R proteins RPM1 and RPS2, PBS1

kinase binding to RPS5, NIP1(N-interacting protein 1) binding to N and Pto binding to Prf (Ade et al. 2007; Burch-Smith et al. 2007; Dinseh-Kumar, personal communication; Mackey et al. 2002; Mucyn et al. 2006). These virulence targets are either cleaved (RIN4 and PBS-1), phosphorylated (RIN4) or modified in an unknown way (NIP1 and Pto) by their attacking effector proteins (AvrRpm1/AvrB and AvrRpt2 for RIN4, AvrPphB for PBS-1, P50 for NIP1 and AvrPto/AvrPtoB for Pto; Axtell et al. 2003; Burch-Smith et al. 2007; Mackey et al. 2002; Mucyn et al. 2006; Shao et al. 2003; Dinesh-Kumar, personal communication). The modification of the guardees is believed to trigger the guarding NB-LRR protein to activate ETI. In the case of R proteins that directly recognise pathogen effectors it is not known whether the effectors also bind to the N-terminal domain or whether they bind to other domains.

Besides the observed interactions with guardees, the N-terminal domain can be involved in homotypic TIR-TIR interactions as shown for the tobacco R protein N resulting in NB-LRR oligomerisation upon activation (Mestre and Baulcombe 2006; see below).

The overall 3D structure of plant TIR domains is not known. However, crystal structures are available for human Toll-like receptors, which could represent appropriate modelling templates as essential residues are conserved in both metazoan and plant TNL proteins (Dinesh-Kumar et al. 2000). The TIR structure is predicted to form a five-stranded parallel β-sheet surrounded by five α-helices (Xu et al. 2000).

The CNL class of R proteins obtained their name because some non-TIR members contain predicted coiled-coil (CC) motifs, consisting of an α-helix-rich domain containing seven residue repeat sequences at their N-terminus (Pan et al. 2000). However, for the CC domain no information about its structure is available, and it is not clear how it folds or even whether it truly represents a coiled coil structure.

A recently recognised domain in the N-terminus of some NB-LRR proteins is the BED-finger (Aravind 2000; Tuskan et al. 2006). This domain is characterized by two motifs; one consisting of a pattern of cysteines and histidines that together might form a metal-chelating zinc finger, and the other containing a conserved tryptophane (Aravind 2000). Besides the name-giving *Drosophila* BEAF and DREF proteins

that function as transcriptional regulators and chromatin insulators, this domain has been found in a subset of plant NB-LRR proteins and in DNA-binding proteins from tomato and tobacco (Aravind 2000). The presence of this BED-finger in DNA-binding domains and the prediction that it forms a zinc finger make it plausible that this structure represents a true DNA-binding domain. The observation of a DNA-binding domain in multi-domain STAND proteins (Leipe et al. 2004) perfectly fits their involvement in signal transduction and transcriptional regulation (see below).

The NB-ARC domain

As the 'N' in NB-ARC indicates, this domain has been predicted to bind nucleotides. This property is based on the presence of several conserved motifs characteristic for P-loop ATPases. Based on these motifs NB-LRRs, and many other proteins, could be classified as signal transduction ATPases with numerous domains (STAND) proteins (Leipe et al. 2004). The STAND protein family consists of five clades, the NB-ARC and NACHT proteins representing two of them (Leipe et al. 2004). All STAND proteins are multi-domain molecules that can contain DNA- or protein-binding domains, and super-repeat structures by which adaptor, regulatory switch, scaffolding, and, in some cases, signal-generating moieties are combined in a single protein. It was predicted that the STAND ATPase domain transmits conformational changes, induced by nucleotide exchange or hydrolysis, to the other domains of the protein thereby allowing it to generate a signal (Leipe et al. 2004). Biochemical studies on I-2, Mi-1.2 and N have indeed confirmed that the NB-ARC of these R proteins is a functional ATPase domain (Tameling et al. 2002; Ueda et al. 2006). The hydrolysis of ATP is likely to be accompanied by a conformational change of the NB-ARC domain, as after ATP-hydrolysis ADP-binding affinity increased dramatically, and because accumulation of mutant I-2 proteins in the ATP-bound state is likely to cause their autoactive phenotype (Tameling et al. 2006). These results are consistent with those obtained with the human NB-ARC protein Apaf-1, for which the various nucleotide binding states (either ADP or ATP) also represent different conformations. Cytochrome c binding to Apaf-1 triggers hydrolysis of the bound dATP and exchange of the formed dADP by dATP subsequently results in formation of the apoptosome that is able to trigger downstream signalling. Low (d)ATP levels result in the inability to exchange dADP and result in the formation of an inactive dADP-bound aggregate (Kim et al. 2005). The crystal structure of Apaf-1 revealed that the NB-ARC domain actually consists of four clearly distinguishable sub-domains (Riedl et al. 2005). These are the core P-loop NTPase fold, forming a five-stranded β-sheet flanked by α-helices, the ARC-1 domain, forming a four-helix bundle, the ARC-2 subdomains forming a winged-helix domain, and the ARC-3 subdomains, also forming a helical bundle. Specific ADP-binding is achieved through eight direct, and four H_2O-mediated, interactions with various conserved residues present in the NB, ARC-1 and ARC-2 subdomains. These three subdomains are also conserved in R proteins, whereas the ARC-3 is lacking there (Albrecht and Takken 2006; Takken et al. 2006). As most of the residues involved in the interaction with the nucleotide as well as several peptide motifs are conserved in R proteins, this suggests a similar fold and possibly a similar molecular mechanism underlying their function. For an overview of the conserved motifs and domain structures we refer to recent reviews (McHale et al. 2006; Rairdan and Moffett 2007; Takken et al. 2006; van Ooijen et al. 2007).

Biochemical analysis of two auto-activating mutants of I-2, which induce plant defence responses in the absence of the pathogen, revealed that these mutants are affected in their ability to hydrolyse ATP, while the binding affinity for this nucleotide is not altered (Tameling et al. 2006). These data support a model in which there is a dynamic equilibrium between the ATP- and ADP-bound state of an NB-LRR R protein. In this model the ATP-bound state represents the active state and hydrolysis of the nucleotide flips the protein back to its inactive, 'resting' state (Takken et al. 2006; Tameling et al. 2006; van Ooijen et al. 2007). In this model at least two conformational changes of the protein are predicted to take place: exchange of ADP for ATP, resulting in the formation of the activated state, and subsequently hydrolysis of ATP whereby the protein returns to its resting state.

The first part of this model is analogous to that proposed for Apaf-1, in which exchange of bound dADP for dATP seems to be sufficient to allow apoptosome formation required for the initiation of apoptosis (Bao et al. 2007). However, it is not clear whether the proposed mechanism is generic for STAND proteins, as for instance ATPase activity has not been observed for the *C. elegans* analogue of APAF-1, CED-4, (Yan et al. 2005). Also for the plant TNL protein N, it has been suggested that not the exchange of ADP by ATP, but rather the hydrolysis of bound ATP is required for the protein to reach its active state (Ueda et al. 2006). Clearly there is a need for more biochemical data on the nucleotide-binding status of different and preferably full-length NB-LRR proteins to further explore the function of the NB-ARC domain. Although the structures are conserved, there will be differences in the underlying molecular mechanisms by which the various STAND proteins perform their function as molecular switches.

The carboxy-terminal extensions of NB-LRRs

Size and composition of the carboxy-terminal extensions differ between TNLs and CNLs. The latter often have short extensions of 40–80 amino acid residues, whereas the TNLs can have extensions of up to 300 amino acid residues (Meyers et al. 2003). In some cases these longer extensions have similarity to other proteins, the one example being *Arabidopsis* RRS1-R containing a WRKY domain and a nuclear localisation signal (NLS) at its carboxy terminus (Deslandes et al. 2003). A WRKY domain is also found in zinc-finger transcription factors and its name is derived from the conserved W-R-K-Y amino acid motif. For the majority of NB-LRRs, however, no recognizable domains have been observed in their C-terminal extensions and there are no known interactors of this domain.

Intramolecular interactions in NB-LRR R proteins

Activation of ETI, which is often accompanied by a cell death response, is costly for a plant and its proliferation could be fatal. Therefore this type of immunity has to be tightly regulated. One way of keeping NB-LRR proteins in check is by auto-inhibition, which seems to be accomplished by intramolecular interactions between the various domains. Deletion of the LRR domain of some NB-

LRR proteins results in a weak auto-activation phenotype, indicative for a negative regulatory role of this domain (Bendahmane et al. 2002; Michael Weaver et al. 2006; Zhang et al. 2004). However, the LRRs can clearly also have a positive regulatory role, as expression of the N-terminal half of a CNL containing auto-activation mutations in the NB domain does not result in the activation of ETI unless the LRR is co-expressed (Moffett et al. 2002; Rairdan and Moffett 2006). Auto-activation mutants can not only be obtained by introducing specific point mutations in R proteins, but also by domain swaps between closely related paralogues as has been shown for Mi-1.2, Rx, Rp1 and L6 (Howles et al. 2005; Hwang et al. 2000; Rairdan and Moffett 2006; Sun et al. 2001). As these chimeras are combinations of wild-type domains, the observed auto-activation phenotype is likely to be due to incompatibility between regulatory subunits. These observations together support a model in which NB-LRR proteins are held in an auto-inhibited state by many (weak) interactions scattered over the various domains. Disturbance or misalignment of these interactions will release the auto-inhibition and allow the protein to proceed to its activated state. Evidence for such intramolecular interactions is provided by the observed association between the CC and NB-ARC-LRR domains of Rx, and between the LRR and the CC-NB-ARC domains in Bs-2 and Rx (Leister et al. 2005; Moffett et al. 2002; Rairdan and Moffett 2006). The first of the above interactions is dependent on a functional NB domain, supporting the model in which nucleotide exchange is required to release the signalling potential of the N-terminus. The latter interaction does not require a functional NB domain and appears to be mediated mainly by the ARC-1 sub-domain (Rairdan and Moffett 2006). One model for the activation of CNLs like Rx is based on the observed interaction between the LRR and the NB-ARC domain. Upon direct/indirect effector recognition, the interaction interface between the LRR and ARC-2 changes, allowing nucleotide exchange by the NB-ARC domain. This nucleotide exchange results in a conformational change of the NB-ARC and the N-terminal domain (Rairdan and Moffett 2007; Rairdan and Moffett 2006) thereby providing the means to convert recognition into signalling. How the activated NB-LRR protein subsequently activates defence signalling will be discussed below.

NB-LRRs and their putative function as transcriptional co-regulators in the nucleus

For a long time it has been thought that plant NB-LRR proteins would localise solely to the cytoplasm as no obvious nuclear localisation signal (NLS) was identified in these proteins. The finding that the atypical NB-LRR protein RRS1-R was present in the nucleus upon co-expression with its cognate effector potein PopP2, from *Ralstonia solanacearum*, was remarkable, but not totally surprising as this protein has a WRKY DNA binding domain and contains a predicted NLS. The two proteins interact in a yeast two-hybrid assay and co-localized exclusively in the nucleus when co-expressed. Co-expression with a PopP2 deletion mutant that lacked its bipartite NLS resulted in cytoplasmic localisation of both proteins. It is not yet clear whether the predicted NLS in RRS1-R is functional and required for nuclear localisation of RRS1-R (Deslandes et al. 2003; see below).

Two recent papers show that surprisingly also typical NB-LRR R proteins (tobacco TNL protein N and barley CNL protein MLA10) localise to the nuclear compartment (in addition to the cytoplasm) and that this localisation is required for activation of ETI (Burch-Smith et al. 2007; Shen et al. 2007). The potato CNL Rx also localises to both the cytoplasm and the nucleus, although it is currently unknown whether this localisation is required for Rx-mediated extreme resistance to potato virus X (PVX) (J. Bakker, personal communication). MLA10 confers resistance to *Blumeria graminis* f. sp. *hordeii* (*Bgh*) races that express the Avr$_{A10}$ effector. Expression of Avr$_{A10}$ in plant cells induces a physical association of MLA10 with WRKY transcription factor *Hv*WRKY2 in the nucleus. In an earlier yeast two-hybrid screen *Hv*WRKY1 and *Hv*WRKY2 were identified as interactors of the CC domain of MLA10 and other MLA proteins containing identical N-termini. Both interactors belong to the WRKY family of transcription factors that bind specific W-box elements present in the promoters of many pathogen-responsive genes (Ulker and Somssich 2004). Shen et al. (2007) showed that *Hv*WRKY1 and *Hv*WRKY2 act as suppressors of PTI, as silencing of these genes resulted in an increased resistance whereas overexpression resulted in hyper-susceptibility to virulent *Bgh* races. Analysis of a double knock-out of the most closely related *Arabidopsis* WRKY genes (*Atwrky18*/

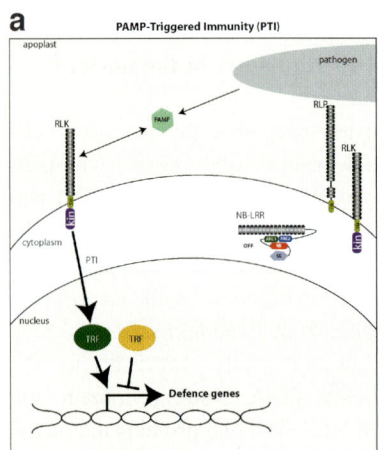 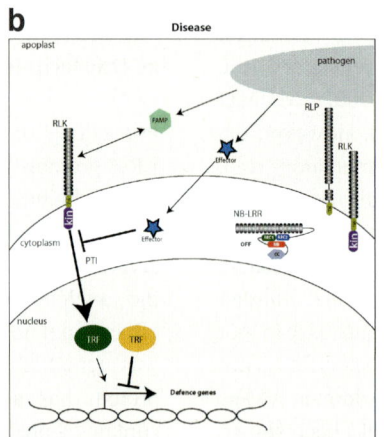 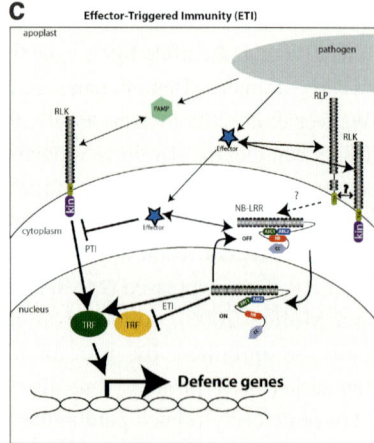

Fig. 2 PAMP and effector-triggered immunity (PTI and ETI, respectively). **a** PAMPs/MAMPs can be recognised by receptor-like proteins (*RLPs*) that subsequently induce defence gene expression through positive (*green*) and negative (*yellow*) regulatory transcription factors (*TRFs*) eventually leading to PTI. **b** Many pathogens interfere with PTI by the production of effector molecules resulting in a diminished defence response.

c Some of those effectors can be recognised by corresponding resistance proteins from the NB-LRR, RLP or RLK family resulting in an amplified form of the defence response termed ETI. For some NB-LRR R proteins nuclear localization is needed to trigger ETI. In that case NB-LRR proteins might facilitate the transcriptional reprogramming leading to ETI by direct interaction with transcriptional regulators in the nucleus

Atwrky40) also revealed a role for these WRKYs as suppressors of PTI. Based on these data, a model was proposed that explains why the transcriptional reprogramming of ETI differs only quantitatively and kinetically from that of PTI (Shen et al. 2007) (Fig. 2c), as was suggested earlier (Tao et al. 2003). The difference in amplitude of the resistance response could also explain why ETI, but not PTI, is often associated with cell death, as proposed by Jones and Dangl 2006. The model presented in Fig. 2a implies negative regulation by transcriptional repressor proteins (e.g. WRKY transcription factors) to dampen the PTI response in order to prevent cell death. As described above specific microbes apparently evolved effector proteins by which they were able to suppress PTI and cause disease (Fig. 2b). When these or other effectors are recognised by host R proteins, ETI is induced. One way to achieve this is by relieving the negative regulatory system of PTI, which will result in faster and higher expression of PTI-triggered genes (Fig. 2c). This direct interaction and manipulation of transcriptional regulators by nuclear NB-LRR proteins could be a generic function of these proteins.

In the example above, MLA10 inhibits transcriptional repressors (*Hv*WRKY1/2) upon pathogen perception. However, stimulation of positive transcriptional regulators is also a possibility as illustrated by the N

protein that confers resistance to tobacco mosaic virus by recognition of viral P50 (Burch-Smith et al. 2007). Using yeast two-hybrid assays an SPL-type transcription factor was found to interact with N and to be required for N-mediated ETI (Dinesh-Kumar, personal communication). This implies that N could trigger ETI through direct interaction with a positive regulatory transcription factor (Fig. 2c). The *Arabidopsis* CNL R proteins RPM1 and RPP5 were also found to interact with a transcriptional regulator, called AtTIP49a, of which the animal homologue interacts with the TATA-binding protein complex. Silencing of this gene enhanced the weak ETI triggered by RPP5 and RPP2, but not the strong ETI mediated by RPM1 (Holt et al. 2002). As silencing of *AtTIP49a* did not lead to enhanced resistance to virulent pathogens, this protein might be a negative regulator of ETI, but not PTI. Whether other NB-LRRs could also function as transcriptional coregulators remains to be investigated, but the presence of DNA binding domains in some NB-LRR proteins such as a WRKY or BED-finger domain (Fig. 1) supports this idea. In this respect, plant NB-LRRs could function similarly to the mammalian NACHT-LRR protein CIITA that translocates into the nucleus to regulate the expression of major histocompatibility complex (MHC) class II genes that are important for antigen presentation (Ting et al. 2006).

Nuclear NB-LRRs and nucleo-cytoplasmic trafficking in PTI and ETI

As no clear NLS sequences have been detected in most NB-LRRs it is unknown how they are translocated into the nucleus. One possibility is that they carry complex NLS sequences that are not easily predicted and deviate strongly from the classical NLS or the bipartite NLS (Gorlich and Kutay 1999). Another mechanism for import could be binding to a co-factor with an NLS sequence, a mechanism termed 'piggyback'. This might be the case for RRS1-R, MLA10 and N, as they all interact with proteins that contain a functional NLS: PopP2, *Hv*WRKY2 and an SPL-type transcription factor, respectively (Deslandes et al. 2003; Shen et al. 2007). However, the atypical RRS1-R protein could also be directly imported through recognition of its own NLS. For Rx, a Ran GTPase-activating protein 2 (RanGAP2) could serve as carrier according to the 'piggyback' mechanism. This protein has recently been identified as an Rx-associated protein by two different research groups (Sacco et al. 2007; Tameling and Baulcombe 2007). RanGAPs are highly conserved in eukaryotes and regulate the activity of the small GTPase Ran that is required for nucleo-cytoplasmic trafficking (Merkle 2003; Rose and Meier 2001). In interphase cells, localisation to the nuclear envelope (NE) is a feature of RanGAPs in both mammals and plants, which in the latter is mediated by the plant-specific WPP domain of RanGAP (Pay et al. 2002; Rose and Meier 2001). This is also the domain responsible for the interaction with the CC domain of Rx (Tameling and Baulcombe 2007). Specific silencing of *RanGAP2* in *Nicotiana benthamiana* plants transgenic for *Rx* resulted in a loss of Rx-mediated extreme resistance to PVX and in local and systemic spread of the virus. *RanGAP2* silencing did not affect N-mediated resistance to TMV or Pto/Prf-mediated resistance to *Pseudomonas syringae* pv. *tabaci* carrying the AvrPto effector, indicating that this protein might be specifically required for Rx (Tameling and Baulcombe 2007). Whether RanGAP2 indeed serves as a carrier or perhaps stimulates Rx import by recruiting yet another protein that serves as a carrier, remains to be investigated. An alternative hypothesis is that Rx activation modulates RanGAP2 activity in order to increase the nucleo-cytoplasmic trafficking of resistance co-factors involved in the induction of ETI.

Transport between the cytoplasm and the nucleus occurs exclusively through the nuclear pore complexes (NPCs) that are inserted in the nuclear envelope (Meier 2007; Merkle 2003). The NPCs are formed by large protein complexes containing nucleoporins. The precise composition of plant NPCs is not known, as most homologues of animal and yeast nucleoporins have not been identified in plants. Proteins of up to 40 kDa in molecular weight are able to diffuse through the NPCs, albeit much slower than the active transport mediated by the import and export receptors. NLS sequences are recognised by the import receptor importin (Imp) α, a member of the karyopherin family. Several karyopherin proteins have been identified in plants (Meier 2007). Docking of the importin cargo complex to the NE is mediated by another karyopherin family member, Imp β. Recently, in a mutagenesis screen, AtImpα3 has been shown to be required for the constitutive ETI mediated by *suppressor of npr1–1 constitutive 1 (snc1)*, a TNL mutant that carries an auto-activating mutation (Palma et al. 2005). AtImpα3 is also required in wild-type *Arabidopsis* for PTI against virulent pathogens. A mutation in a second gene that was implicated in nucleo-cytoplasmic trafficking and required for *snc1*-mediated defence was identified and encodes an *Arabidopsis* nucleoporin 96 homolog that is important for both PTI and ETI (Zhang et al. 2004). These results indicate that nucleo-cytoplasmic trafficking plays an important role in both PTI and ETI, although further research is needed to identify which resistance co-factors are inhibited in their translocation to the nucleus when AtImpα3 and nucleoporin 96 function is abolished. It will be interesting to determine whether the tested NB-LRRs indeed localise to the nucleus and whether translocation of these might be inhibited in the described mutants. Although we are just starting to explore the role of nucleo-cytoplasmic trafficking in plant defence, it might turn out to play a crucial role in this process.

Crosstalk between R proteins classes

As mentioned above, some NB-LRR R proteins may initiate ETI by functioning as transcriptional co-regulators. How extracellular R proteins belonging to the RLP and RLK families (Fig. 1) could activate ETI is unclear. The founding members of the RLP class of R protein are the Cf proteins (Rivas and Thomas 2005). Due to their homology to the *Arabidopsis* RLP CLAVATA2 (CLV2) it was proposed that Cf proteins might relay downstream signalling by a similar mechanism (Joosten and De Wit 1999). CLV proteins control meristem development and function in a complex consisting of an RLP (CLV2), an RLK (CLV1) and an extracellular ligand (CLV3; Doerner 2003). Analogously, Cf proteins might depend on a plasma membrane-localised RLK for downstream signalling, with which it would form a heterodimer upon (indirect) perception of an extracellular effector protein of *C. fulvum* (indicated by a question mark in Fig. 2c). Although initial attempts to identify such Cf-protein complexes suggested the existence of ±400 kDA complexes for Cf-4 and Cf-9 in size exclusion chromatography experiments (Rivas et al. 2002a, b), later experiments revealed that the fast migration in the column is an intrinsic property of the Cf proteins (Van Der Hoorn et al. 2003). Therefore, it is currently unclear whether R proteins from either RLP or RLK class form heterodimers similar to the CLAVATA proteins.

An alternative approach to elucidate how RLP R proteins trigger downstream resistance signalling is via the identification of putative signal transduction components that are transcriptionally regulated upon the activation of Cf-mediated ETI. Gabriëls and associates performed a transcriptional profiling of tomato seedlings mounting Cf-4-mediated ETI (Gabriëls et al. 2006). A subset of these differentially expressed genes were silenced in transgenic *N. benthamiana* plants expressing *Cf-4* to investigate their function in Cf-4-mediated HR. Interestingly this screen identified a gene coding for a CNL (Gabriëls et al. 2006). Since this protein was shown to be required for the Cf-4-mediated HR it was named NRC1 for NB-LRR protein required for HR-associated cell death 1 (see Fig. 2c, dotted arrow). NRC1 was not only required for RLP R proteins, but also for LeEIX (Gabriëls et al. 2007), an RLP that mediates recognition of ethylene-inducing xylanase (EIX), a potent elicitor of plant defence (Ron and Avni 2004). The genetic dependence of both RLP and RLK proteins on the same CNL suggests that they can trigger plant defences via a similar signalling pathway. It will be interesting to identify the cellular localisation of NRC1, since that would provide a clue on how RLP and RLKs affect defence gene expression.

Interestingly, NRC1 was also shown to be required for HR mediated by the CNL R proteins Prf, Rx and Mi-1.2 suggesting that also the signalling pathways of NB-LRR proteins are interwoven (Gabriëls et al. 2007). Additional support for such cross-talk is provided by the discovery of NRG1 (N-requirement gene 1), a CNL that is specifically required for the function of the TNL R protein N (Peart et al. 2005). Future studies should reveal whether crosstalk between CNLs, TNLs, RLPs and RLKs is a general phenomenon. If so, it could explain why the responses induced by the various R proteins are largely overlapping and depend on a limited number of conserved downstream signalling components (Martin et al. 2003; Tao et al. 2003).

Another major challenge for future studies is to solve the 3D structure of R proteins and to visualise their dynamics both at the subcellular as well as at the conformational level. A major bottleneck for biochemical and structural analyses of R proteins is the great difficulty to produce and purify sufficient amounts of intact and soluble native protein. A recent paper however, demonstrated the feasibility to use the methylotrophic yeast *Pichia pastoris* to produce and purify a relatively large amount of the almost full-length NB-LRR flax rust R protein M (Schmidt et al. 2007). If this protocol can also be applied for purification of other NB-LRR proteins it could provide the basis for experiments aimed to further our understanding of the mechanism by which these proteins operate in plant defence.

Acknowledgements The authors would like to acknowledge Savithramma Dinesh-Kumar and Jaap Bakker for sharing data before publication. We are grateful to Matthieu Joosten, Gerben van Ooijen and Martijn Rep for providing a critical review of the manuscript and helpful comments. Wladimir Tameling is supported by the EU-funded Integrated Project Bioexploit.

References

Ade, J., DeYoung, B. J., Golstein, C., & Innes, R. W. (2007). Indirect activation of a plant nucleotide binding site-leucine-rich repeat protein by a bacterial protease. *Proceedings of the National Academy of Sciences of the United States of America, 104*, 2531–2536.

Albrecht, M., & Takken, F. L. W. (2006). Update on the domain architectures of NLRs and R proteins. *Biochem-*

ical and Biophysical Research Communications, *339*, 459–462.

Aravind, L. (2000). The BED finger, a novel DNA-binding domain in chromatin-boundary-element-binding proteins and transposases. *Trends in Biochemical Sciences*, *25*, 421–423.

Ausubel, F. M. (2005). Are innate immune signaling pathways in plants and animals conserved. *Nauret Immunoogyl*, *6*, 973–979.

Axtell, M. J., Chisholm, S. T., Dahlbeck, D., & Staskawicz, B. J. (2003). Genetic and molecular evidence that the *Pseudomonas syringae* type III effector protein AvrRpt2 is a cysteine protease. *Molecular Microbiology*, *49*, 1537–1546.

Azevedo, C., Betsuyaku, S., Peart, J., Takahashi, A., Noel, L., Sadanandom, A., et al. (2006). Role of SGT1 in resistance protein accumulation in plant immunity. *EMBO Journal*, *25*, 2007–2016.

Bao, Q., Lu, W., Rabinowitz, J. D., & Shi, Y. (2007). Calcium blocks formation of apoptosome by preventing nucleotide exchange in Apaf-1. *Molecular Cell*, *25*, 181–192.

Bendahmane, A., Farnham, G., Moffett, P., & Baulcombe, D. C. (2002). Constitutive gain-of-function mutants in a nucleotide binding site- leucine rich repeat protein encoded at the *Rx* locus of potato. *Plant Journal*, *32*, 195–204.

Bieri, S., Mauch, S., Shen, Q. H., Peart, J., Devoto, A., Casais, C., et al. (2004). RAR1 positively controls steady state levels of barley MLA resistance proteins and enables sufficient MLA6 accumulation for effective resistance. *Plant Cell*, *16*, 3480–3495.

Birch, P. R., Rehmany, A. P., Pritchard, L., Kamoun, S., & Beynon, J. L. (2006). Trafficking arms: Oomycete effectors enter host plant cells. *Trends in Microbiology*, *14*, 8–11.

Burch-Smith, T. M., Schiff, M., Caplan, J. L., Tsao, J., Czymmek, K., & Dinesh-Kumar, S. P. (2007). A novel role for the TIR domain in association with pathogen-derived elicitors. *PLoS Biol*, *5*, e68.

Catanzariti, A. M., Dodds, P. N., & Ellis, J. G. (2007). Avirulence proteins from haustoria-forming pathogens. *FEMS Microbiology Letters*, *269*, 181–188.

Chen, X., Shang, J., Chen, D., Lei, C., Zou, Y., Zhai, W., et al. (2006). A B-lectin receptor kinase gene conferring rice blast resistance. *Plant Journal*, *46*, 794–804.

Chisholm, S. T., Coaker, G., Day, B., & Staskawicz, B. J. (2006). Host-microbe interactions: shaping the evolution of the plant immune response. *Cell*, *124*, 803–814.

De la Fuente van Bentem, S., Vossen, J. H., de Vries, K., van Wees, S. C., Tameling, W. I. L., Dekker, H., et al. (2005). Heat shock protein 90 and its co-chaperone protein phosphatase 5 interact with distinct regions of the tomato I-2 disease resistance protein. *Plant Journal*, *43*, 284–298.

Deslandes, L., Olivier, J., Peeters, N., Feng, D. X., Khounlothan, M., Boucher, C., et al. (2003). Physical interaction between RRS1-R, a protein conferring resistance to bacterial wilt, and PopP2, a type III effector targeted to the plant nucleus. *Proceedings of the National Academy of Sciences of the United States of America*, *100*, 8024-8029

Di Matteo, A., Federici, L., Mattei, B., Salvi, G., Johnson, K. A., Savino, C., et al. (2003). The crystal structure of polygalacturonase-inhibiting protein (PGIP), a leucine-rich repeat protein involved in plant defense. *Proceedings of the National Academy of Sciences of the United States of America*, *100*, 10124–10128.

Dinesh-Kumar, S. P., Tham, W. H., & Baker, B. J. (2000). Structure-function analysis of the tobacco mosaic virus resistance gene N. *Proceedings of the National Academy of Sciences of the United States of America*, *97*, 14789–14794.

Doerner, P. (2003). Plant meristems: A merry-go-round of signals. *Current Biology*, *13*, R368–R374.

Ellis, J. G., Dodds, P. N., & Lawrence, G. J. (2007). Flax rust resistance gene specificity is based on direct resistance-avirulence protein interactions. *Annual Review of Phytopathology*, *45*, 289–306.

Feys, B. J., & Parker, J. E. (2000). Interplay of signaling pathways in plant disease resistance. *Trends in Genetics*, *16*, 449–455.

Gabriëls, S. H. E. J., Takken, F. L. W., Vossen, J. H., de Jong, C. F., Liu, Q., Turk, S. C., et al. (2006). CDNA-AFLP combined with functional analysis reveals novel genes involved in the hypersensitive response. *Molecular Plant-Microbe Interactions*, *19*, 567–576.

Gabriëls, S. H. E. J., Vossen, J. H., Ekengren, S. K., van Ooijen, G., Abd-El-Haliem, A. M., van den Berg, G. C. M., et al. (2007). An NB-LRR protein required for HR signalling mediated by both extra- and intracellular resistance proteins. *The Plant Journal*, *50*, 14–28.

Gorlich, D., & Kutay, U. (1999). Transport between the cell nucleus and the cytoplasm. *Annual Review of Cell and Developmental Biology*, *15*, 607–660.

Grant, S. R., Fisher, E. J., Chang, J. H., Mole, B. M., & Dangl, J. L. (2006). Subterfuge and Manipulation: Type III Effector Proteins of Phytopathogenic Bacteria. *Annual Review of Microbiology*, *60*, 425–449.

Holt 3rd, B. F., Belkhadir, Y., & Dangl, J. L. (2005). Antagonistic control of disease resistance protein stability in the plant immune system. *Science*, *309*, 929–932.

Holt 3rd, B. F., Boyes, D. C., Ellerstrom, M., Siefers, N., Wiig, A., Kauffman, S., et al. (2002). An evolutionarily conserved mediator of plant disease resistance gene function is required for normal *Arabidopsis* development. *Developmental Cell*, *2*, 807–817.

Howles, P., Lawrence, G., Finnegan, J., McFadden, H., Ayliffe, M., Dodds, P., et al. (2005). Autoactive alleles of the flax *L6* rust resistance gene induce non-race-specific rust resistance associated with the hypersensitive response. *Molecular Plant-Microbe Interactions*, *18*, 570–582.

Hubert, D. A., Tornero, P., Belkhadir, Y., Krishna, P., Takahashi, A., Shirasu, K., et al. (2003). Cytosolic HSP90 associates with and modulates the *Arabidopsis* RPM1 disease resistance protein. *EMBO Journal*, *22*, 5679–5689.

Hwang, C. F., Bhakta, A. V., Truesdell, G. M., Pudlo, W. M., & Williamson, V. M. (2000). Evidence for a role of the N terminus and leucine-rich repeat region of the Mi gene product in regulation of localized cell death. *Plant Cell*, *12*, 1319–1329.

Jones, J. D. G., & Dangl, J. L. (2006). The plant immune system. *Nature, 444*, 323–329.

Joosten, M. H. A. J., & De Wit, P. J. G. M. (1999). The tomato-*Cladosporium fulvum* interaction: A versatile experimental system to study plant-pathogen interactions. *Annual Review of Phytopathology, 37*, 355–367.

Kajava, A. V. (1998). Structural diversity of leucine-rich repeat proteins. *Journal of Molecular Biology, 277*, 519–527.

Kim, H. E., Du, F., Fang, M., & Wang, X. (2005). Formation of apoptosome is initiated by cytochrome c-induced dATP hydrolysis and subsequent nucleotide exchange on Apaf-1. *Proceedings of the National Academy of Sciences of the United States of America, 102*, 17545–17550.

Kobe, B., & Deisenhofer, J. (1994). The leucine-rich repeat: A versatile binding motif. *Trends in Biochemical Sciences, 19*, 415–421.

Leipe, D. D., Koonin, E. V., & Aravind, L. (2004). STAND, a class of P-loop NTPases including animal and plant regulators of programmed cell death: Multiple, complex domain architectures, unusual phyletic patterns, and evolution by horizontal gene transfer. *Journal of Molecular Biology, 343*, 1–28.

Leister, R. T., Dahlbeck, D., Day, B., Li, Y., Chesnokova, O., & Staskawicz, B. J. (2005). Molecular genetic evidence for the role of SGT1 in the intramolecular complementation of Bs2 protein activity in *Nicotiana benthamiana*. *Plant Cell, 17*, 1268–1278.

Liu, Y., Burch-Smith, T., Schiff, M., Feng, S., & Dinesh-Kumar, S. P. (2004). Molecular chaperone Hsp90 associates with resistance protein N and its signaling proteins SGT1 and Rar1 to modulate an innate immune response in plants. *Journal of Biological Chemistry, 279*, 2101–2108.

Liu, Y., Schiff, M., Serino, G., Deng, X. W., & Dinesh-Kumar, S. P. (2002). Role of SCF Ubiquitin-Ligase and the COP9 Signalosome in the *N* Gene- Mediated Resistance Response to Tobacco mosaic virus. *Plant Cell, 14*, 1483–1496.

Mackey, D., Holt, B. F., Wiig, A., & Dangl, J. L. (2002). RIN4 interacts with *Pseudomonas syringae* type III effector molecules and is required for *RPM1*-mediated resistance in *Arabidopsis*. *Cell, 108*, 743–754.

Martin, G. B., Bogdanove, A. J., & Sessa, G. (2003). Understanding the functions of plant disease resistance proteins. *Annual Review of Plant Biology, 54*, 23–61.

McHale, L., Tan, X., Koehl, P., & Michelmore, R. W. (2006). Plant NBS-LRR proteins: Adaptable guards. *Genome Biology, 7*, 212.

Meier, I. (2007). Composition of the plant nuclear envelope: Theme and variations. *Journal of Experimental Botany, 58*, 27–34.

Merkle, T. (2003). Nucleo-cytoplasmic partitioning of proteins in plants: Implications for the regulation of environmental and developmental signalling. *Current Genetics, 44*, 231–260.

Mestre, P., & Baulcombe, D. C. (2006). Elicitor-mediated oligomerization of the tobacco N disease resistance protein. *Plant Cell, 18*, 491–501.

Meyers, B. C., Dickerman, A. W., Michelmore, R. W., Sivaramakrishnan, S., Sobral, B. W., & Young, N. D. (1999). Plant disease resistance genes encode members of an ancient and diverse protein family within the nucleotide-binding superfamily. *Plant Journal, 20*, 317–332.

Meyers, B. C., Kozik, A., Griego, A., Kuang, H., & Michelmore, R. W. (2003). Genome-wide analysis of NBS-LRR-encoding genes in *Arabidopsis*. *Plant Cell, 15*, 809–834.

Moffett, P., Farnham, G., Peart, J., & Baulcombe, D. C. (2002). Interaction between domains of a plant NBS-LRR protein in disease resistance-related cell death. *EMBO Journal, 21*, 4511–4519.

Monosi, B., Wisser, R. J., Pennill, L., & Hulbert, S. H. (2004). Full-genome analysis of resistance gene homologues in rice. *Theoretical & Applied Genetics, 109*, 1434–1447.

Mucyn, T. S., Clemente, A., Andriotis, V. M. E., Balmuth, A. L., MucynOldroyd, G. E. D., Staskawicz, B. J., et al. (2006). The tomato NBARC-LRR protein Prf Interacts with Pto kinase in vivo to regulate specific plant immunity. *Plant Cell, 18 (10)*, 2792–806.

Nürnberger, T., & Kemmerling, B. (2006). Receptor protein kinases–pattern recognition receptors in plant immunity. *Trends in Plant Scencei, 11*, 519–522.

Palma, K., Zhang, Y., & Li, X. (2005). An importin alpha homolog, MOS6, plays an important role in plant innate immunity. *Current Biology, 15*, 1129–1135.

Pan, Q., Wendel, J., & Fluhr, R. (2000). Divergent evolution of plant NBS-LRR resistance gene homologues in dicot and cereal genomes. *Journal of Molecular Evolution, 50*, 203–213.

Pay, A., Resch, K., Frohnmeyer, H., Fejes, E., Nagy, F., & Nick, P. (2002). Plant RanGAPs are localized at the nuclear envelope in interphase and associated with microtubules in mitotic cells. *Plant Journal, 30*, 699–709.

Peart, J. R., Mestre, P., Lu, R., Malcuit, I., & Baulcombe, D. C. (2005). NRG1, a CC-NB-LRR Protein, together with N, a TIR-NB-LRR Protein, Mediates Resistance against Tobacco Mosaic Virus. *Current Biology, 15*, 968–973.

Rairdan, G. J., & Moffett, P. (2006). Distinct domains in the ARC region of the potato resistance protein Rx mediate LRR binding and inhibition of activation. *Plant Cell, 18*, 2082–2093.

Rairdan, G., & Moffett, P. (2007). Brothers in arms? Common and contrasting themes in pathogen perception by plant NB-LRR and animal NACHT-LRR proteins. *Microbes and Infection, 9*, 677–686.

Riedl, S. J., Li, W., Chao, Y., Schwarzenbacher, R., & Shi, Y. (2005). Structure of the apoptotic protease-activating factor 1 bound to ADP. *Nature, 434*, 926–933.

Rivas, S., Mucyn, T., van den Burg, H. A., Vervoort, J., & Jones, J. D. G. (2002a). An approximately 400 kDa membrane-associated complex that contains one molecule of the resistance protein Cf-4. *Plant Journal, 29*, 783–796.

Rivas, S., Romeis, T., & Jones, J. D. G. (2002b). The Cf-9 disease resistance protein is present in an approximately 420- kilodalton heteromultimeric membrane-associated complex at one molecule per complex. *Plant Cell, 14*, 689–702.

Rivas, S., & Thomas, C. M. (2005). Molecular interactions between tomato and the leaf mold pathogen *Cladosporium fulvum*. *Annual Review of Phytopathology, 43*, 395–436.

Ron, M., & Avni, A. (2004). The receptor for the fungal elicitor ethylene-inducing xylanase is a member of a resistance-like gene family in tomato. *Plant Cell, 16*, 1604–1615.

Rose, A., & Meier, I. (2001). A domain unique to plant RanGAP is responsible for its targeting to the plant nuclear rim. *Proceedings of the National Academy of Sciences of the United States of America, 98*, 15377–15382.

Sacco, M., Mansoor, S., & Moffett, P. (2007). A RanGAP protein physically interacts with the NB-LRR protein Rx and is required for Rx-mediated viral resistance. *Plant Journal*, DOI 10.111/j.1365-313x.2007.03213.x.

Schmidt, S. A., Williams, S. J., Wang, C. I., Sornaraj, P., James, B., Kobe, B., et al. (2007). Purification of the M flax-rust resistance protein expressed in *Pichia pastoris*. *Plant Journal, 50*, 1107–1117.

Shao, F., Golstein, C., Ade, J., Stoutemyer, M., Dixon, J. E., & Innes, R. W. (2003). Cleavage of *Arabidopsis*. PBS1 by a bacterial type III effector. *Science, 301*, 1230–1233.

Shen, Q. H., Saijo, Y., Mauch, S., Biskup, C., Bieri, S., Keller, B., et al. (2007). Nuclear activity of MLA immune receptors links isolate-specific and basal disease-resistance responses. *Science, 315*, 1098–1103.

Sun, X., Cao, Y., & Wang, S. (2006). Point mutations with positive selection were a major force during the evolution of a receptor-kinase resistance gene family of rice. *Plant Physiology, 140*, 998–1008.

Sun, X., Cao, Y., Yang, Z., Xu, C., Li, X., Wang, S., et al. (2004). *Xa26*, a gene conferring resistance to *Xanthomonas oryzae* pv. *oryzae*. in rice, encodes an LRR receptor kinase-like protein. *Plant Journal, 37*, 517–527.

Sun, Q., Collins, N. C., Ayliffe, M., Smith, S. M., Drake, J., Pryor, T., et al. (2001). Recombination between paralogues at the *rp1* rust resistance locus in maize. *Genetics, 158*, 423–438.

Takahashi, A., Casais, C., Ichimura, K., & Shirasu, K. (2003). HSP90 interacts with RAR1 and SGT1 and is essential for RPS2-mediated disease resistance in *Arabidopsis*. *Proceedings of the National Academy of Sciences of the United States of America, 100*, 11777–11782.

Takken, F. L. W., Albrecht, M., & Tameling, W. I. L. (2006). Resistance proteins: Molecular switches of plant defence. *Current Opinion in Plant Biology, 9*, 383–390.

Tameling, W. I. L., & Baulcombe, D. C. (2007). Physical association of the NB-LRR resistance protein Rx with a Ran GTPase-activating protein is required for extreme resistance to potato virus X. *Plant Cell, 19*, 1682–694.

Tameling, W. I. L., Elzinga, S. D., Darmin, P. S., Vossen, J. H., Takken, F. L. W., Haring, M. A., et al. (2002). The tomato *R* gene products I-2 and Mi-1 are functional ATP binding proteins with ATPase activity. *Plant Cell, 14*, 2929–2939.

Tameling, W. I. L., Vossen, J. H., Albrecht, M., Lengauer, T., Berden, J. A., Haring, M. A., et al. (2006). Mutations in the NB-ARC domain of I-2 that impair ATP hydrolysis cause autoactivation. *Plant Physiology, 140*, 1233–1245.

Tan, X., Calderon-Villalobos, L. I., Sharon, M., Zheng, C., Robinson, C. V., Estelle, M., et al. (2007). Mechanism of auxin perception by the TIR1 ubiquitin ligase. *Nature, 446*, 640–645.

Tao, Y., Xie, Z., Chen, W., Glazebrook, J., Chang, H. S., Han, B., et al. (2003). Quantitative nature of *Arabidopsis* responses during compatible and incompatible interactions with the bacterial pathogen Pseudomonas syringae. *Plant Cell, 15*, 317–330.

Ting, J. P., Kastner, D. L., & Hoffman, H. M. (2006). CATERPILLERs, pyrin and hereditary immunological disorders. *Natures Review. Immunology, 6*, 183–195.

Tuskan, G. A., Difazio, S., Jansson, S., Bohlmann, J., Grigoriev, I., Hellsten, U., et al. (2006). The genome of black cottonwood, *Populus trichocarpa* (Torr. & Gray). *Science, 313*, 1596–1604.

Ueda, H., Yamaguchi, Y., & Sano, H. (2006). Direct interaction between the tobacco mosaic virus helicase domain and the ATP-bound resistance protein, N factor during the hypersensitive response in tobacco plants. *Plant Molecular Biology, 61*, 31–45.

Ulker, B., & Somssich, I. E. (2004). WRKY transcription factors: From DNA binding towards biological function. *Current Opinion in Plant Biology, 7*, 491–498.

van der Biezen, E. A., & Jones, J. D. G. (1998). The NB-ARC domain: A novel signalling motif shared by plant resistance gene products and regulators of cell death in animals. *Current Biology, 8*, R226–227.

Van Der Hoorn, R. A., Rivas, S., Wulff, B. B., Jones, J. D., & Joosten, M. H. (2003). Rapid migration in gel filtration of the Cf-4 and Cf-9 resistance proteins is an intrinsic property of Cf proteins and not because of their association with high-molecular-weight proteins. *Plant Journal, 35*, 305–315.

van Ooijen, G., van den Burg, H. A., Cornelissen, B. J. C., & Takken, F. L. W. (2007). Structure and function of Resistance proteins in solanaceous plants. *Annual Review of Phytopathology, 45*, 43–72.

Weaver, M. L., Swiderski, M. R., Li, Y., & Jones, J. D. (2006). The *Arabidopsis thaliana* TIR-NB-LRR R-protein, RPP1A; protein localization and constitutive activation of defence by truncated alleles in tobacco and *Arabidopsis*. *Plant Journal, 47*, 829–840.

Xu, Y., Tao, X., Shen, B., Horng, T., Medzhitov, R., Manley, J. L., et al. (2000). Structural basis for signal transduction by the Toll/interleukin-1 receptor domains. *Nature, 408*, 111–115.

Yan, N., Chai, J., Lee, E. S., Gu, L., Liu, Q., He, J., et al. (2005). Structure of the CED-4-CED-9 complex provides insights into programmed cell death in *Caenorhabditis elegans*. *Nature, 437*, 831–837.

Zhang, Y., Dorey, S., Swiderski, M., & Jones, J. D. (2004). Expression of RPS4 in tobacco induces an AvrRps4-independent HR that requires EDS1, SGT1 and HSP90. *Plant Journal, 40*, 213–224.

Eur J Plant Pathol (2008) 121:257–266
DOI 10.1007/s10658-008-9271-8

How can we exploit functional genomics approaches for understanding the nature of plant defences? Barley as a case study

David B. Collinge · Michael K. Jensen ·
Michael F. Lyngkjaer · Jesper Rung

Received: 22 May 2007 / Accepted: 14 January 2008
© KNPV 2008

Abstract The concept 'functional genomics' refers to the methods used for the functional characterisation of genomes. The methods utilised provide new opportunities for studying the nature and role of defence mechanisms in plants. Unlike *Arabidopsis*, poplar and rice, the full genomic sequence of barley is not available. In this case, the analysis of barley gene expression data plays a pivotal role for obtaining insight into the functional characterisation of individual gene products. Many genes are activated transcriptionally following attack by pathogens and these often contribute to the defence mechanisms which underlie disease resistance. The use of large-scale complementary DNA library constructions and genome-wide transcript profiles of plants exposed to biotic stress provide the data required to drive hypotheses concerning the function of newly identified genes. In this paper, we illustrate how publicly available gene expression data has proved valid for studies of plant defence responses; enabling a cost-effective workflow starting from isolated gene transcripts to elucidation of biological function upon biotic stress.

Keywords Barley · Functional genomics · Plants · NAC transcription factors · *Hordeum vulgare* · *Blumeria graminis* f.sp. *hordei* · Pathogen

Introduction

Plants are constantly under attack by microorganisms. However, only a few of these are potential pathogens capable of causing disease on a particular plant species. Even for host pathogen species, only few pathogen attacks actually develop to cause a successful infection. More often than not, the plant succeeds in repelling attack through deployment of its defences, with disease resistance as the result. Before we describe the means by which functional genomic approaches can be deployed for the elucidation of molecular components engaged in plant defences, a brief summary of plant defensive strategies is presented.

Plant defences comprise the production of antimicrobial compounds (Field et al. 2006; Hammerschmidt 1999) and proteins (van Loon et al. 2006), chemical and physical changes to secondary cell walls (Mörschbacher and Mendgen 2000), and the induction of programmed cell death, known as the hypersensitive response (HR; Jabs and Slusarenko 2000). Some defence mechanisms are essentially constitutive, that is, they are always produced at a particular stage in the host's development.

D. B. Collinge (✉) · M. K. Jensen · J. Rung
Department of Plant Biology, Faculty of Life Sciences,
University of Copenhagen,
1871 Frederiksberg, Denmark
e-mail: dbc@life.ku.dk

M. F. Lyngkjaer
Biosystems Department, Risoe National Laboratory,
Technical University of Denmark,
4000 Roskilde, Denmark

Others are first induced or activated when a pathogen attacks the host. Many of the same defences are activated in a particular host by different pathogen species (Collinge et al. 2002). Even in compatible interactions, plant defences comprise an effective barrier to pathogens, often limiting the rate at which the pathogen invades the host tissues (Trujillo et al. 2004). However, the particular mechanisms which are effective against a specific pathogen will depend on physiology of the pathogen, which in part reflects its taxonomic group, e.g., fungus, bacteria or virus.

The employment of different life style strategies by different pathogens also plays a role in the efficacy of specific defence mechanisms used by the attacked host plant. The extremes are represented by necrotrophy, in which the pathogen destroys and consumes the host tissues, and biotrophy, in which the pathogen parasitizes living tissue. Hemibiotrophs in essence utilise both strategies at different phases of their life cycles. Defence mechanisms, for example, the HR, differ in their effectiveness against pathogens using these different strategies, and the regulation of the activation of these mechanisms differs too; thus salicylic acid signalling and jasmonic acid signalling are associated, at least in *Arabidopsis*, with defence against biotrophs and necrotrophs, respectively (Glazebrook 2005; Zimmerli et al. 2004). This generalisation highlights important correlations, though some degree of stimuli-dependent discrepancies have been observed (Zimmerli et al. 2004). Effective resistance thus depends on successful recognition by the host that it is being attacked by the pathogen and deployment of the appropriate defence mechanisms at the right place and at the right time.

In this article, we will look at the means by which plant defences are studied and the tools which can be used for determining whether a particular defence mechanism has a role in disease resistance towards a specific pathogen. As a case study, we will focus on defence against a biotrophic fungal pathogen and use the interaction between barley, *Hordeum vulgare*, and the barley powdery mildew fungus, *Blumeria graminis* f.sp. *hordei* (*Bgh*). Functional genomics encompass a wealth of scientific disciplines not covered in this case study. Thus, high-throughput bioimaging, proteomics and metabolomics approaches are also excellent tools which are being utilised in various biological systems for an improved understanding of genome function, even for those not yet sequenced.

Why barley and *Blumeria graminis*?

Bgh is the causal agent of powdery mildew; one of the most important diseases of barley worldwide. The barley–*Bgh* interaction has evolved as a model system for several reasons, both biological and practical (Collinge et al. 2002). Firstly, a very large number of race-specific resistance genes have been described (Jørgensen 1994)–and many of these have been incorporated into near-isogenic lines of barley (Kølster et al. 1986). The resistant phenotype for the majority of these disease resistance genes is associated with the HR. Recently, allelic diversity has become accessible with ecoTILLING lines (Mejlhede et al. 2006) and single-nucleotide polymorphisms (SNP) populations (Rostoks et al. 2005), which offer great potential for exploitation of the natural variation of disease resistance. Thus, ecoTILLING is a polymerase chain reaction (PCR)-based technique that allows identification of allelic variants in known genes of interest, for example R-genes (Mejlhede et al. 2006). A new microarray-based technique has been developed for identifying SNP in specific genes (Rostoks et al. 2005). These methods represent new tools for identification of the allelic variation in known genes, which can be tested phenotypically for their influence on disease resistance. Secondly, the development of the fungus on the host is synchronised, facilitating meaningful experiments where physiological and molecular responses of the barley host can be correlated perfectly with the development of the fungus, using bioimaging analyses (Shen et al. 2007) and transcript profiles among other methods (Caldo et al. 2006; Gjetting et al. 2007; Gregersen et al. 1997; Zierold et al. 2005). Collectively, this has made the barley–*Bgh* interaction among the best-studied pathosystems for investigating plant responses towards pathogen attack.

Historical perspective of transcriptomics in the barley–*Bgh* interaction

The first molecular studies to assay responses in the barley transcriptome to *Bgh* used in vitro translation products by two-dimensional polyacrylamide gel electrophoresis (Collinge et al. 2002; Gregersen et al. 1990; Manners and Scott 1985). The next phase, in the 1990s, was the utilisation of various differential and subtractive hybridisation techniques to

isolate complementary DNA (cDNA) clones (Collinge et al. 2002; Gregersen et al. 1997; Hein et al. 2004), and the differential expression suggested by the screening method was confirmed by northern blotting. In some cases, sequence-based identification was supported by biochemical evidence (see Collinge et al. 2002). For the barley–*Bgh* interaction, these approaches for gene discovery have been superseded largely by the use of expressed sequence tag (EST) libraries which provide a vast open resource of partial–and even full-length–cDNA sequences, representing roughly 500,000 individual cDNA clones, which reflect gene expression in specific tissues and physiological states. We illustrate this in Table 1 with the NAC transcription factor family of barley (see below). EST databases are now providing a corroborative effort to assemble contigs (as sequences assembled from smaller overlapping fragments) encoding full-length or near full-length gene products. For instance, the UniGene database (http://www.ncbi.nlm.nih.gov/sites/entrez?db=unigene) is an in silico experimental system for automatically partitioning GenBank sequences into a non-redundant set of gene-oriented clusters. It uses part of a coding sequence to extract all sequences including transcripts exhibiting homologies to a given query sequence (Boguski and Schuler 1995). From such homology searches, sequences including ESTs with high sequence similarity are grouped into clusters. Subsequently, these UniGene clusters can be used to (1) obtain an indication of the level of transcript accumulation for a given UniGene member for a specific tissue and (2) perform intercluster comparisons for the possible discovery of expressed genes responding to a particular environmental stress factor, physiological state or developmental stage (Zhang et al. 2004). However, it should be stressed that UniGene clusters are dynamic entities, and represent the current best model for interpreting the coding sequences of genes in the databases expressing these 'tags' at the time of sampling.

EST data, (e.g. Zhang et al. 2004) also provided the basis for the design of an Affymetrix GeneChip®, which in barley carries 22,792 gene sequences (Close et al. 2004; Shen et al. 2005). Parallel to the development of this Affymetrix Barley1 GeneChip®, dotted filter array technology has also been utilised for barley–*Bgh* interactions (Zierold et al. 2005). Each of these technologies offers its advantages and disadvantages which we will discuss below. However, all hybridization-based transcriptome techniques suf-

fer the limitation that a specific gene will not be present unless the cDNA is prepared from a tissue in the physiological state where it is expressed. This problem is essentially solved once the entire genomic sequence is available for the species in question. Although the barley genome is large (5,000 Mb), it is predicted that a draft sequence for barley will be available within a few years.[1]

Transcriptomics in barley today

The Affymetrix Barley1 GeneChip® microarray has been used for several studies of gene expression in barley after *Bgh* inoculation (Caldo et al. 2004, 2006) as well as for other interactions in barley involving biotic, namely *Fusarium graminearum* (Boddu et al. 2006), and abiotic (Svensson et al. 2006) stress, namely cold. Likewise the dotted filter array has been utilised for barley–*Bgh* interactions (Eichmann et al. 2006; Gjetting et al. 2007; Zierold et al. 2005). The advantage of the dotted filter array compared to the microarray is that it is straightforward to add new sequences to the study as they are discovered. A disadvantage lies in reduced sensitivity and specificity. This makes it difficult to distinguish closely related and weakly expressed gene sequences from each other. The advantage of both array technologies is that it is possible to study the expression of a large number of genes simultaneously. Thus detailed time course studies with appropriate biological replicates using array technologies have been conducted (Caldo et al. 2004, 2006) and much of the data from these and other studies can be accessed through the public Affymetrix-specific data repository PLEXdb (www.plexdb.org).

Case study: EST libraries and their exploitation for studying the NAC transcription factors of barley

Members of the large plant-specific gene family encoding NAC transcription factors share a common N-terminal domain, comprised of five highly conserved motifs. The domain is termed NAC from its

---

[1] http://pgrc.ipk-gatersleben.de/etgi/publications/whitepaper_barley_physmap_and_sequence.pdf; http://www.ars.usda.gov/research/projects/projects.htm?ACCN_NO=411452

Table 1 Gene-oriented UniGene clusters of NAC transcript sequences

Barley NAC UniGenes	Transcripts[a]	Rice homologue (% iden.)	*Arabidopsis* homologue (% iden.)	cDNA source
Hv.6550 (*HvNAC4*)	12[a]	Os01g0816100 (79.4%)	ATAF2 (At5g63790) (62%)	*B. graminis* inoc. leaf, seed, stem, root
Hv.1425	88[a]	Os03g0815100 (85.4%)	ATAF2 (At5g63790) (65.6%)	*B. graminis* inoc. leaf, seed, callus, root
Hv.984	3	Os03g0624600 (55.9%)	–	Root
Hv.6308	38	Os02g0822400 (71.4%)	ANAC051/ANAC052 (At3g10490) (58%)	*B. graminis* inoc. leaf, seed, callus, flower
Hv.5295	28[a]	Os08g0562200 (81.4%)	ANAC053 (At3g10500) (49.3%)	*B. graminis* inoc. leaf, callus, flower
Hv.5097	8[a]	Os03g0327800 (76.2%)	ATAF2 (At5g63790) (56%)	Seed, leaf, root
Hv.5147	7	Os07g0683200 (85.5%)	ANAC032 (At1g77450) (71.6%)	Leaf, flower, root
Hv.877	10	Os04g0460600 (80.5%)	ANAC092 (At5g39610) (69.8%)	Root, leaf
Hv.17199	13	Os09g0493700 (68.2%)	ANAC051/ANAC052 (At3g10490) (48.4%)	*B. graminis* inoc. leaf, seed
Hv.2154	5	–	–	Seed
Hv.4825	26[a]	Os08g0157900 (69.8%)	–	*B. graminis* inoc. leaf, stem, callus
Hv.2154	14	Os08g0200600 (69.9%)	(At3g12977) (64.7%)	Seed
Hv.15755 (*HvNAC6*)	47[a]	Os01g0884300 (86.5%)	ATAF1 (At1g01720) (76.5%)	*B. sorokiniana* inoc. leaf, root, callus
Hv.19392	10	Os07g0684800 (61.1%)	ANAC092 (At5g39610) (67.6%)	Seed
Hv.19815	6	Os04g0536500 (56.4%)	ANAC092 (At5g39610) (72.4%)	Seed, flower, *F. graminearum* inoc. leaf
Hv.6910	3	Os06g0131700 (46.1%)	ANAC092 (At5g39610) (64.4%)	Seed
Hv.18811	3	Os04g0536500 (58%)	ANAC092 (At5g39610) (61.2%)	Seed
Hv.18323	7	Os07g0684800 (75.7%)	ANAC092 (At5g39610) (67.6%)	*B. graminis* inoc. leaf, root, callus
Hv.17687	3	Os11g0294400 (57.1%)	–	Flower, meristem
Hv.21351	7[a]	Os11g0126900 (73%)	RD26 (At4g27410) (64.7%)	*B. graminis* inoc. leaf, callus, seed
Hv.21779 (*HvNAC1*)	8[a]	Os12g0610600 (81.8%)	NAC1 (At1g56010) (48.8%)	*B. graminis* inoc. leaf, stem, root
Hv.19852	24	Os11g0184900 (80.9%)	ANAC032 (At1g77450) (75.3%)	*B. graminis* inoc. leaf, stem, root
Hv.25370	6	Os06g0675600 (84.2%)	–	Leaf, seed, root
Hv.24230	3	Os03g0624600 (71.2%)	ANAC058 (At3g18400) (75.7%)	Root, stem
Hv.16999	24	Os03g0327100 (84.3%)	ANAC092 (At5g39610) (75.8%)	Root, leaf, seed, callus
Hv.16783	3	–	ANAC074 (At4g28540) (63.7%)	Root, seed

[a] Full-length mRNA available

first identified members *NAM*, *ATAF1,2* and *CUC2* (Aida et al. 1997; Souer et al. 1996). Genes encoding NAC transcription factors have been reported to be induced by both abiotic and biotic stresses. Furthermore, modulation of the expression of individual members has resulted in improved salt and drought tolerance, and enhanced resistance towards *Fusarium oxysporum*, in rice and *Arabidopsis* (Hu et al. 2006; Lu et al. 2006). We have isolated several NAC gene members from barley, using differential display and cDNA library screening techniques for transcripts expressed in barley upon *Bgh* inoculation (Gregersen and Collinge 2001; Jensen et al. 2007). We have shown subsequently that *HvNAC6* (*H. vulgare NAC6*) has a positive role in penetration resistance against *Bgh* (Jensen et al. 2007). In the following, we will present how public transcript data repositories can be used in a data-driven approach for developing

hypotheses on the functionality of specific genes of interest. We will use expression profiles from NAC gene members as a case study, but any gene of interest can be exploited, as long as a transcript sequence originating from the gene of interest is present on the array platform to be analyzed.

A Basic Local Alignment Search Tool search of European Molecular Biology Laboratory and GenBank databases using a nucleotide sequence encoding the conserved NAC domain yields approximately 400 putative barley NAC derived transcript sequences. However, most of these are partial sequences which can be grouped nevertheless into 26 UniGene clusters based on sequence similarities (Table 1). Each UniGene cluster comprises several partial transcript sequences, ideally making up a contig (i.e. contributing to a composite and potentially complete gene sequence), deciphering the full-length mRNA sequence of the individual gene. Table 1 show that approximately 50% of the current (UniGene Build #48) NAC domain containing UniGene clusters include ESTs originating from *Bgh*-inoculated barley cDNA libraries. As transcripts from biotically stressed barley cDNA libraries are included in approximately 35% of the total number of barley UniGene clusters, NAC members could be over represented in biotically stressed barley cDNA libraries, making them interesting candidates for the understanding of the regulatory mechanisms involved in the barley–*Bgh* interaction.

Apart from the Barley1 Gene Chip®, one such platform is the ~3 k cDNA microarray from the *Bgh*-infected barley epidermis cDNA library developed by Institute of Plant Genetics and Crop Plant Research (Zierold et al. 2005). In barley, the recessive loss-of-function alleles of the *Mlo* gene mediate durable and race-nonspecific resistance towards *Bgh* associated with the rapid formation of large epidermal cell wall-associated structures termed papillae (Jørgensen 1992). By comparing the large-scale transcript responses of *mlo5* mutant plants with wild-type *Mlo* plants upon *Bgh* inoculation, Zierold et al. (2005) aimed at identifying candidate genes mediating durable resistance towards *Bgh* in barley. Investigating the origin of the approximately 400 transcripts representing 26 NAC UniGene clusters, for transcripts originating from the *Bgh*-inoculated DNA library used for spotting the nylon filter used by Zierold and co-workers, we identified 11 gene-oriented NAC clones, of which eight had been successfully spotted

on the cDNA array (Fig. 1). Among the spotted clones, two belonged to UniGene clusters *Hv.21779* and *Hv.15755*, of which we have isolated full-length cDNA clones (*HvNAC1* and *HvNAC6*, respectively, see Jensen et al. 2007). Interestingly, from our data-mining, we observed *mlo5*-specific up-regulation of *HvNAC6* upon *Bgh* inoculation. In the susceptible *Mlo* wild-type background, no *HvNAC6* induction was observed (Fig. 1), possibly due to a *Mlo*-dependent negative control of *HvNAC6* transcription (Zierold et al. 2005). Another interesting transcript profile is depicted by the HO13D12 cDNA clone (UniGene cluster Hv.1425). HO13D12 abundance shows delayed accumulation in the wild-type plants compared to *mlo5* plants upon *Bgh* attack. As the outcome of race non-specific lines of defence are believed to depend on the timing of most responses towards attacking pathogens (Caldo et al. 2006) the observed delayed induction of HO13D12-specific transcripts in wild-type plants could affect the delicate timing of effective race non-specific resistance. However, the *HvNAC6* expression profile shown in Fig. 1 was not identical to another member of the Hv.15755 UniGene cluster, which did not show any genotype- or treatment-specific regulations (HO05F04, data not shown). Hence caution and thorough investigation of the repositories available should be performed when mining transcripts originating from genomes not yet sequenced. Finally, the generation of working hypotheses should include expression profiling by an alternate approach to the one analysed in silico (e.g., quantitative real-time (qRT)-PCR).

From an independent transcript analysis, we supplemented the *HvNAC6* expression pattern presented in Fig. 1 by using qRT-PCR on a pattern of a *Bgh*-challenged Pallas near-isogenic line (Kølster et al. 1986) and continued with functional studies to examine the possible importance *HvNAC6* for resistance towards *Bgh* (Jensen et al. 2007). For this purpose, we made use of the particle bombardment transformation assay of barley epidermal cells (Shirasu et al. 1999). Individual NAC gene constructs for in vivo gene silencing or over expression were co-transformed with *uidA* the β-glucuronidase reporter gene, thereby providing a reverse genetics tool to study the cell-autonomous interaction outcomes between barley and *Bgh* of transformed cells (Fig. 2). Our studies showed that *HvNAC6* transcript abundance indeed affects the defence responses in barley by

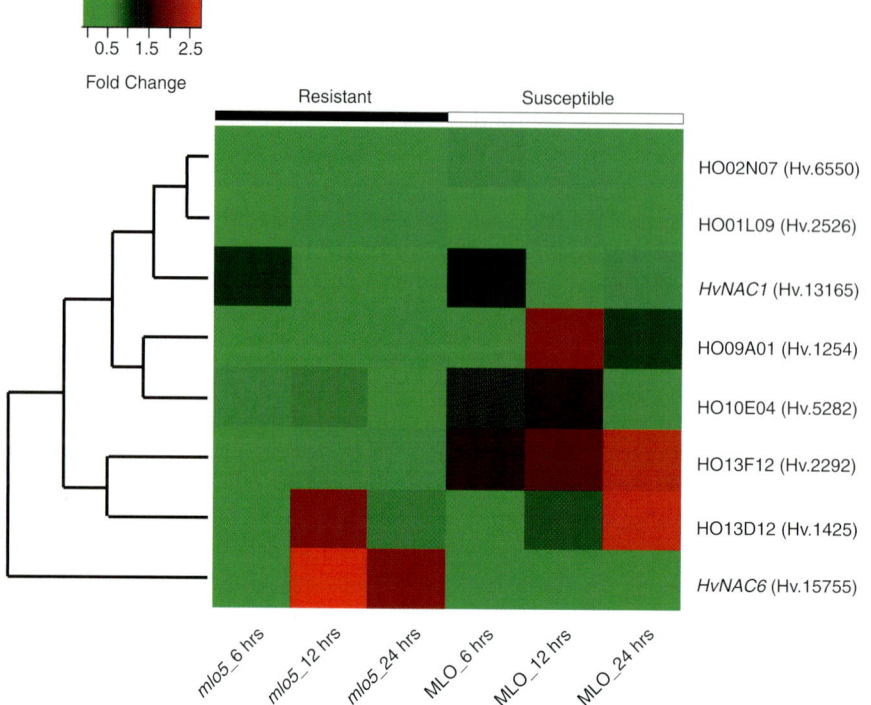

Fig. 1. Transcript accumulation of eight barley NAC genes and UniGene members upon *Bgh* inoculation and modulation of *Mlo*. An analysis of the epidermis-specific *Bgh*-inoculated cDNA library spotted onto nylon membranes (Zierold et al. 2005) reveals eight NAC encoding transcripts. The *bottom panel* displays the experimental conditions; genotype and hours after *Bgh* inoculation. Gene names and UniGene cluster gene-oriented names are given to the *right*. The *colours* refer to mean ratios of gene centred signal intensities of inoculated samples versus corresponding control samples (Eisen et al. 1998). Hierarchical clustering was performed using unscaled correlation and complete linkage clustering. *Colour key* displays correlation between colour and fold changes of *Bgh*-inoculated vs. control samples (non-inoculated)

positively regulating the formation of papilla and effective penetration resistance (Jensen et al. 2007).

To summarize, the wealth of data deposited in publicly available repositories provide a cost-effective tool for 'desk-top-to-bench-top' analyses of tran-

scripts of interest (Fig. 3). Though data should be thoroughly inspected with respect to their origin and relevance to the research in question, it can accommodate new hypothesis to be tested in the laboratory or field. In the case of barley NAC transcription

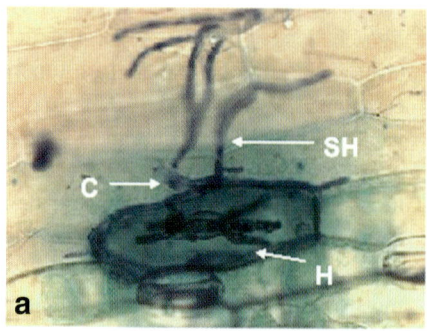
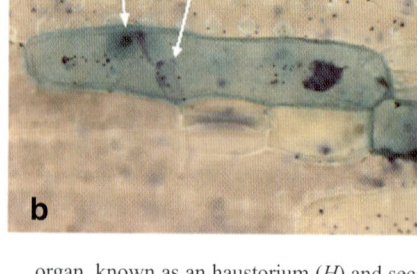

Fig. 2. Barley epidermal single-cell interaction outcomes with *Bgh* provides a well-established system for transient expression studies of genes of interest using GUS as a transformation control. (**a**) Susceptibility. An epidermal cell penetrated by a *Bgh* conidium (*C*) and subsequent development a feeding organ, known as an haustorium (*H*) and secondary hyphae (*SH*) elongation. (**b**) Race-non-specific resistance. A penetration resistant epidermal cell showing race non-specific resistance towards *Bgh* penetration attempts by formation of a papilla (*P*)

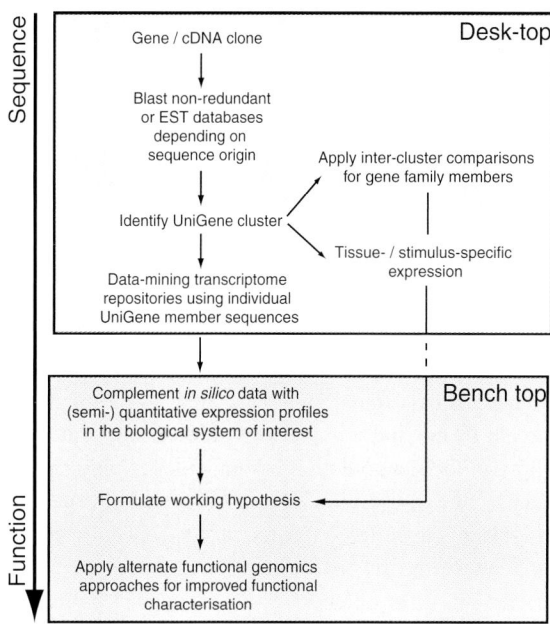

Fig. 3 Diagram showing a cost-effective strategy for the generation of working hypotheses aiming at functional characterisation of isolated genes or gene products

factors, interesting hypothesis have been tested and verified, partially based on publicly available EST resources and relatively simple analyses of expression platforms and reverse genetics approaches.

The role of individual genes in defence

Much of the effort to understand the defence mechanisms of plants concerns the identification of components of defences rather than understanding the role of the individual genes expressed in the defence response. It is abundantly clear from the literature that mutational approaches aimed at identifying genes necessary for disease resistance rarely lead to the identification of defence genes per se, i.e. those encoding antimicrobial proteins or enzymes involved in biosynthesis of antimicrobial phytoalexins (Field et al. 2006; Hammerschmidt 1999; van Loon et al. 2006). Instead, mutations affecting resistance are generally in genes involved in the regulation of defence mechanisms, including race-specific resistance genes themselves, and are often associated with signal transduction pathways (Glazebrook 2005; Panstruga and Schulze-Lefert 2002; Takken et al. 2008). They therefore fall outside the subject of this review. The lack of mutants in defence genes which

exhibit compromised resistance implies that individual components of the defence response have an incremental, rather than determinative, role on the outcome of an interaction with a pathogen. The approach which to date has given the most extensive data set for understanding the impact that individual genes encoding components of defence mechanisms have on resistance to *Bgh*, is the use of transient RNA interference, itself a defence mechanism which operates against viruses (Lindbo and Dougherty 2005; MacDiarmid 2005). An inverse repeat RNA structure is produced in the host cell (Waterhouse et al. 2001). This RNA folds to make a double-stranded RNA molecule which stimulates the host cell's defence against viruses with the result that both extraneous and endogenous copies of the transcript are essentially eliminated, and, as a consequence, the host gene product is no longer produced. This can be achieved by direct particle bombardment using plasmid constructs which contain an inverted repeat of the sequence of interest, or by infecting with a virus containing the sequence, termed virus-induced gene silencing (VIGS). Most studies have used the former approach (Christensen et al. 2004; Douchkov et al. 2005; Jensen et al. 2007; Schweizer et al. 2000; Shen et al. 2007), and this includes the use of high-throughput technologies for studying large groups of genes (Douchkov et al. 2005). To date, few studies have used VIGS successfully to investigate the role of individual defence and disease resistance genes in the barley–*Bgh* interaction (Hein et al. 2005; Shen et al. 2007). A common feature of the results obtained is that the effect of silencing is usually partial. This is in accordance with results obtained with mutational studies where mutations in regulators provide the major phenotypic effects and no defence-related genes per se have been identified (Glazebrook 2005).

Concluding remarks

In this review, we have demonstrated the use of the genomics resources available in barley, a species where the genomic sequence is not yet available, for the identification and validation of the roles of individual genes involved in the regulation of, or encoding components of, specific defence mechanisms, and have illustrated the process with the NAC transcription factor HvNAC6. *Arabidopsis*, poplar and rice have

the advantage of full sequenced genomes and many tools have been developed which are not available in barley. For example, T-DNA insertion lines (i.e. tagged mutant collections) are available for many *Arabidopsis* genes, as are whole genome (TILLING) arrays. Furthermore, well-characterised mutants are available in barley, and several alleles of key (regulatory) genes have been described in *Arabidopsis*. A large body of knowledge about genes and encoded proteins and regulatory networks have been elucidated. Nevertheless, there is a wealth of well-documented natural genetic variation in barley, especially with respect to disease resistance, and this material is supplemented with many good mutants (Jørgensen 1994; Panstruga and Schulze-Lefert 2002).

Although microarrays are available for many species, it is a major undertaking to design and prepare microarrays carrying a significant part of a genome. What do you do if there is no microarray in your biological system? In these cases, the best approach is to use more classical gene discovery techniques such as the use of subtractive libraries (van den Berg et al. 2004) or differential display techniques, such as cDNA-amplified fragment length polymorphism (Liang and Pardee 1997; Ramonell and Somerville 2002). An alternative, though more costly approach would be to prepare a customized array, designed from the cDNA library of interest, for example from subtractive libraries, reflecting the transcriptome of the interaction of interest or for specific families of genes. One technology uses the Affymetrix GeneChip® platform, for example, through the collaboration between NimbleGen Systems and Affymetrix. Another very promising platform is the Agilent® multiplex custom arrays, which brings down the price for 15,000 features to roughly €100. Moreover laboratories worldwide are set up to make dotted oligonucleotide arrays, which may be a cheaper option. However, with a price tag of up to €800 per replicate, depending on platform, the microarray approach is still prohibitively expensive for many laboratories which instead make a more limited array experiment and support this with more detailed expression studies of individual candidate genes using northern blotting or quantitative real-time PCR.

Finally, for what can the knowledge gained from these studies be used for in the context of developing sustainable agriculture? The answer is currently not much–yet! Studies made to date have made it clear

that the idea of taking an antimicrobial protein and using it to make a transgenic plant which has gained effective disease resistance is now largely discredited as the effect observed is at best, partial resistance (Collinge et al. 2008). This, however, is in itself valuable knowledge. A second achievement from the study of defence mechanisms is the realisation that the regulation of plant defence mechanisms is more complex than previously imagined. It is now becoming clear that antagonistic regulatory and interlinked signalling pathways are involved in both biotic and abiotic stress signalling. This may mean that the dream of making a universally disease resistant plant through the manipulation of single genes may remain that for the foreseeable future. The real challenge, and therefore our efforts, needs to be concentrated on understanding the nature of the regulatory networks underlying host defence responses against biotic and abiotic stress and through this be able to manipulate them to achieve resistance.

Acknowledgements   Authors would like to thank Drs. Patrick Schweizer for fruitful discussions and access to data. MKJ was supported by a Ph.D. scholarship from the University of Copenhagen, Faculty of Life Sciences (formerly the Royal Veterinary and Agricultural University) and research financed by a Danish Research Council grant 'Cell specific analysis of host plant responses to pathogens using a functional genomic approach' SJVF 23-03-0167 (to MFL and DBC). This paper is based on two oral contributions held at the EFPP conference held in Copenhagen in August 2006.

References

Aida, M., Ishida, T., Fukaki, H., Fujisawa, H., & Tasaka, M. (1997). Genes involved in organ separation in *Arabidopsis*: An analysis of the cup-shaped cotyledon mutant. *The Plant Cell, 9*, 841–857.

Boddu, J., Cho, S. G., Kruger, W. M., & Muehlbauer, G. J. (2006). Transcriptome analysis of the barley–*Fusarium graminearum* interaction. *Molecular Plant–Microbe Interactions, 19*, 407–417.

Boguski, M. S., & Schuler, G. D. (1995). ESTablishing a human transcript map. *Nature Genetics, 10*, 369–371.

Caldo, R. A., Nettleton, D., Peng, J., & Wise, R. P. (2006). Stage-specific suppression of basal defense discriminates barley plants containing fast- and delayed-acting Mla powdery mildew resistance alleles. *Molecular Plant–Microbe Interactions, 19*, 939–947.

Caldo, R. A., Nettleton, D., & Wise, R. P. (2004). Interaction-dependent gene expression in Mla-specified response to barley powdery mildew. *The Plant Cell, 16*, 2514–2528.

Christensen, A. B., Thordal-Christensen, H., Zimmermann, G., Gjetting, T., Lyngkjær, M. F., Dudler, R., et al. (2004). The

germinlike protein GLP4 exhibits superoxide dismutase activity and is an important component of quantitative resistance in wheat and barley. *Molecular Plant–Microbe Interactions, 17*, 109–117.

Close, T. J., Wanamaker, S. I., Caldo, R. A., Turner, S. M., Ashlock, D. A., Dickerson, J. A., et al. (2004). A new resource for cereal genomics: 22K barley GeneChip comes of age. *Plant Physiology, 134*, 960–968.

Collinge, D. B., Gregersen, P. L., & Thordal-Christensen, H. (2002). The nature and role of defence response genes in cereals. In *The powdery mildews: A comprehensive treatise* (pp. 146–160). St. Paul: APS.

Collinge, D. B., Lund O. S., Thordal-Christensen, H. (2008) What are the prospects for genetically engineered, disease resistant plants? *European Journal of Plant Pathology*, (in press).

Douchkov, D., Nowara, D., Zierold, U., & Schweizer, P. (2005). A high-throughput gene-silencing system for the functional assessment of defense-related genes in barley epidermal cells. *Molecular Plant–Microbe Interactions, 18*, 755–761.

Eichmann, R., Biemelt, S., Schäfer, P., Scholz, U., Jansen, C., Felk, A., et al. (2006). Macroarray expression analysis of barley susceptibility and nonhost resistance to *Blumeria graminis. Journal of Plant Physiology, 163*, 657–670.

Eisen, M. B., Spellman, P. T., Brown, P. O., & Botstein, D. (1998). Cluster analysis and display of genome-wide expression patterns. *Proceedings of the National Academy of Sciences of the United States of America, 95*, 14863–14868.

Field, B., Jordan, F., & Osbourn, A. (2006). First encounters–Deployment of defence-related natural products by plants. *New Phytologist, 172*, 193–207.

Gjetting, T., Hagedorn, P. H., Schweizer, P., Thordal-Christensen, H., Carver, T. L. W., & Lyngkjær, M. F. (2007). Single-cell transcript profiling of barley attacked by the powdery mildew fungus. *Molecular Plant–Microbe Interactions, 20*, 235–246.

Glazebrook, J. (2005). Contrasting mechanisms of defense against biotrophic and necrotrophic pathogens. *Annual Review of Phytopathology, 43*, 205–227.

Gregersen, P. L., & Collinge, D. B. (2001). Penetration attempts by the powdery mildew fungus into barley leaves are accompanied by increased gene transcript accumulation in the epidermal cell layer. *Journal of Plant Pathology, 83*, 257–260.

Gregersen, P. L., Collinge, D. B., & Smedegaard-Petersen, V. (1990). Early induction of new mRNAs accompanies the resistance reaction of barley to the wheat pathogen, *Erysiphe graminis* f.sp. *tritici. Physiological and Molecular Plant Pathology, 36*, 471–481.

Gregersen, P. L., Thordal-Christensen, H., Forster, H., & Collinge, D. B. (1997). Differential gene transcript accumulation in barley leaf epidermis and mesophyll in response to attack by *Blumeria graminis* f.sp. *hordei* (syn. *Erysiphe graminis* f.sp. *hordei*). *Physiological and Molecular Plant Pathology, 51*, 85–97.

Hammerschmidt, R. (1999). Phytoalexins: What have we learned after 60 years? *Annual Review of Phytopathology, 37*, 285–306.

Hein, I., Campbell, E. I., Woodhead, M., Hedley, P. E., Young, V., Morris, W., et al. (2004). Characterisation of early transcriptional changes involving multiple signalling pathways in the Mla13 barley interaction with powdery mildew (Blumeria graminis f.sp. hordei). *Planta, 218*, 803–813.

Hein, I., Barciszewska-Pacak, M., Hrubikova, K., Williamson, S., Dinesen, M., Soenderby, I. E., et al. (2005). Virus-induced gene silencing-based functional characterization of genes associated with powdery mildew resistance in barley. *Plant Physiology, 138*, 2155–2164.

Hu, H., Dai, M., Yao, J., Xiao, B., Li, X., Zhang, Q., et al. (2006). Overexpressing a NAM, ATAF, and CUC (NAC) transcription factor enhances drought resistance and salt tolerance in rice. *Proceedings of the National Academy of Sciences of the United States of America, 103*, 12987–12992.

Jabs, T., & Slusarenko, A. (2000). The hypersensitive response. In A. Slusarenko, R. S. S. Fraser, & L. C. Loon (Eds.) *Mechanisms of resistance to plant diseases* (pp. 279–323). Dordrecht: Kluwer.

Jensen, M. K., Rung, J. H., Gregersen, P. L., Gjetting, T., Fuglsang, A. T., Hansen, M., et al. (2007). The HvNAC6 transcription factor: A positive regulator of penetration resistance in barley and *Arabidopsis. Plant Molecular Biology, 65*, 137–150.

Jørgensen, J. H. (1992). Discovery, characterization and exploitation of Mlo powdery mildew resistance in barley. *Euphytica, 63*, 141–152.

Jørgensen, J. H. (1994). Genetics of powdery mildew resistance in barley. *Critical Reviews in Plant Sciences, 13*, 97–119.

Kølster, P., Munk, L., Stølen, O., & Løhde, J. (1986). Near-isogenic barley lines with genes for resistance to powdery mildew. *Crop Science, 26*, 903–907.

Liang, P., & Pardee, A. B. (1997). Differential display. A general protocol. *Methods Molecular Biology, 85*, 3–11.

Lindbo, J. A., & Dougherty, W. G. (2005). Plant pathology and RNAi: A brief history. *Annual Review of Phytopathology, 43*, 191–204.

Lu, P. L., Chen, N. Z., An, R., Su, Z., Qi, B. S., Ren, F., et al. (2006). A novel drought-inducible gene, ATAF1, encodes a NAC family protein that negatively regulates the expression of stress-responsive genes in *Arabidopsis. Plant Molecular Biology, 63*, 289–305.

MacDiarmid, R. (2005). RNA silencing in productive virus infections. *Annual Review of Phytopathology, 43*, 523–544.

Manners, J. M., & Scott, K. J. (1985). Reduced translatable messenger RNA activities in leaves of barley infected with *Erysiphe graminis* f.sp. *hordei. Physiological Plant Pathology, 26*, 297–308.

Mejlhede, N., Kyjovska, Z., Backes, G., Burhenne, K., Rasmussen, S. K., & Jahoor, A. (2006). EcoTILLING for the identification of allelic variation in the powdery mildew resistance genes mlo and Mla of barley. *Plant Breeding, 125*, 461–467.

Mörschbacher, B., & Mendgen, K. W. (2000). Structural aspects of defense. In A. J. Slusarenko, R. S. S. Fraser, & L. C. van Loon (Eds.) *Mechanisms of resistance to plant diseases* (pp. 231–227). Dordrecht: Kluwer.

Panstruga, R., & Schulze-Lefert, P. (2002). Live and let live: Insights into powdery mildew disease and resistance. *Molecular Plant Pathology, 3*, 495–502.

Ramonell, K. M., & Somerville, S. (2002). The genomics parade of defense responses: To infinity and beyond. *Current Opinion in Plant Biology, 5*, 291–294.

Rostoks, N., Borewitz, J., Hedley, P. E., Russell, J., Mudie, S., Morris, J., et al. (2005). Single-feature polymorphism discovery in the barley transcriptome. *Genome Biology, 6*, R54.

 Springer

Schweizer, P., Pokorny, J., Schulze-Lefert, P., & Dudler, R. (2000). Double-stranded RNA interferes with gene function at the single-cell level in cereals. *The Plant Journal*, *24*, 895–903.

Shen, L. H., Gong, J., Caldo, R. A., Nettleton, D., Cook, D., Wise, R. P., et al. (2005). BarleyBase–An expression profiling database for plant genomics. *Nucleic Acids Research*, *33*, D614–D618.

Shen, Q. H., Saijo, Y., Mauch, S., Biskup, C., Bieri, S., Keller, B., et al. (2007). Nuclear activity of MLA immune receptors links isolate-specific and basal disease-resistance responses. *Science*, *315*, 1098–1103.

Shirasu, K., Nielsen, K., Piffanelli, P., Oliver, R., & Schulze-Lefert, P. (1999). Cell-autonomous complementation of mlo resistance using a biolistic transient expression system. *The Plant Journal*, *17*, 293–299.

Souer, E., van Houwelingen, A., Kloos, D., Mol, J., & Koes, R. (1996). The no apical meristem gene of Petunia is required for pattern formation in embryos and flowers and is expressed at meristem and primordia boundaries. *Cell*, *85*, 159–170.

Svensson, J. T., Crosatti, C., Campoli, C., Bassi, R., Stanca, A. M., Close, T. J., et al. (2006). Transcriptome analysis of cold acclimation in barley albina and xantha mutants. *Plant Physiology*, *141*, 257–270.

Takken, F. L. W., Tameling, W. I. L., & Joosten, M. H. A. J. (2008). Molecular basis of plant disease/resistance and disease resistance proteins. *European Journal of Plant Pathology*, (in press).

Trujillo, M., Troeger, M., Niks, R. E., Kogel, K.-H., & Hückelhoven, R. (2004). Mechanistic and genetic overlap of barley host and non-host resistance to *Blumeria graminis*. *Molecular Plant Pathology*, *5*, 389–396.

van den Berg, N., Crampton, B. G., Hein, I., Birch, P. R. J., & Berger, D. K. (2004). High-throughput screening of suppression subtractive hybridization cDNA libraries using DNA microarray analysis. *Biotechniques*, *37*, 818–824.

van Loon, L. C., Rep, M., & Pieterse, C. M. J. (2006). Significance of inducible defense-related proteins in infected plants. *Annual Review of Phytopathology*, *44*, 135–162.

Waterhouse, P. M., Wang, M. B., & Lough, T. (2001). Gene silencing as an adaptive defence against viruses. *Nature*, *411*, 834–842.

Zhang, H. N., Sreenivasulu, N., Weschke, W., Stein, N., Rudd, S., Radchuk, V., et al. (2004). Large-scale analysis of the barley transcriptome based on expressed sequence tags. *Plant Journal*, *40*, 276–290.

Zierold, U., Scholz, U., & Schweizer, P. (2005). Transcriptome analysis of mlo-mediated resistance in the epidermis of barley. *Molecular Plant Pathology*, *6*, 139–151.

Zimmerli, L., Stein, M., Lipka, V., Schulze-Lefert, P., & Somerville, S. (2004). Host and non-host pathogens elicit different jasmonate/ethylene responses in *Arabidopsis*. *Plant Journal*, *40*, 633–646.

Eur J Plant Pathol (2008) 121:267–280
DOI 10.1007/s10658-008-9302-5

Roles of reactive oxygen species in interactions between plants and pathogens

Nandini P. Shetty · Hans J. Lyngs Jørgensen · Jens Due Jensen · David B. Collinge · H. Shekar Shetty

Received: 23 May 2007 / Accepted: 3 March 2008
© KNPV 2008

Abstract The production of reactive oxygen species (ROS) by the consumption of molecular oxygen during host–pathogen interactions is termed the oxidative burst. The most important ROS are singlet oxygen ($^1O_2$), the hydroxyperoxyl radical ($HO_2\cdot$), the superoxide anion $\left(O_2^-\right)$, hydrogen peroxide (H_2O_2), the hydroxyl radical (OH^-) and the closely related reactive nitrogen species, nitric oxide (NO). These ROS are highly reactive, and therefore toxic, and participate in several important processes related to defence and infection. Furthermore, ROS also play important roles in plant biology both as toxic by-products of aerobic metabolism and as key regulators of growth, development and defence pathways. In this review, we will assess the different roles of ROS in host–pathogen interactions with special emphasis on fungal and Oomycete pathogens.

Keywords Antimicrobial · Cell wall cross-linking · Hypersensitive response · Signal transduction · Gene expression · Successful pathogenesis · Hydrogen peroxide

Abbreviations
ROS   reactive oxygen species
SA   salicylic acid
ET   ethylene
MAPK   mitogen-activated protein kinase
SOD   superoxide dismutase
CWA   cell wall appositions
NO   nitric oxide
JA   jasmonic acid
HR   hypersensitive response
PCD   programmed cell death

N. P. Shetty (✉) · H. J. L. Jørgensen · J. D. Jensen ·
D. B. Collinge
Department of Plant Biology, Faculty of Life Sciences,
University of Copenhagen,
Thorvaldsensvej 40,
DK-1871 Frederiksberg, Denmark
e-mail: nps@life.ku.dk

H. S. Shetty
Department of Studies in Applied Botany,
Seed Pathology and Biotechnology, University of Mysore,
Manasagangotri,
Mysore 570 006, India

Introduction

Generation of ROS, especially hydrogen peroxide (H_2O_2), has been recorded in interactions with a variety of pathogens (Mellersh et al. 2002; Shetty et al. 2003; Thordal-Christensen et al. 1997; Unger et al. 2005). Avirulent pathogens often induce a biphasic ROS accumulation with a small, transient first phase, followed by a continuous phase of much higher intensity that correlates with disease resistance (Lamb and Dixon 1997; Torres et al. 2006). However, three phases of ROS accumulation have been observed in

some cases, e.g., for *Blumeria graminis* f. sp. *hordei* infecting barley (Hückelhoven and Kogel 2003) and *Septoria tritici* infecting wheat (Shetty et al. 2003). These differences can be attributed to the more complicated development of these fungal pathogens and the influence of the host genotype, which presumably determine whether two or three phases of ROS accumulation occur. Virulent pathogens that avoid or suppress host recognition induce only the transient, first phase of this response (Bolwell et al. 2002). Elicitors of defence responses, often now referred to as microbe or pathogen-associated molecular patterns (PAMPs), also trigger an oxidative burst (Chisholm et al. 2006). There are several potential sources of ROS in plants and different sources of ROS may be activated within a species in different situations depending on the type of stress (Bolwell et al. 2002; Lamb and Dixon 1997). A variety of enzyme systems have been implicated in ROS generation following pathogen recognition, i.e., reduced form of nicotinamide adenine dinucleotide phosphate (NADPH) oxidase (Bedard et al. 2007; Carter et al. 2007; Grant et al. 2000), SOD (Auh and Murphy 1995; Deepak et al. 2006), oxalate oxidases (Hu et al. 2003; Zimmermann et al. 2006), peroxidases (Bindschedler et al. 2006; Bolwell et al. 2002), lipoxygenases (Babitha et al. 2004) and amine oxidases (Allan and Fluhr 1997; Cona et al. 2006; Walters 2003). Stress on of ROS-producing organelles during pathogenesis may also contribute to ROS production during host–pathogen interactions. Mitochondria are normally considered relatively unimportant ROS generators in photosynthesising tissue (Apel and Hirt 2004; Kuźniak and Skłodowska 2005). However, a recent review by Amirsadeghi et al. (2007) discusses evidence that mitochondria are a potential source of ROS in response to biotic stress. Chloroplasts (Kariola et al. 2005) and peroxisomes (Kuźniak and Skłodowska 2005) have also been shown to be important. In this review, we present recent knowledge on the roles of ROS during host–pathogen interactions with special emphasis on fungal and Oomycete pathogens.

Roles of ROS in host–pathogen interactions

ROS have been implicated in many different processes related to pathogen interactions with their hosts. In the initial phases of the interactions, this essentially means involvement in defence processes, whereas at the later stages, during pathogen colonisation, the role of ROS may be more ambiguous.

ROS as antimicrobial agents

ROS, especially H_2O_2, was suggested as an antimicrobial agent during the plant defence response (Apostol et al. 1989; Custers et al. 2004; Legendre et al. 1993; Walters 2003). However, the actual toxicity of ROS in a given plant–pathogen interaction will depend on the sensitivity of the pathogen to the concentration of ROS present (Levine et al. 1994). The amount of extracellular H_2O_2 produced depends on several factors including the nature of the elicitor, the plant species, and age or developmental stages of the plant cells (Legendre et al. 1993; Małolepsza 2005; Nurnberger et al. 1994). Micromolar concentrations of H_2O_2 inhibited spore germination of a number of fungal pathogens *in vitro* (Peng and Kuc 1992). Thus, a concentration of 0.1 mM H_2O_2 completely inhibited the growth of cultured bacteria *Pectobacterium carotovorum* subsp. *carotovorum* (formerly *Erwinia carotovora* pv. *carotovora*), and resulted in >95% inhibition of *Phytophthora infestans* growth (Wu et al. 1995). Shetty et al. (2007) demonstrated by in vitro experiments that 5 mM H_2O_2 inhibited the development of inoculum from 4 day-old *S. tritici* cultures whereas a concentration of about 50 mM was required to inhibit inoculum from 16 day-old cultures. This reflects the ability of the pathogen to tolerate H_2O_2 during the different stages of its life-cycle. Shetty et al. (2007) also demonstrated that in the wheat–*S. tritici* interaction, infiltration of 4 mM H_2O_2 into a susceptible cultivar made it more resistant, symptoms appearing 6 days later than in control plants, whereas infiltration of catalase resulted in symptoms appearing 4 days earlier. It is currently not known whether the effect of H_2O_2 was direct, i.e., by toxicity of ROS, or indirect by affecting signal transduction or defence gene expression.

It is difficult to determine which H_2O_2 concentrations actually inhibit pathogens in planta since the necessary manipulation of the host tissue may itself trigger the production of ROS and/or antioxidants. However, ROS are also toxic to the plant. Thus, soybean suspension-cultured cells remained viable with up to 4 mM H_2O_2, whereas slightly higher levels (6–8 mM) resulted in extensive cell death (Levine et

al. 1994). Therefore, ROS accumulation is tightly regulated by the plant to avoid high concentrations, which could damage the plant tissue (Torres et al. 2006).

Involvement of ROS in signal transduction and gene expression

ROS are involved in different signalling pathways for defence mechanisms, such as triggering of the HR, accumulation of phytoalexins and a number of other defence-response genes (cf. Fig. 1). It has been suggested that ROS are sensed by plants via three mechanisms (Mittler et al. 2004): unidentified receptor proteins; redox-sensitive transcription factors such as natriuretic peptide receptor 1 (NPR1) or heat-shock transcription factors (HSFs); and direct inhibition of phosphatase (Apel and Hirt 2004; Mittler et al. 2004; Neill et al. 2002). ROS signalling is the subject of intense studies, but the role of ROS in signalling is poorly understood. Here, we will give an overview of the initial events involved in signalling in plant–pathogen interactions.

Protein phosphorylation, changes in ion fluxes and the oxidative burst, leading to either HR or defence gene expression, or both, are important events taking place after pathogen infection (Chandra et al. 1996; Jabs et al. 1997; Lamb and Dixon 1997; Sasabe et al. 2000). The earliest reactions of plant cells include changes in plasma membrane permeability, which leads to Ca^{2+} and proton influx and K^+ and Cl^- efflux (McDowell and Dangl 2000). Ion fluxes subsequently induce extracellular production of ROS catalysed by

enzymes that act as secondary messengers for the HR and defence gene expression (Lamb and Dixon 1997). Calcium has been shown to be important in signalling. Heteromeric guanosine triphosphate (GTP)-binding proteins and protein phosphorylation/dephosphorylation events are probably involved in transferring the signals from the receptor to calcium channels that activate downstream processes (Legendre et al. 1992). Furthermore, elevation of the cytosolic calcium concentration has been shown to occur during most biotic and abiotic stresses (Price et al. 1996). For example, oxidative stress increased the cytosolic calcium concentration in tobacco (Price et al. 1994) and H_2O_2 induced calcium influx-mediated stomatal closure in *Commelina communis* and *Arabidopsis thaliana* (McAinsh et al. 1996; Pei et al. 2000). In further support of the involvement of calcium in signalling, lanthanide ions (calcium channel blocker) inhibited bacterial elicitor-induced ROS production in tobacco (Baker et al. 1993). Moreover, Urquhart et al. (2007) showed that transient expression of the chimeric cyclic nucleotide-gated ion channel gene ATCNGC11/12 in *Nicotiana benthamiana* gave rise to cell death with characteristics of the HR. Furthermore, it was shown that this gene could function as a Ca^{2+}-conducting channel and that calcium ions were important for the observed cell death. Recently, Ashtamker et al. (2007) showed that nuclei isolated from tobacco were capable of producing H_2O_2. This was dependent on calcium, suggesting that nuclei can be a source of ROS production.

Different models for the action of calcium in the regulation of ROS have been proposed. One model

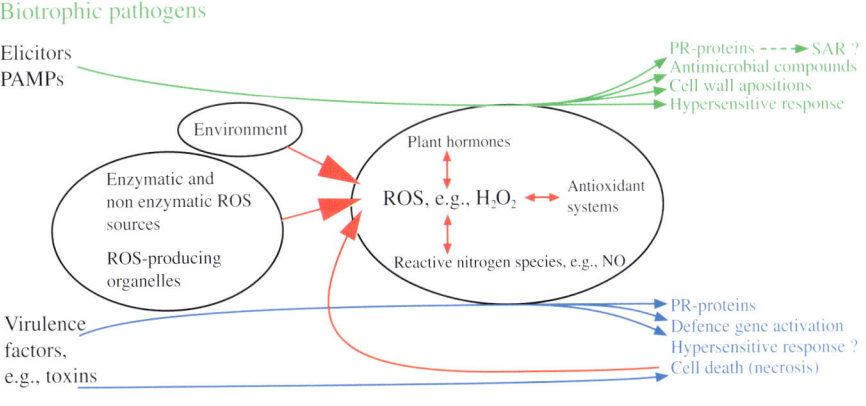

Fig. 1 Putative sources and functions of reactive oxygen species (ROS) in host–pathogen interactions of biotrophic and necrotrophic organisms

suggests that an elicitor interacts with a receptor coupled with a G-protein, which leads to Ca^{2+} influx that activates a Ca^{2+}-dependent protein kinase and ultimately NADPH oxidase (Blumwald et al. 1998). Another model, based on studies of innate immunity in *Arabidopsis*, suggests that pathogens or PAMPs are recognised by (unknown) receptors which trigger an ion (calcium) channel, leading to increases in cytosolic Ca^{2+} and subsequent NO generation (Ali et al. 2007). NO generation, together with other required factors such as an avirulent pathogen and an oxidative burst, could lead to the HR and potentially, diffusion of NO to neighbouring cells could act as a signal that thereby activates further calcium channels.

Activation of the oxidative burst is governed by phosphorylation/dephosphorylation (Lamb and Dixon 1997). Thus, a protein phosphorylation cascade that has been shown to be activated by H_2O_2 is a MAPK cascade, which has an important role in signal transduction (Zwerger and Hirt 2001). H_2O_2 has been shown to activate MAPK in *Arabidopsis* suspension cultures (Desikan et al. 1999). Furthermore, Petersen et al. (2000) showed that mutation of the MAPK gene *MPK4* in *Arabidopsis* altered plant defence activation.

H_2O_2 mediates the transcription of specific genes, though the exact mechanism is as yet unknown. Neill et al. (2002) suggested that it could be due to oxidation of cysteine residues of transcription factors. Activation of MAPKs is a common reaction of plant cells in defence-related signal transduction pathways (Neill et al. 2002). Perception of an extracellular signal activates a MAPK, which in turn can facilitate translocation of the signal to the nucleus where it can phosphorylate and activate transcription factors, thereby modulating gene expression (Apel and Hirt 2004; Hirt 1997; Zhou et al. 2004). For example, it has been reported that two tobacco MAPKs, namely salicylic acid-induced protein kinase (SIPK) and wound-induced protein kinase (WIPK), are regulated by a common upstream MAPK, which is involved in signalling for PCD (Zhang and Klessig 1998; Zhang et al. 2000). Ren et al. (2006) showed that there was another MAPK, Ntf4, with a similar function to SIPK and WIPK, which, when expressed in transgenic tobacco plants, accelerated the PCD when treated with the elicitin cryptogen from *Phytophthora cryptogea*. This indicates a role in signalling for PCD. Recently, Liu et al. (2007) showed that the combined activation of SIPK, Ntf4 and WIPK induced an HR-like PCD.

Activation of signal transduction could lead to increased ROS accumulation and activation of defence genes coding for PR-proteins, enzymes involved in the generation of phytoalexins, enzymes involved in oxidative stress protection, lignification and other defence responses (Alvarez et al. 1998; Apel and Hirt 2004; Lamb and Dixon 1997). Further evidence for a role of ROS in signalling has come from the fact that addition of low doses of ROS inducers stimulates the induction of detoxification mechanisms, such as SOD and glutathione-*S*-transferase, and activation of other defence mechanisms in neighbouring cells (Levine et al. 1994). Mittler et al. (2004) suggested that NADPH oxidase could be involved in ROS signalling by creating a loop where a small enhancement of ROS production and amplification of the ROS signals occurs in specific cellular locations. Pharmacological and genetic studies (Dat et al. 2003) support the existence of positive amplification loops involving NADPH oxidases in ROS signalling. These loops might be activated by low levels of ROS and result in enhanced production and amplification of the ROS signals. It has been reported that a small GTP-binding protein, Rac, regulates ROS production in rice, most likely through an NADPH oxidase, and induces cell death in rice cells with biochemical and morphological features similar to apoptosis in mammalian cells (Kawasaki et al. 1999). Together, MAPK and calcium-dependent protein kinases seem to play central roles in the regulation of pathogen-responsive NADPH oxidases at the transcriptional and post-transcriptional levels, respectively (Kobayashi et al. 2007).

It has been suggested that the HR is triggered only by balanced production of NO and ROS (Delledonne et al. 1998, 2002)-see also below. More specifically, dismutation of O_2^- to H_2O_2 is required to activate cell death, which depends on synergistic interactions between NO, H_2O_2 and SA (Delledonne et al. 1998; Mur et al. 2006). Scavenging of O_2^- by surplus NO (or vice versa) disturbs the NO/H_2O_2 ratio, resulting in reduced cell death (Mur et al. 2006). Little is known about signalling pathways downstream of NO/H_2O_2. Nevertheless, it has been shown that NO signalling during both PCD and defence responses requires cyclic GMP and cyclic ADP ribose, two molecules that can serve as secondary messengers for NO signalling in mammals (Van Breusegem and Dat 2006).

SA has been shown to be an important signalling molecule involved in defence responses to pathogen attack in many plant–pathogen interactions. Thus, Enyedi et al. (1992) showed that SA levels increased dramatically in tobacco cells surrounding infection sites when infected with *Tobacco mosaic virus*. Torres et al. (2006) suggested that ROS acted synergistically in a signal amplification loop with SA to drive the HR and the establishment of systemic defences. SA accumulation can also down-regulate those ROS-scavenging systems that, in turn, can contribute to increased overall ROS levels following pathogen recognition (Klessig et al. 2000; Shah 2003). In addition to SA, ET and JA are also involved in signalling (Thatcher et al. 2005). The activation of a redox-signalling pathway possessing a MAPK module has also been reported in response to infection by avirulent pathogens in *Arabidopsis* (Suzuki 2002). This signalling network functions independently of the plant hormones ET, SA and JA (Thatcher et al. 2005). Additionally, when some mutants, which develop spontaneous lesions mimicking HR cell death, are placed in a *NahG* background to degrade SA (Shah 2003), lesion formation is suppressed but can be restored by SA treatment (Lorrain et al. 2003). However, this is not the case for all lesion-mimic mutants; some show intensified lesions in plants defective in JA signalling, while others have delayed lesion formation in plants defective in ET signalling (Lorrain et al. 2003). These differences in lesion formation can be due to synergistic or antagonistic effects between SA, JA and ET signalling pathways (Lorrain et al. 2003), but also indicates that mutants displaying the same phenotype could be mutated in widely different genes.

Involvement of ROS in oxidative cross-linking of cell walls

Barriers operating at the cell periphery to prevent invasion represent the first line of defence against pathogens that penetrate plant cells directly (Schulze-Lefert 2004). These barriers can, for example, depend on the nature and thickness of the epicuticular wax layer and cuticle or the composition and physical properties of the cell wall. Alternatively, they may occur by reinforcement of the cell wall, e.g., by deposition of callose-rich papillae and lignin at attempted penetration sites (Heitefuss 1997). ROS

production has been associated with the formation of physical defensive barriers (Hückelhoven and Kogel 2003; Lamb and Dixon 1997)-see also Fig. 2b. Association of H_2O_2 with lignification during plant development has been shown in several systems (Olson and Varner 1993; Repka 2002). Thus, H_2O_2 accumulation resulted in lignification in wounded *Zinnia* stem sections (Olson and Varner 1993). Furthermore, Thordal-Christensen et al. (1997) showed that the H_2O_2 production in barley infected with *B. graminis* f.sp. *hordei* led to cell wall cross-linking. Collins et al. (2003) showed that H_2O_2 was associated with vesicles containing cell wall components which were in transit to CWA, suggesting that H_2O_2 may play a role upstream of CWA or that compounds of CWA are oxidatively cross-linked on the way to the site of deposition. These findings have been confirmed by An et al. (2006), who showed that multivesicular bodies, intravacuolar vesicle aggregates and paramural bodies, which might participate in the secretion of building blocks for CWA, are associated with H_2O_2 accumulation. Another conspicuous role of the CWA, besides arresting fungal penetration, is blockage of all plasmodesmata between intact cells and those undergoing the HR, thereby containing the hypersensitive cell death. Also, Iwano et al. (2002) showed that in suspension-cultured rice cells infected with *Acidovorax avenae* (formerly *Pseudomonas avenae*), callose synthesis occurred at the H_2O_2 generation site.

Studies of pearl millet infected with *Sclerospora graminicola* have indicated that cell wall protein cross-linking is induced by enhanced H_2O_2 production at the time of pathogen attack (Kumudini and Shetty 2002). Possibly, hyroxyproline-rich glycoproteins (HRGPs) accumulate and contribute to disease resistance involving cross-linking between HRGP monomers to form a network which provide anchorage for lignification. This might also lead to obstruction of haustorial formation and nutrient shortage, which may be particularly unfavourable for biotrophic pathogens that use specific organs, e.g., haustoria for feeding (Bradley et al. 1992; Shailashree et al. 2004). Likewise, studies on the interaction between wheat and *B. graminis* f.sp. *tritici* showed that H_2O_2 plays important roles in defence, by driving among others the cross-linking to strengthen the cell wall (in effective papillae), and in association with HR (Li et al. 2005).

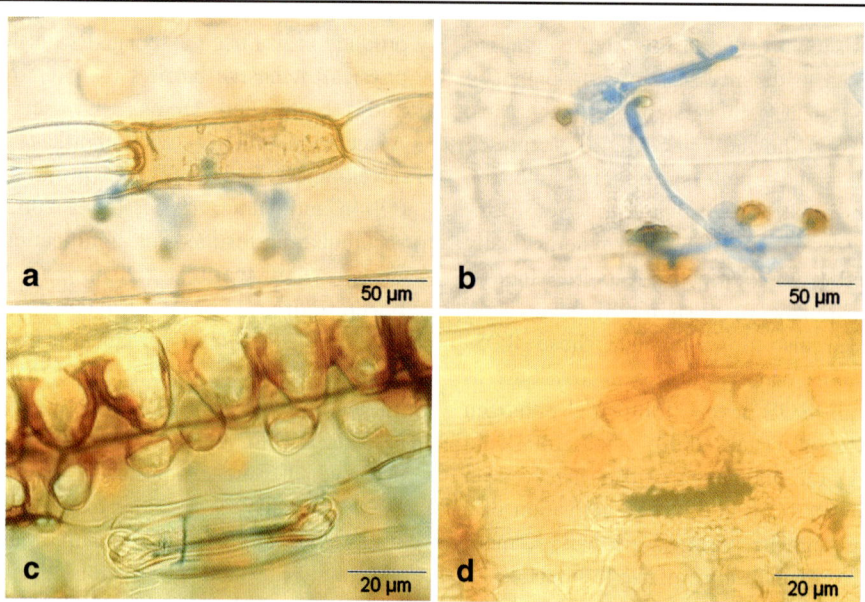

Fig. 2 Accumulation of H$_2$O$_2$ as seen by DAB-staining (Thordal-Christensen et al. 1997) in the barley–*B. graminis* f. sp. *hordei* interaction (**a**, **b**) and the wheat–*S. tritici* interaction (**c**, **d**). **a** shows barley isoline P-01 inoculated with isolate c15. A cell is undergoing HR as a response to penetration and is completely stained with DAB. **b** shows barley isoline P-02 inoculated with isolate c15, 2 days after inoculation. Note red–brown staining in the papillae and that the papilla in the cell containing an haustorium is not stained with DAB whereas the other papillae are stained. **c** shows wheat cv. Stakado inoculated with isolate IPO323 of *S. tritici* (incompatible interaction) at 5 days after inoculation. H$_2$O$_2$ is accumulating in the apoplast of the substomatal cavity of a stoma penetrated by the pathogen. **d** shows wheat cv. Sevin inoculated with isolate IPO323 of *S. tritici* (compatible interaction) at 15 days after inoculation. H$_2$O$_2$ is accumulating throughout the tissue in which fungal sporulation occurs

Involvement of ROS in the hypersensitive response (HR)

The HR is a rapid host response occurring in a host cell, which is infected by a pathogen (Lam et al. 2001; Lam 2004). The cells die shortly after penetration Fig. 2a, often together with some of the surrounding cells (Greenberg 1997; Van Breusegem and Dat 2006). The HR occurs in order to restrict pathogen growth and is highly effective against biotrophic pathogens, since, with the death of host cells, the nutrient supply is removed (Greenberg and Yao 2004; Mellersh et al. 2002; Thordal-Christensen et al. 1997). In addition, toxic substances like ROS and phytoalexins produced in these cells apparently help to kill the pathogen (Lamb and Dixon 1997; Li et al. 2006; Moerschbacher and Reisener 1997). ROS may originate primarily from chloroplasts, mitochondria and peroxisomes (Amirsadeghi et al. 2007; Lam 2004; Op den Camp et al. 2003; Van Breusegem and Dat 2006). The HR is often not effective against necrotrophic pathogens because these usually kill host cells to feed on them (Govrin and Levine 2000; Mayer et al.

2001). Thus, for true necrotrophic pathogens, such as *Botryotinia fuckeliana* (formerly *Botrytis cinerea*), it has been suggested that plant cell death is beneficial for infection, leading to enhanced colonisation (Govrin and Levine 2000; Greenberg and Yao 2004). However, the general nature of this conclusion has been questioned even for *B. fuckeliana* (Unger et al. 2005). In addition, there is a group of pathogens, often considered to be necrotrophic, which are in fact inhibited to some extent by HR, e.g., *Pyrenophora teres* (anamorph *Drechslera teres*; Jørgensen et al. 1998) and *Magnaporthe grisea* (Iwai et al. 2007). Collectively, these findings raise the question whether these pathogens are, in fact, necrotrophic, or should be considered as hemibiotrophic or whether some necrotrophic pathogens may also be inhibited by HR under some circumstances. In our view, the term 'necrotrophic' includes a diverse group of pathogens with quite different modes of pathogenicity.

The HR is a type of active PCD (Greenberg and Yao 2004; Lam 2004; Li et al. 2006; Sasabe et al. 2000; Van Breusegem and Dat 2006), which is often characterised by discrete cellular lesions and preceded

by an oxidative burst (Baker and Orlandi 1995; Dat et al. 2003; Levine et al. 1994; Sasabe et al. 2000). The process of HR may involve several steps including chromatin condensation, DNA cleavage and membrane blebbing, eventually leading to membrane disruption and release of cell contents (Dat et al. 2003; Hoeberichts and Woltering 2003; Lam 2004; Li et al. 2006; Sasabe et al. 2000;). The cell death process thus shares some features with mammalian apoptosis (Greenberg and Yao 2004; Hoeberichts et al. 2003; Hoeberichts and Woltering 2003; Lam 2004). Further similarities include a group of proteins, termed metacaspases, in the genome of *Arabidopsis* with homology to the specific cysteine proteases, termed caspases, which play a key role in execution of mammalian apoptosis (Hoeberichts et al. 2003).

Involvement of ROS in HR has been studied by several different tools, including infiltration of antioxidants (Li et al. 2006) and ROS inhibitors or scavengers (Levine et al. 1994; Li et al. 2006; Sasabe et al. 2000). Furthermore, specific lines of *Arabidopsis* and other plants mutated in their ability to accumulate ROS or express antioxidants (such as SOD, catalase and ascorbate peroxidase) and subsequently activate HR have been studied (Dat et al. 2003; Hoeberichts and Woltering 2003; Jabs et al. 1996; Lorrain et al. 2003; Mateo et al. 2004; Mittler et al. 1999; Montillet et al. 2005; Op den Camp et al. 2003; Torres et al. 2005; Van Breusegem and Dat 2006). Also ROS accumulation has been studied following treatment with pathogen elicitors of HR (Greenberg and Yao 2004; Levine et al. 1994; Montillet et al. 2005; Sasabe et al. 2000). The use of such a diversity of approaches and systems strengthens and substantiates our knowledge on the role of ROS in HR, since broad background information is obtained. However, it also reveals that this process is highly complex, and not yet understood in detail (Van Breusegem and Dat 2006). For example, several studies have shown a correlation between accumulation of ROS (H_2O_2, $^1O_2$, O_2^-), NO and HR (Dat et al. 2003; Floryszak-Wieczorek et al. 2007; Jabs et al. 1996; Levine et al. 1994; Mittler et al. 1999; Montillet et al. 2005; Op den Camp et al. 2003). In this respect, NADPH oxidase has been found to be an important generator of ROS (Lamb and Dixon 1997; Li et al. 2006; Torres et al. 2002). On the other hand, lack of correlation has also been reported in some cases (Dorey et al. 1999; Glazner et

al. 1996; Repka 2002; Torres et al. 2005). This discrepancy in results illustrate that, although there is overwhelming evidence that ROS accumulation plays a central role for the HR, we do not yet understand the process or processes in detail and different pathways may operate in different systems or under different conditions. Elucidation of the causal relation between ROS and HR is further complicated by the fact that, for example, plant hormones such as SA, JA, ET and abscisic acid also influence the elicitation and expression of HR (Hoeberichts and Woltering 2003; Torres et al. 2005; Van Breusegem and Dat 2006).

Different models have been proposed to explain how ROS (H_2O_2, O_2^-, $^1O_2^-$) and NO may elicit and regulate HR (e.g., Delledonne et al. 2002; Torres et al. 2005; Van Breusegem and Dat 2006). Thus, when plants are subjected to stress and ROS accumulate at levels insufficient to kill the cell (as opposed to necrosis which is passive, accidental cell death), signalling events lead to PCD represented by HR. Initiation of HR may further lead to activation of other defence responses and systemic acquired resistance (Greenberg 1997; Van Breusegem and Dat 2006).

The exact role and mechanism of ROS in elicitation of the HR remains unclear. For example, it has been suggested from studies of soybean suspension-cultured cells that elicitation of HR requires tightly balanced production of H_2O_2 and NO where NO reacts with H_2O_2 (generated from O_2^- by SOD) and elicits the HR (Delledonne et al. 2001), although it has been questioned whether this is a general phenomenon (Greenberg and Yao 2004). In wheat infected by the hemibiotrophic pathogen *S. tritici*, there was a strong accumulation of H_2O_2 in a resistant cultivar, coinciding with the restriction of pathogen growth and expression of defence genes, but there is no classical HR in the host (Shetty et al. 2003, 2007). Likewise, Sasabe et al. (2000) also found that elicitor treatment of tobacco cell suspension cultures resulted in an oxidative burst, but not in cell death or defence gene activation. This indicates that signalling pathways leading to the oxidative burst, cell death and defence gene activation may branch at an early stage. Also Montillet et al. (2005) found that an HR could be elicited in tobacco by different pathways in light and darkness. There are also reports where elicitors and pathogens have been shown to trigger a strong oxidative burst without causing an HR but activate other defence mechanisms involved with the oxidative burst (Glazner et al. 1996; Jabs et al. 1997; Repka 2002). For example, Glazner et al. (1996) showed that ROS accumulation in tobacco leaves and cultured cells in response to an incompatible strain of *Pseudomonas syringae* pv. *syringae* was not sufficient to cause HR. Likewise, in parsley suspension-cultured cells inoculated with fungal elicitor, ROS production and activation of defence genes was observed, but no HR, indicating that ROS accumulation could play other roles, such as to act as a direct antimicrobial agent, for induction of defence gene expression and phytoalexin accumulation in the absence of the HR (Jabs et al. 1997).

Clearly, the role of ROS for triggering and executing the HR is complicated and influenced by many factors. Already Levine et al. (1994) suggested that ROS at interaction sites may have different roles in the elicitation or prevention of cell death depending on their concentration, sub-cellular localisation and the duration of their production.

Involvement of ROS in successful pathogenesis

Biotrophic pathogens obtain their nutrition from living host cells (Oliver and Ipcho 2004), and H_2O_2 has been reported as an effective factor in stopping growth of biotrophic pathogens such as *B. graminis* f. sp. *hordei* (Mellersh et al. 2002; Thordal-Christensen et al. 1997; Trujillo et al. 2006). For example, Trujillo et al. (2006) found in the barley–*Blumeria* interaction that H_2O_2 is produced in non-penetrated CWA (see Fig. 2b). However, they also showed that superoxide (O_2^-) was produced locally at the site of penetration and appeared to enhance infection, thus suggesting that ROS can also be important for the pathogenesis of biotrophic pathogens. Biotrophic pathogens may suppress the host defence responses during infection (Ferreira et al. 2007). For example, the fungal metabolite mannitol, which can suppress ROS-related defence mechanisms by scavenging ROS, was found in apoplastic fluids of *Vicia faba* leaves infected with *Uromyces viciae-fabae* (formerly *Uromyces fabae*; Link et al. 2005; Voegele et al. 2005). On the other hand, HR and ROS such as H_2O_2 have been reported to benefit infection by necrotrophic pathogens, which may even be able to produce ROS themselves or stimulate the host to do so (Govrin and Levine 2000; Von Gönner and Schlösser 1992). For example, Van

der Vlugt-Bergmans et al. (1997) studied *B. fuckeliana* infection in *V. faba* and reported that the fungus released H_2O_2 which could destroy host membrane lipids, thereby facilitating penetration. *B. fuckeliana* was also reported to enhance ROS production to aid its tissue colonization by triggering changes in the host (tomato) peroxisomal antioxidant system, leading to a collapse of regulatory mechanisms (Kuźniak and Skłodowska 2005). In vitro studies by Gil-ad and Mayer (1999) showed that *B. fuckeliana* spores could germinate at a concentration of about 180 mM H_2O_2 and that the mycelium could reduce the H_2O_2 present in the culture medium, thus clearly demonstrating that the pathogen could cope with this high H_2O_2 concentration. In contrast, Unger et al. (2005) performed studies in bean leaves and suspension-cultured cells, which indicated that a non-aggressive isolate of *B. fuckeliana* was in fact inhibited by ROS (O_2^- and partly H_2O_2) and HR. On the other hand, an aggressive isolate induced HR-like necrosis and was able to complete its life cycle. The aggressive isolate produced high amounts of a suppressor of O_2^-, i.e., 2-methyl succinate. When this suppressor was added to the non-aggressive isolate, enhanced tissue necrosis occurred. These results demonstrate a situation equivalent to biotrophic pathogens suppressing the host defence responses and indicate that ROS may also inhibit the necrotrophic pathogen *B. fuckeliana* in some cases (c.f. Małolepsza and Urbanek 2002), contrary to previous conclusions regarding this organism.

The so-called hemibiotrophic pathogens are a diverse group of organisms with an initial biotrophic phase where infection is established, followed by a necrotrophic phase where the pathogen completes its life-cycle (Oliver and Ipcho 2004). A correlation between pathogen growth at the late stages of their life-cycle and large quantities of H_2O_2 has also been reported in such host-pathogen systems. Thus, Able (2003) reported that in barley infected by *Rhynchosporium secalis* and *P. teres*, there was a large accumulation of H_2O_2 in compatible interactions in the later stage of infection, and it was concluded that H_2O_2 was necessary for successful infection as for *B. fuckeliana* (see above), however, based only on correlative evidence. Recently, doubts have been raised whether this is a valid conclusion. Shetty et al. (2003) observed a similar correlation in wheat

infected by the hemibiotrophic pathogen *S. tritici*. During the biotrophic phase of the interaction, H_2O_2 accumulation occurred as a defence response only in an incompatible interaction (Fig. 2c). On the other hand, in a compatible interaction, large amounts of H_2O_2 accumulated after extensive tissue colonization just before the symptoms appeared and the pathogen sporulated (Fig. 2d). However, Shetty et al. (2007) found that, even though sporulation of *S. tritici* in wheat coincided with a massive accumulation of H_2O_2, removal of this H_2O_2 by infiltration of catalase resulted in increased pathogen growth, indicating both that it can survive and tolerate the presence of H_2O_2, but also that this H_2O_2 inhibits the pathogen.

The ability to colonise and proliferate in an environment with high concentrations of ROS shows that the pathogens have efficient systems enabling them to protect themselves against the harsh environment in the host. Thus, for the necrotrophic pathogen *B. fuckeliana*, Van der Vlugt-Bergmans et al. (1997) found that it could protect itself by expressing genes encoding catalase which could scavenge H_2O_2. In accordance with this, Goodwin et al. (2001) cloned a catalase gene from the hemibiotrophic pathogen *Colletotrichum gloesporioides* f.sp. *malvae* (pathogen of *Malva pusilla*). Catalase genes have also been reported from *S. tritici* (Levy et al. 1992), but their expression was not studied. However, it can be predicted that catalases will be activated during the necrotrophic phase of *S. tritici* growth to help protect the pathogen from the deleterious effect of H_2O_2. In agreement with this, Shetty et al. (2003) found high levels of catalase activity in a susceptible host during pycnidial formation, but were unable to determine whether this was of host or pathogen origin.

Other types of hemibiotrophic pathogens might benefit from ROS accumulation at some stage. Thus, Kumar et al. (2001) reported that barley with *mlo* resistance against *B. graminis* f.sp. *hordei* was very susceptible to *Bipolaris sorokiniana*. Toxins from *B. sorokiniana* killed the host cells, leading to massive H_2O_2 accumulation, and it was hypothesised that the *mlo*-resistant plants were very susceptible because cell death occurred more easily compared to other barley genotypes. After cell death, the pathogen could grow unhindered in the dead tissue and was not inhibited by H_2O_2 accumulation (cf. Fig. 1). However, in the initial stages of penetration, it was concluded that *B. sorokiniana* was inhibited by H_2O_2 accumulation just beneath appressoria from which penetration was attempted, i.e., before cells died.

Most conclusions regarding the role of ROS in successful pathogenesis are based solely on correlative data and come from rather few pathosystems. *B. fuckeliana* is most often used as a representative necrotrophic pathogen. The influence of ROS on this pathogen is fairly well studied even though conflicting results have been reported, but generalisations to other pathogens should be made with caution. Thus, as pointed out by Shetty et al. (2007), even if there is a correlation between ROS accumulation and pathogen colonisation, the fact that the pathogen can tolerate the presence of large amounts of ROS does not necessarily mean that it benefits from ROS. Furthermore, there is disagreement as to which categories different pathogens belong (see, e.g., Oliver and Ipcho 2004). For example, *R. secalis* was reported to be necrotrophic (Able 2003), whereas previous research has shown this pathogen to have a long symptomless phase (Jørgensen et al. 1993), suggesting that it should be considered as hemibiotrophic. Therefore, caution should also be taken when concluding about a definite role of ROS for specific types or even species of pathogens before thorough studies have been conducted.

Discussion

Although our understanding of the oxidative burst in plant–pathogen interactions has advanced considerably since the first reports, there are still several unanswered questions. Thus, Fig. 1 shows an overview of our current knowledge of the different roles of ROS in host–pathogen interactions, but also indicates some of the areas where there are unanswered questions and gaps in our knowledge.

The rapid production of ROS in the apoplast in response to pathogens has been proposed to orchestrate the establishment of different defensive barriers against pathogens (Torres et al. 2006). Our understanding of sources and roles of ROS has been greatly enhanced by the identification of defence-associated mutants in the model plant *A. thaliana* (Lorrain et al. 2003). These mutants have not only allowed the identification of important signalling intermediates but also allowed the dissection of ROS-mediated

signalling pathways. However, the influence of other factors such as environment, plant hormones and activation of different signalling pathways (Op den Camp et al. 2003; Montillet et al. 2005; Sasabe et al. 2000; Torres et al. 2005) plays an important role for the accumulation of ROS, and this needs to be taken into consideration and studied in detail. Likewise, external factors such as different types of pathogens and elicitors may vary in their ability to trigger ROS production (cf. Fig. 1), and are therefore possible reasons for conflicting results. This emphasises the need for studies of several different host–pathogen systems in order to clarify if and when different pathways are activated in different situations. It is therefore essential to study several different hosts infected by taxonomically different pathogens which represent different life-style strategies before making general conclusions and thus avoid over-simplification. There are profound differences between monocots and dicots as well as in the biology of biotrophic, hemi-biotrophic and necrotrophic pathogens. Caution should therefore be exercised before stating that processes occur in a similar way in totally different systems.

It is also important to adopt different approaches to increase the robustness of conclusions as illustrated by the example of involvement of ROS in HR. Besides a genetic approach, using mutants, gene silencing, gene knock-outs and/or over-expression, careful physiological and biochemical characterisation of different host–pathogen interactions and defence responses activated should be carried out followed by studies of the role of proteins encoded by ROS genes in the different systems. This approach provides insight into their precise function in defence, cell death, and/or pathogen development, through determination of their sub-cellular localisation and biochemical function. In particular, in relation to the evaluation of the role of ROS in successful pathogenesis, it is important to try to inhibit the cell death machinery selectively and simultaneously to monitor other defence and pathogenesis-related events. Using this approach, it should be possible to determine whether cell death can be uncoupled from other defence responses and if so, the specific contribution to resistance or susceptibility in the interaction in question. Of particular interest in this context is to determine which role ROS plays in HR/necrosis against necrotrophic pathogens (cf. Fig. 1). Thus, do these pathogens all benefit from

ROS accumulation or are some of them actually inhibited to some extent by ROS or other substances in the dying cells? Another important question regarding necrotrophic pathogens relates to potential toxins produced (cf. Fig. 1). Thus, will those toxins which kill host cells always cause the release of ROS, which in turn causes an HR (cf. *B. sorokiniana*)? It is also important to study the interplay between ROS and SA/NO, in order to gain further insights into the regulation of resistance, as these are important defence response regulators that interact with ROS signalling in response to pathogens (Mur et al. 2006). Thus, ROS may be part of many signalling pathways and provide a crucial link in the cross-talk to different responses (Apel and Hirt 2004).

The flux of information between different cell compartments needs to be elucidated to further understand the regulatory capabilities of ROS. Previously, genetic engineering for improved disease resistance has mainly targeted genes involved in the recognition of the pathogen or in the over-expression of defence molecules like phytoalexins (Jalali et al. 2006). An interesting alternative approach would be to target key molecules like ROS that act at points of convergence of different signalling pathways. Engineering plants with such genes using a pathogen-inducible promoter would enable expression of different downstream genes simultaneously in the host, ensuring that plants develop an array of effective responses, which will ultimately secure a sustainable and durable resistance against a range of plant pathogens.

References

Able, A. J. (2003). Role of reactive oxygen species in the response of barley to necrotrophic pathogens. *Protoplasma, 221*, 137–143.

Ali, R., Ma, W., Lemtiri-Chlieh, F., Tsaltas, D., Leng, Q., von Bodman, S., et al. (2007). Death don't have no mercy and neither does calcium: *Arabidopsis* CYCLIC NUCLEOTIDE GATED CHANNEL2 and innate immunity. *The Plant Cell, 19*, 1081–1095.

Allan, A. C., & Fluhr, R. (1997). Two distinct sources of elicited reactive oxygen species in tobacco epidermal cells. *The Plant Cell, 9*, 1559–1572.

Alvarez, M. E., Pennell, R. I., Meijer, P.-J., Ishikawa, A., Dixon, R. A., & Lamb, C. (1998). Reactive oxygen intermediates mediate a systemic signal network in the establishment of plant immunity. *Cell, 92*, 773–784.

Amirsadeghi, S., Robson, C. A., & Vanlerberghe, G. C. (2007). The role of the mitochondrion in plant responses to biotic stress. *Physiologia Plantarum, 129*, 253–266.

An, Q., Ehlers, K., Kogel, K.-H., van Bel, A. J. E., & Hükelhoven, R. (2006). Multivesicular compartments proliferate in susceptible and resistant *MLA12*-barley leaves in response to infection by the biotrophic powdery mildew fungus. *New Phytologist, 172*, 563–576.

Apel, K., & Hirt, H. (2004). Reactive oxygen species: Metabolism, oxidative stress, and signal transduction. *Annual Review of Plant Biology, 55*, 373–399.

Apostol, I., Heinstein, P. F., & Low, P. S. (1989). Rapid induction of an oxidative burst during elicitation of cultured plant cells. *Plant Physiology, 90*, 109–116.

Ashtamker, C., Kiss, V., Sagi, M., Davydov, O., & Fluhr, R. (2007). Diverse subcellular locations of cryptogein-induced reactive oxygen species production in tobacco Bright Yellow-2 cells. *Plant Physiology, 143*, 1817–1826.

Auh, C. K., & Murphy, T. M. (1995). Plasma-membrane redox enzyme is involved in the synthesis of O_2^- and H_2O_2 by *Phytophthora* elicitor-stimulated rose cells. *Plant Physiology, 107*, 1241–1247.

Babitha, M. P., Prakash, H. S., & Shetty, H. S. (2004). Purification and properties of lipoxygenase induced in downy mildew resistant pearl millet seedlings due to infection with *Sclerospora graminicola*. *Plant Science, 166*, 31–39.

Baker, C. J., & Orlandi, E. W. (1995). Active oxygen in plant pathogenesis. *Annual Review of Phytopathology, 33*, 299–321.

Baker, C. J., Orlandi, E. W., & Mock, N. M. (1993). Harpin, an elicitor of the hypersensitive response in tobacco caused by *Erwinia amylovora*, elicits active oxygen production in suspension cells. *Plant Physiology, 102*, 1341–1344.

Bedard, K., Lardy, B., & Krause, K.-H. (2007). NOX family NADPH oxidases: Not just in mammals. *Biochimie, 89*, 1107–1112.

Bindschedler, L. V., Dewdney, J., Blee, K. A., Stone, J. M., Asai, T., Plotnikov, J., et al. (2006). Peroxidase-dependent apoplastic oxidative burst in *Arabidopsis* required for pathogen resistance. *The Plant Journal, 47*, 851–863.

Blumwald, E., Aharon, G. S., & Lam, B. C.-H. (1998). Early signal transduction pathways in plant–pathogen interactions. *Trends in Plant Science, 3*, 342–346.

Bolwell, G. P., Bindschedler, L. V., Blee, K. A., Butt, V. S., Davies, D. R., Gardner, S. L., et al. (2002). The apoplastic oxidative burst in response to biotic stress in plants: A three-component system. *Journal of Experimental Botany, 53*, 1367–1376.

Bradley, D. J., Kjellbom, P., & Lamb, C. J. (1992). Elicitor- and wound-induced oxidative cross-linking of a proline-rich plant cell wall structural protein: A novel, rapid plant defense response. *Cell, 70*, 21–30.

Carter, C., Healy, R., O'Tool, N. M., Naqvi, S. M. S., Ren, G., Park, S., et al. (2007). Tobacco nectaries express a novel NADPH oxidase implicated in the defense of floral reproductive tissues against microorganisms. *Plant Physiology, 143*, 389–399.

Chandra, S., Martin, G. B., & Low, P. S. (1996). The Pto kinase mediates a signaling pathway leading to the oxidative burst in tomato. *Proceedings of the National Academy of Sciences of the United States of America, 93*, 13393–13397.

Chisholm, S. T., Coaker, G., Day, B., & Staskawicz, B. J. (2006). Host–microbe interactions: Shaping the evolution of the plant immune response. *Cell, 124*, 803–814.

Collins, N. C., Thordal-Christensen, H., Lipka, V., Bau, S., Kombrink, E., Qiu, J. L., et al. (2003). SNARE-protein-mediated disease resistance at the plant cell wall. *Nature, 425*, 973–977.

Cona, A., Rea, G., Angelini, R., Frederico, R., & Tavaldorak, P. (2006). Functions of amine oxidases in plant development and defence. *Trends in Plant Science, 11*, 80–89.

Custers, J. H. H. V., Harrison, S. J., Sela-Buurlage, M. B., van Deventer, E., Lageweg, W., Howe, P. W., et al. (2004). Isolation and characterisation of a class of carbohydrate oxidases from higher plants, with a role in active defence. *Plant Journal, 39*, 147–160.

Dat, J. F., Pellinen, R., Beeckman, T., Van de Cotte, B., Langebartels, C., Kangasjärvi, J., et al. (2003). Changes in hydrogen peroxide homeostasis trigger an active cell death process in tobacco. *The Plant Journal, 33*, 621–632.

Deepak, S. A., Ishii, H., & Park, P. (2006). Acibenzolar-*S*-methyl primes cell wall strengthening genes and reactive oxygen species forming/scavenging enzymes in cucumber after fungal pathogen attack. *Physiological and Molecular Plant Pathology, 69*, 52–61.

Delledonne, M., Murgia, I., Ederle, D., Sbicego, P. F., Biondani, A., Polverari, A., et al. (2002). Reactive oxygen intermediates modulate nitric oxide signaling in the plant hypersensitive disease-resistance response. *Plant Physiology and Biochemistry, 40*, 605–610.

Delledonne, M., Xia, Y., Dixon, R. A., & Lamb, C. (1998). Nitric oxide functions as a signal in plant disease resistance. *Nature, 394*, 585–588.

Delledonne, M., Zeier, J., Marocco, A., & Lamb, C. (2001). Signal interactions between nitric oxide and reactive oxygen intermediates in the plant hypersensitive disease resistance response. *Proceedings of the National Academy of Sciences of the United States of America, 98*, 13454–13459.

Desikan, R., Clarke, A., Hancock, J. T., & Neill, S. J. (1999). H_2O_2 activates a MAP kinase-like enzymes in *Arabidopsis thaliana* suspension cultures. *Journal of Experimental Botany, 50*, 1863–1866.

Dorey, S., Kopp, M., Geoffroy, P., Fritig, B., & Kauffmann, S. (1999). Hydrogen peroxide from the oxidative burst is neither necessary nor sufficient for hypersensitive cell death induction, phenylalanine ammonia lyase stimulation, salicylic acid accumulation or scopoletin consumption in cultured tobacco cells treated with elicitin. *Plant Physiology, 121*, 163–171.

Enyedi, A. J., Yalpani, N., Silverman, P., & Raskin, I. (1992). Localization, conjugation, and function of salicylic acid in tobacco during the hypersensitive reaction to tobacco mosaic virus. *Proceedings of the National Academy of Sciences of the United States of America, 89*, 2480–2484.

Ferreira, R. B., Monteiro, S., Freitas, R., Santos, C. N., Chen, Z., Batista, L. M., et al. (2007). The role of plant defence proteins in fungal pathogenesis. *Molecular Plant Pathology, 8*, 677–700.

Floryszak-Wieczorek, J., Arasimowicz, M., Milczarek, G., Jelen, H., & Jackowiak, H. (2007). Only an early nitric oxide burst and the following wave of secondary nitric oxide generation enhanced effective defence responses of pelargonium to a necrotrophic pathogen. *New Phytologist*, *175*, 718–730.

Gil-ad, N. L., & Mayer, A. M. (1999). Evidence for rapid breakdown of hydrogen peroxide by *Botrytis cinerea*. *FEMS Microbiology Letters*, *176*, 455–461.

Glazner, J. A., Orlandi, E. W., & Baker, C. J. (1996). The active oxygen response of cell suspensions to incompatible bacteria is not sufficient to cause hypersensitive cell death. *Plant Physiology*, *110*, 759–763.

Goodwin, P. H., Li, J., & Jin, S. (2001). A catalase gene of *Colletotrichum gloeosporioides* f. sp. *malvae* is highly expressed during the necrotrophic phase of infection of round-leaved mallow, *Malva pusilla*. *FEMS Microbiology Letters*, *202*, 103–107.

Govrin, E. M., & Levine, A. (2000). The hypersensitive response facilitates plant infection by the necrotrophic pathogen *Botrytis cinerea*. *Current Biology*, *10*, 751–757.

Grant, J. J., Yun, B.-W., & Loake, G. J. (2000). Oxidative burst and cognate redox signalling reported by luciferase imaging: Identification of a signal network that functions independently of ethylene, SA and Me-JA but is dependent on MAPKK activity. *The Plant Journal*, *24*, 569–582.

Greenberg, J. T. (1997). Programmed cell death in plant–pathogen interactions. *Annual Review of Plant Physiology and Plant Molecular Biology*, *48*, 525–545.

Greenberg, J. T., & Yao, N. (2004). The role and regulation of programmed cell death in plant–pathogen interactions. *Cellular Microbiology*, *6*, 201–211.

Heitefuss, R. (1997). Cell wall modification in relation to resistance. In H. Hartleb, R. Heitefuss, & H.-H. Hoppe (Eds.), *Resistance of crop plants against fungi* (pp. 100–125). Jena: Gustav Fischer.

Hirt, H. (1997). Multiple roles of MAP kinases in plant signal transduction. *Trends in Plant Science*, *2*, 11–15.

Hoeberichts, F. A., ten Have, A., & Woltering, E. J. (2003). A tomato metacaspase gene is upregulated during programmed cell death in *Botrytis cinerea*-infected leaves. *Planta*, *217*, 517–522.

Hoeberichts, F. A., & Woltering, E. J. (2003). Multiple mediators of plant programmed cell death: Interplay of conserved cell death mechanisms and plant-specific regulators. *BioEssays*, *25*, 47–57.

Hu, X., Bidney, D. L., Yalpani, N., Duvick, J. P., & Crasta, O. (2003). Overexpression of a gene encoding hydrogen peroxide-generating oxalate oxidase evokes defense responses in sunflower. *Plant Physiology*, *133*, 170–181.

Hückelhoven, R., & Kogel, K.-H. (2003). Reactive oxygen intermediates in plant microbe interactions: Who is who in powdery mildew resistance? *Planta*, *216*, 891–902.

Iwai, T., Seo, S., Mitsuhara, I., & Ohashi, Y. (2007). Probenazole-induced accumulation of salicylic acid confers resistance to *Magnaporthe grisea* in adult rice plants. *Plant and Cell Physiology*, *48*, 915–924.

Iwano, M., Che, F.-S., Goto, K., Tanaka, N., Takayama, S., & Isogai, A. (2002). Electron microscopic analysis of the H_2O_2 accumulation preceding hypersensitive cell death induced by an incompatible strain of *Pseudomonas avenae* in cultured rice cells. *Molecular Plant Pathology*, *3*, 1–8.

Jabs, T., Dietrich, R. A., & Dangl, J. L. (1996). Initiation of runaway cell death in an *Arabidopsis* mutant by extracellular superoxide. *Science*, *273*, 1853–1856.

Jabs, T., Tschöpe, M., Colling, C., Hahlbrock, K., & Scheel, D. (1997). Elicitor-stimulated ion fluxes and O_2^- from the oxidative burst are essential components in triggering defense gene activation and phytoalexin synthesis in parsley. *Proceedings of the National Academy of Sciences of the United States of America*, *94*, 4800–4805.

Jalali, B. L., Bhargava, S., & Kamble, A. (2006). Signal transduction and transcriptional regulation of plant defence responses. *Journal of Phytopathology*, *154*, 65–74.

Jørgensen, H. J. L., de Neergaard, E., & Smedegaard-Petersen, V. (1993). Histological examination of the interaction between *Rhynchosporium secalis* and susceptible and resistant cultivars of barley. *Physiological and Molecular Plant Pathology*, *42*, 345–358.

Jørgensen, H. J. L., Lübeck, P. S., Thordal-Christensen, H., de Neergaard, E., & Smedegaard-Petersen, V. (1998). Mechanisms of induced resistance in barley against *Drechslera teres*. *Phytopathology*, *88*, 698–707.

Kariola, T., Brader, G., Li, J., & Palva, E. T. (2005). Chlorophyllose 1, a damage control enzyme, affects the balance between defense pathways in plants. *The Plant Cell*, *17*, 282–294.

Kawasaki, T., Henmi, K., Ono, E., Hatakeyama, S., Iwano, M., Satoh, H., et al. (1999). The small GTP-binding protein Rac is a regulator of cell death in plants. *Proceedings of the National Academy of Sciences of the United States of America*, *96*, 10922–10926.

Klessig, D. F., Durner, J., Noad, R., Navarre, D. A., Wendehenne, D., Kumar, D., et al. (2000). Nitric oxide and salicylic acid signaling in plant defense. *Proceedings of the National Academy of Sciences of the United States of America*, *97*, 8849–8855.

Kobayashi, M., Ohura, I., Kawakita, K., Yokota, N., Fujiwara, M., Shimamoto, K., et al. (2007). Calcium-dependent protein kinases regulate the production of reactive oxygen species by potato NADPH oxidase. *The Plant Cell*, *19*, 1065–1080.

Kumar, J., Hückelhoven, R., Beckhove, U., Nagarajan, S., & Kogel, K.-H. (2001). A compromised Mlo pathway affects the response of barley to the necrotrophic fungus *Bipolaris sorokiniana* (teleomorph: *Cochliobolus sativus*) and its toxins. *Phytopathology*, *91*, 127–133.

Kumudini, B. S., & Shetty, H. S. (2002). Association of lignification and callose deposition with host cultivar resistance and induced systemic resistance of pearl millet to *Sclerospora graminicola*. *Australasian Plant Pathology*, *32*, 157–164.

Kuźniak, E., & Skłodowska, M. (2005). Fungal pathogen-induced changes in the antioxidant systems of leaf peroxisomes from infected tomato plants. *Planta*, *222*, 192–200.

Lam, E. (2004). Controlled cell death, plant survival and development. *Nature Reviews in Molecular Cell Biology*, *5*, 305–315.

Lam, E., Kato, N., & Lawton, M. (2001). Programmed cell death, mitochondria and the plant hypersensitive response. *Nature*, *411*, 848–853.

Lamb, C., & Dixon, R. A. (1997). The oxidative burst in plant disease resistance. *Annual Review of Plant Physiology and Plant Molecular Biology, 48*, 251–275.

Legendre, L., Heinstein, P. F., & Low, P. S. (1992). Evidence for the participation of GTP-binding proteins in the elicitation of rapid oxidative burst in cultured soybean cells. *Journal of Biological Chemistry, 267*, 20140–20147.

Legendre, L., Rueter, S., Heinstein, P. S., & Low, P. S. (1993). Characterisation of the oligogalacturonide-induced oxidative burst in cultured soybean (*Glycine max*) cells. *Plant Physiology, 102*, 233–240.

Levine, A., Tenhaken, R., Dixon, R., & Lamb, C. (1994). H_2O_2 from the oxidative burst orchestrates the plant hypersensitive disease resistance response. *Cell, 79*, 583–593.

Levy, E., Eyal, Z., & Hochman, A. (1992). Purification and characterization of a catalase–peroxidase from the fungus *Septoria tritici*. *Archives of Biochemistry and Biophysics, 296*, 321–327.

Li, A. L., Wang, M. L., Zhou, R. H., Kong, X. Y., Huo, N. X., Wang, W. S., et al. (2005). Comparative analysis of early H_2O_2 accumulation in compatible and incompatible wheat–powdery mildew interactions. *Plant Pathology, 54*, 308–316.

Li, J., Zhang, Z.-G., Ji, R., Wang, Y.-C., & Zheng, X.-B. (2006). Hydrogen peroxide regulates elicitor PB90-induced cell death and defense in non-heading Chinese cabbage. *Physiological and Molecular Plant Pathology, 67*, 220–230.

Link, T., Lohaus, G., Heiser, I., Mendgen, K., Hahn, M., & Voegele, R. T. (2005). Characterization of a novel $NADP^+$-dependent D-arabitol dehydrogenase from the plant pathogen *Uromyces fabae*. *Biochemical Journal, 389*, 289–295.

Liu, G., Greenshields, D. L., Sammynaiken, R., Hirji, R. N., Selvaraj, G., & Wei, Y. (2007). Targeted alterations in iron homeostasis underlie plant defense responses. *Journal of Cell Science, 120*, 596–605.

Lorrain, S., Vailleau, F., Balagué, C., & Roby, D. (2003). Lesion mimic mutants: Keys for deciphering cell death and defense pathways in plants? *Trends in Plant Science, 8*, 263–271.

Małolepsza, U. (2005). Spatial and temporal variation of reactive oxygen species and antioxidant enzymes in *o*-hydroxyethylorutin-treated tomato leaves inoculated with *Botrytis cinerea*. *Plant Pathology, 54*, 317–324.

Małolepsza, U., & Urbanek, H. (2002). *o*-Hydroxyethylorutin-mediated enhancement of tomato resistance to *Botrytis cinerea* depends on a burst of reactive oxygen species. *Journal of Phytopathology, 150*, 616–624.

Mateo, A., Mühlenbock, P., Rustérucci, C., Chang, C. C.-C., Miszalski, Z., Karpinska, B., et al. (2004). *LESION SIMULATING DISEASE 1* is required for acclimation to conditions that promote excess excitation energy. *Plant Physiology, 136*, 2818–2830.

Mayer, A. M., Staples, R. C., & Gil-ad, N. L. (2001). Mechanisms of survival of necrotrophic fungal plant pathogens in hosts expressing the hypersensitive response. *Phytochemistry, 58*, 33–41.

McAinsh, M. R., Clayton, H., Mansfield, T. A., & Hetherington, A. M. (1996). Changes in stomatal behavior and guard cell cytosolic free calcium in response to oxidative stress. *Plant Physiology, 111*, 1031–1042.

McDowell, J. M., & Dangl, J. L. (2000). Signal transduction in the plant immune response. *Trends in Biochemical Science, 25*, 79–82.

Mellersh, D. G., Foulds, I. V., Higgens, V. J., & Heath, M. C. (2002). H_2O_2 plays different roles in determining penetration failure in three diverse plant–fungal interactions. *The Plant Journal, 29*, 257–268.

Mittler, R., Herr, E. H., Orvar, B. L., van Camp, W., Willekens, H., Inzé, D., et al. (1999). Transgenic tobacco plants with reduced capability to detoxify reactive oxygen intermediates are hyperresponsive to pathogen infection. *Proceedings of the National Academy of Sciences of the United States of America, 96*, 14165–14170.

Mittler, R., Vanderauwera, S., Gollery, M., & Van Breusegem, F. (2004). The reactive oxygen gene network in plants. *Trends in Plant Science, 9*, 490–498.

Moerschbacher, B. M., & Reisener, H.-J. (1997). The hypersensitive resistance reaction. In H. Hartleb, R. Heitefuss, & H.-H. Hoppe (Eds.), *Resistance of crop plants against fungi* (pp. 126–158). Jena: Gustav Fischer.

Montillet, J.-L., Chamnongpol, S., Rustérucci, C., Dat, J., Van de Cotte, B., Agnel, J.-P., et al. (2005). Fatty acid hydroperoxides and H_2O_2 in the execution of hypersensitive cell death in tobacco leaves. *Plant Physiology, 138*, 1516–1526.

Mur, L. A. J., Carver, T. L. W., & Prats, E. (2006). NO way to live; the various roles of nitric oxide in plant–pathogen interactions. *Journal of Experimental Botany, 57*, 489–505.

Neill, S. J., Desikan, R., Clarke, A., Hurst, R. D., & Hancock, J. T. (2002). Hydrogen peroxide and nitric oxide as signalling molecules in plants. *Journal of Experimental Botany, 53*, 1237–1247.

Nurnberger, T. M., Nennsteil, O., Jabs, T., Sacks, W. R., Hahlbrock, K., & Scheel, D. (1994). High affinity binding of a fungal oligopeptide elicitor to parsley plasma membranes triggers multiple defense responses. *Cell, 78*, 449–460.

Oliver, R. P., & Ipcho, S. V. S. (2004). *Arabidopsis* pathology breathes new life into the necrotrophs-vs.-biotrophs classification of fungal pathogens. *Molecular Plant Pathology, 4*, 347–352.

Olson, P. D., & Varner, J. E. (1993). Hydrogen peroxide and lignification. *The Plant Journal, 4*, 887–892.

Op den Camp, R. G. L., Przybyla, D., Ochsenbein, C., Laloi, C., Kim, C., Danon, A., et al. (2003). Rapid induction of distinct stress responses after the release of singlet oxygen in *Arabidopsis*. *The Plant Cell, 15*, 2320–2332.

Pei, Z.-M., Murata, Y., Benning, G., Thomine, S., Klüsener, B., Allen, G. J., et al. (2000). Calcium channels activated by hydrogen peroxide mediate abscisic acid signalling in guard cells. *Nature, 406*, 731–734.

Peng, M., & Kuc, J. (1992). Peroxidase-generated hydrogen peroxide as a source of antifungal activity in vitro and on tobacco leaf disks. *Phytopathology, 82*, 696–699.

Petersen, M., Brodersen, P., Naested, H., Andreasson, E., Lindhart, U., Johansen, B., et al. (2000). *Arabidopsis*

MAP kinase 4 negatively regulates systemic acquired resistance. *Cell*, *103*, 1111–1120.

Price, A., Knight, M., Knight, H., Cuin, T., Tomos, D., & Ashenden, T. (1996). Cytosolic calcium and oxidative plant stress. *Biochemical Society Transactions*, *24*, 479–483.

Price, A. H., Taylor, A., Ripley, S. J., Griffiths, A., Trewavas, A. J., & Knight, M. R. (1994). Oxidative signals in tobacco increase cytosolic calcium. *The Plant Cell*, *6*, 1301–1310.

Ren, D., Yang, K.-Y., Li, G.-J., Liu, Y., & Zhang, S. (2006). Activation of Ntf4, a tobacco mitogen-activated protein kinase, during plant defence response and its involvement in hypersensitive response-like cell death. *Plant Physiology*, *141*, 1482–1493.

Repka, V. (2002). Hydrogen peroxide generated via the octadecanoid pathway is neither necessary nor sufficient for methyl jasmonate-induced hypersensitive cell death in woody plants. *Biologia Plantarum*, *45*, 105–115.

Sasabe, M., Takeuchi, K., Kamoun, S., Ichinose, Y., Govers, F., Toyoda, K., et al. (2000). Independent pathways leading to apoptotic cell death, oxidative burst and defense gene expression in response to elicitin in tobacco cell suspension culture. *European Journal of Biochemistry*, *267*, 5005–5013.

Schulze-Lefert, P. (2004). Knocking on the heaven's wall: Pathogenesis of and resistance to biotrophic fungi at the cell wall. *Current Opinion in Plant Biology*, *7*, 377–383.

Shah, J. (2003). The salicylic acid loop in plant defense. *Current Opinion in Plant Biology*, *6*, 365–371.

Shailashree, S., Kini, K. R., Deepak, S., Kumudini, B. S., & Shetty, H. S. (2004). Accumulation of hydroxyproline-rich glycoproteins in pearl millet seedlings in response to *Sclerospora graminicola* infection. *Plant Science*, *167*, 1227–1234.

Shetty, N. P., Kristensen, B. K., Newman, M.-A., Møller, K., Gregersen, P. L., & Jørgensen, H. J. L. (2003). Association of hydrogen peroxide with restriction of *Septoria tritici* in resistant wheat. *Physiological and Molecular Plant Pathology*, *62*, 333–346.

Shetty, N. P., Mehrabi, R., Lütken, H., Haldrup, A., Kema, G. H. J., Collinge, D. B., et al. (2007). Role of hydrogen peroxide during the interaction between the hemibiotrophic fungal pathogen *Septoria tritici* and wheat. *New Phytologist*, *174*, 637–647.

Suzuki, K. (2002). Map kinase cascade in elicitor signal transduction. *Journal of Plant Research*, *115*, 237–244.

Thatcher, L. F., Anderson, J. P., & Singh, K. B. (2005). Plant defence responses: What have we learnt from *Arabidopsis*? *Functional Plant Biology*, *32*, 1–19.

Thordal-Christensen, H., Zhang, Z., Wei, Y., & Collinge, D. B. (1997). Subcellular localization of H_2O_2 in plants. H_2O_2 accumulation in papillae and hypersensitive response during the barley–powdery mildew interaction. *The Plant Journal*, *11*, 1187–1194.

Torres, M. A., Dangl, J. L., & Jones, J. D. G. (2002). *Arabidopsis gp91^{phox}* homologues, *AtrbohD* and *AtrbohF* are required for accumulation of reactive oxygen intermediates in the plant defense response. *Proceedings of the National Academy of Sciences of the United States of America*, *99*, 517–522.

Torres, M. A., Jones, J. D. G., & Dangl, J. L. (2005). Pathogen-induced, NADPH oxidase-derived reactive oxygen intermediates suppress spread of cell death in *Arabidopsis thaliana*. *Nature Genetics*, *37*, 1130–1134.

Torres, M. A., Jones, J. D. G., & Dangl, J. L. (2006). Reactive oxygen species signaling in response to pathogens. *Plant Physiology*, *141*, 373–378.

Trujillo, M., Altschmeid, L., Schweizer, P., Kogel, K.-H., & Hückelhoven, R. (2006). *Respiratory Burst Oxidase Homologue A* of barley contributes to penetration by the powdery mildew fungus *Blumeria graminis* f. sp. *hordei*. *Journal of Experimental Botany*, *57*, 3781–3791.

Unger, C., Kleta, S., Jandl, G., & v. Tiedemann, A. (2005). Suppression of the defence-related oxidative burst in bean leaf tissue and bean suspension cells by the necrotrophic pathogen *Botrytis cinerea*. *Journal of Phytopathology*, *153*, 15–26.

Urquhart, W., Gunawardena, A. H. L. A. N., Moeder, W., Ali, R., Berkowitz, G. A., & Yoshioka, K. (2007). The chimeric cyclic nucleotide-gated ion channel ATCNGC11/12 constitutively induces programmed cell death in a Ca^{2+} dependent manner. *Plant Molecular Biology*, *65*, 747–761.

Van Breusegem, F., & Dat, J. F. (2006). Reactive oxygen species in plant cell death. *Plant Physiology*, *141*, 384–390.

Van der Vlugt-Bergmans, C. J. B., Wagemakers, C. A. M., Dees, D. C. T., & Van Kan, J. A. L. (1997). Catalase A from *Botrytis cinerea* is not expressed during infection on tomato leaves. *Physiological and Molecular Plant Pathology*, *50*, 1–15.

Voegele, R. T., Hahn, M., Lohaus, G., Link, T., Heiser, I., & Mendgen, K. (2005). Possible roles for mannitol and mannitol dehydrogenase in the biotrophic plant pathogen *Uromyces fabae*. *Plant Physiology*, *137*, 190–198.

Von Gönner, M., & Schlösser, E. (1992). Effect of radical scavengers on pathogenesis in the host–parasite-system *Avena sativa–Drechslera avenae*. *Zeitschrift für Pflanzenkrankheiten und Pflanzenschutz*, *99*, 617–625.

Walters, D. R. (2003). Polyamines and plant disease. *Phytochemistry*, *64*, 97–107.

Wu, G. S., Short, B. J., Lawrence, E. B., Levine, E. B., Fitzsimmons, K. C., & Shah, D. M. (1995). Disease resistance conferred by expression of a gene encoding H_2O_2-generating glucose oxidase in transgenic potato plants. *The Plant Cell*, *7*, 1357–1368.

Zhang, S. Q., & Klessig, D. F. (1998). Resistance gene N-mediated de novo synthesis and activation of a tobacco mitogen-activated protein kinase by *Tobacco mosaic virus*. *Proceedings of the National Academy of Sciences of the United States of America*, *95*, 7433–7438.

Zhang, S., Liu, Y., & Klessig, D. F. (2000). Multiple levels of tobacco WIPK activation during the induction of cell death by fungal elicitins. *The Plant Journal*, *23*, 339–347.

Zhou, F., Menke, F. L. H., Yoshioka, K., Moder, W., Shirano, Y., & Klessig, D. F. (2004). High humidity suppresses *ssi4*-mediated cell death and disease resistance upstream of MAP kinase activation, H_2O_2 production and defense gene expression. *The Plant Journal*, *39*, 920–932.

Zimmermann, G., Baumlein, H., Mock, H. P., Himmelbach, A., & Schweizer, P. (2006). The multigene family encoding germin-like proteins of barley. Regulation and function in basal host resistance. *Plant Physiology*, *142*, 181–192.

Zwerger, K., & Hirt, H. (2001). Recent advances in plant MAP kinase signalling. *Biological Chemistry*, *382*, 1123–1131.

Eur J Plant Pathol (2008) 121:281–289
DOI 10.1007/s10658-007-9257-y

Mechanisms modulating fungal attack in post-harvest pathogen interactions and their control

Dov Prusky · Amnon Lichter

Received: 21 June 2007 / Accepted: 19 November 2007
© KNPV 2007

Abstract As biotrophs, insidious fungal infections by post-harvest pathogens remain quiescent during fruit growth, but at a particular phase, during ripening and senescence, the pathogens transform to necrotrophs and elicit the typical decay symptoms. Exposure of unripe hosts to pathogens initiates defensive signal-transduction cascades that limit fungal growth and development. Exposure to the same pathogens during ripening and storage activates a substantially different signalling cascade that facilitates fungal colonization. This review will focus on modulation of post-harvest host-pathogen interactions by pH and reactive oxygen species (ROS). Modulation of host pH in response to a host signal is bidirectional and includes either alkalinisation by ammonification of the host tissue, or acidification by secretion of organic acids. These changes sensitise the host and activate the transcription and secretion of fungal hydrolases that promote maceration of the host tissue. This sensitisation is further enhanced at various stages by the accumulation of host or fungal ROS that can further weaken host tissue and amplify fungal development. Several specific examples of coordinated responses that conform with this scheme are described, followed by discussion of the means to exploit these mechanisms to establish new approaches to post-harvest disease control.

Keywords Fruit ripening · Fruit senescence · Pectolytic enzymes · Quiescent infections · pH regulation · PELB · PACC

Introduction

Post-harvest fungal pathogens exploit three main routes to penetrate the host tissue: (a) through wounds caused by biotic and/or abiotic agents during growth and storage; (b) through natural openings such as lenticels, stem-ends and the pedicel–fruit interface; and (c) by directly breaching the host cuticle, which can occur throughout fruit growth. An active pathogen can start its attack process immediately after spores land on the wounded tissue, whereas other pathogens can breach the unripe fruit cuticle and then remain inactive for months until the harvested fruit ripens. The penetration process may go unnoticed by the host, or it may result in rapid defence signalling that results in the induction of defence molecules that will limit fungal development. The period from host infection to the activation of fungal development and symptom expression is designated the quiescent stage (Prusky 1996). After harvest, during ripening and storage, the mechanism that protects the fruit from fungal attack becomes non-functional. This transition

D. Prusky (✉) · A. Lichter
Department of Postharvest Science of Fresh Produce,
Agricultural Research Organization,
Bet Dagan 50250, Israel
e-mail: dovprusk@agri.gov.il

Eur J Plant Pathol (2008) 121:281–289

from a resistant to a susceptible state parallels physiological changes that occur during ripening, and which the pathogen senses and responds to. In the present review, we focus on some of the interactions between the fungus and the host environment, that serve the pathogen as a basis for fungal colonization. We also touch upon the signals that may activate the transition from quiescent to necrotrophic mode during fruit ripening.

The quiescent stage

During the colonization of plant hosts, post-harvest fungal pathogens exploit two main modes of nutrition: biotrophy, in which the nutrients are obtained from the living host cells, and necrotrophy, in which nutrients are obtained from dead host cells killed by the fungus (Perfect et al. 1999). Both of these nutritional modes are exhibited by post-harvest pathogens. Opportunistic post-harvest pathogens, in an inactive mode, may also be located within the fruit, awaiting fruit wounding, ripening or senescence. The length of each period varies among pathogens, hosts, and host developmental stages. Pathogens such as *Colletotrichum*, *Monilinia*, *Botrytis* and *Alternaria* may remain quiescent for long periods in developing fruit tissues, but initiate immediate necrotrophic development on ripening and senescing fruits.

Colletotrichum is one of the major post-harvest pathogens in which quiescence has been studied. *Colletotrichum* spores adhere to and germinate on the plant surface, where they produce germ tubes; the tip of the germ tube developing from the appressorium penetrates through the cuticle with an infection peg. Following penetration, *Colletotrichum* initiates subcuticular intramural colonization (Perfect et al. 1999) and spreads rapidly throughout the tissue with both inter- and intracellular hyphae that kill cells as they advance. After colonizing one or more host cells, the infecting hyphae, which can be described as biotrophic (Kramer-Haimovitch et al. 2006), subsequently give rise to secondary necrotrophic hyphae (Bailey and Jeger 1992; Coates et al. 1993; Latunde-Dada et al. 1996; Mendgen and Hahn 2001; O'Connell et al. 1985). *Botrytis* and *Monilinia* can penetrate through wounds and also breach the fruit cuticle by extending an infection peg from an appressorium, which then remains quiescent for long periods (Fourie and Holz

1995; Pezet et al. 2003). Depending on the physiological status of the organ, these hyphae may continue the infective process or remain quiescent.

The close association between the infection peg and the host fruit is likely to involve exchanges of chemical and physical signals that control transport of nutrients and modulate defence responses. This type of interaction has been demonstrated in rust haustorial penetration (Heath 1997) but not in systems involving post-harvest pathogens.

In the light of published reports, three major modes for the activation of quiescent biotrophic pathogens have been hypothesized (Prusky 1996): (a) a lack in the host of the nutritional resources required for pathogen development; (b) the presence of preformed or inducible fungistatic antifungal compounds in resistant unripe fruits; and (c) an unsuitable environment for the activation of fungal pathogenicity factors.

Factors facilitating pathogenicity

Quiescence may result from a localized host response that is often associated with an oxidative burst, i.e., the production of reactive oxygen species (ROS). Localized generation of ROS during quiescence was found to be one of the earliest (within 2–3 h) detectable cytological defence responses to attempted penetration by *Colletotrichum gloeosporioides* into unripe, resistant avocado fruits (Prusky et al. 1988). During quiescent infections of tomato leaves by the hemibiotroph *Colletotrichum coccodes*, a localized accumulation of H_2O_2 could be clearly seen during the initial stages of fungal penetration. In addition, treatment of tomato fruits with the H_2O_2-scavenging protein catalase, prior to inoculation with *C. coccodes*, resulted in a significant increase in fungal penetration efficiency. At the same time, less H_2O_2 was detected in the infection zone, which suggests that ROS production by the host may be important in inducing resistance and maintaining fungal quiescence (Beno-Moualem and Prusky 2000). Similar behaviour was observed when another hemibiotroph, *Septoria tritici* infecting wheat was inhibited by H_2O_2 during the biotrophic phase (Shetty et al. 2007). Decomposition of H_2O_2 by infiltrating catalase increased susceptibility whereas addition of H_2O_2 to the wheat increased tolerance to fungal attack.

What possible factors could affect ROS production in ripening fruits? ROS production is dependent on the physiological stage of fruit ripening. *Colletotrichum gloeosporioides* infection of unripe, resistant avocado fruits (quiescent interaction) activated a threefold increase in the level of oxygen radicals, with a corresponding increase in NADPH oxidase activity (Beno-Moualem and Prusky 2000), whereas no significant enhancement of ROS production was observed in ripened fruits.

ROS production by pathogens that induce quiescent infections may depend on the environmental pH (Mellersh et al. 2002), a factor that changes naturally during fruit ripening and storage. The pH is also changed, or induced to change, by the pathogens around the infection site (Beno-Moualem and Prusky 2000; Wang et al. 2004). Thus, an increase in the natural pH during fruit ripening or as a result of fungal activity can suppress oxidative responses of either the pathogen or the host, and so facilitate activation of the quiescent infection.

ROS can be produced, perceived and detoxified by fungi, whose growth, virulence and differentiation may be profoundly affected by these metabolites (Aguirre et al. 2006). However, little is known about the exact spatial and temporal mechanisms that protect fungal cells from self-produced or host-produced ROS during the interaction *in planta*. Fungal pathogens have also evolved strategies to compromise or delay the activation of efficient host defence responses. Cessna et al. (2000) suggested that secretion of oxalic acid by *Sclerotinia sclerotiorum* suppresses the initiation of the oxidative burst that marks the activation of defence responses. *Botrytis cinerea* possesses an arsenal of genes which are capable of detoxifying ROS in axenic media (Gil-ad et al. 2000), and it is likely that these enzymes can prevent ROS damage to fungal cells *in planta*. Also in *Botrytis*-bean cell interactions, suppression of a second specific ROS burst following the first non-specific ROS production was observed in aggressive isolates of the pathogen (Unger et al. 2005). In this case, the suppressor of ROS production was identified as 2-methyl succinate, and was able to suppress the hypersensitive response-like necrosis that reduces the virulence of *B. cinerea*. Proline was also shown to be an effective scavenger of intracellular ROS in *Colletotrichum trifolii* (Chen and Dickman 2005). This ability of proline to function as a potent

antioxidant and inhibitor of the hypothesized programmed cell death process may represent an important and broad-based function in the cellular response to stress.

The production of ROS by the host has two mutually conflicting effects during the transition of the pathogen from quiescent to active infection. On the one hand, the accumulation of ROS is a potential trigger of defence mechanisms that confine the pathogen during the biotrophic phase. On the other hand, cell death induced by ROS can lead to the formation of necrotic tissue from which quiescent pathogens obtain the nutrients that will fuel their development during the necrotrophic phase of either the hembiotrophic or the necrotrophic pathogen (van Baarlen et al. 2004). Which of the two effects is more important in modulating the activation of a quiescent infection is still unknown. For example, in *Botrytis*, modulation of ROS was correlated with necrosis and pathogenicity: deletion of a Cu–Zn superoxide dismutase gene retarded the development of necrosis (Rolke et al. 2004). Govrin and Levin (2000) used another approach to show that inhibitors of NADPH oxidase reduced the level of ROS in *Arabidopsis*, and restricted necrosis caused by *B. cinerea*: infiltration of solutions that increase ROS production consistently stimulated leaf necrosis caused by *Botrytis* and *S. sclerotiorum*.

Effect of pH on the expression of fungal virulence factors

During fruit ripening and senescence, pH levels change as part of the ripening process: for instance, the pH of avocado fruit increased from 5.2 to 6.0 during ripening (Yakoby et al. 2000). Pathogens can also alter the pH in the vicinity of the infection site, and the change in pH can modulate the expression of pathogenicity factors (Denison 2000; Yakoby et al. 2000; Prusky et al. 2001b; Eshel et al. 2002; Prusky and Yakoby 2003). Expression of the endoglucanase gene *AAK*1 in *Alternaria alternata* was found to be maximal at pH levels above 6.0, which are characteristic of decayed tissue; it was not expressed at the lower pH values at which the pathogen was quiescent (Eshel et al. 2002). In the pathogen *C. gloeosporioides*, the gene *PEL*B was expressed when the pH was above 5.7, a value similar to that of decaying tissue (Yakoby et al. 2000, 2001). The transcription factor PacC, which is involved in pH regulation, exhibits a similar expression pattern to that of *PEL*B, which suggests that they are co-regulated or that *PAC1* regulates the expression of *PELB* (Fig. 1) (Drori et al. 2003). Disruption of *PAC1*, the orthologue of *PACC*, in *C. gloeosporioides* resulted in loss-of-function mutants with severely attenuated virulence, suggesting that pH responsiveness is critical for the activation of the pathogenicity of this fungus (Miyara et al., unpublished). In another example, expression of the

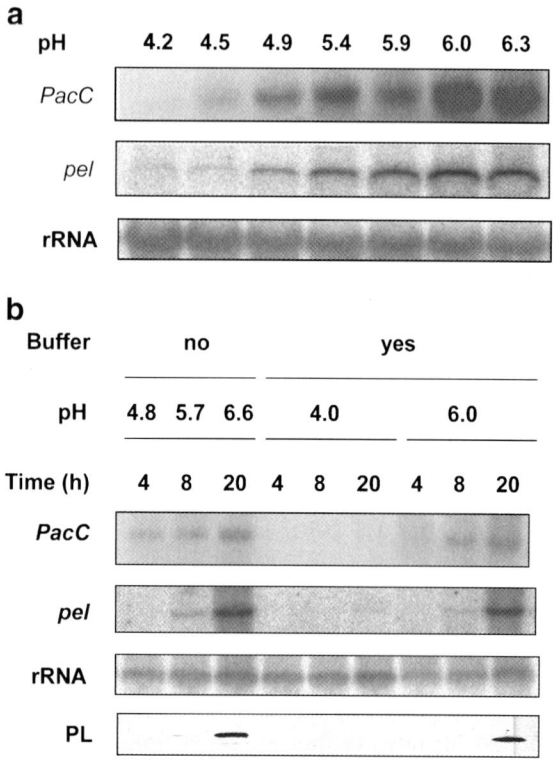

Fig. 1 Transcriptional activation of *pel*B and *PacC* by *C. gloeosporioides* as a function of pH. **a** Expression as a function of pH. Northern analysis of total RNA isolated from *C. gloeosporioides* mycelia 16 after transfer to fresh secondary cultures buffered with phthalate to the indicated pH values. Blots were probed with *pelB* (*middle panel*), and then sequentially stripped and reprobed with *pac1* (*top panel*) and ribosomal DNA (rDNA – *bottom panel*) probes as indicated. **b** Expression as a function of time and pH. Growth medium was buffered with phthalate to pH 4.0 or 6.0, or was unbuffered. At different times the culture media were harvested, hyphae were subjected to RNA extraction and blots were probed with *pelB*, and then sequentially stripped and reprobed with *pac1* and ribosomal DNA (rDNA) probes as indicated. Western analysis of secreted PL was performed on concentrated and dialysed culture medium

*Fusarium oxysporum PG*1 and *PG*5 genes was enhanced in an acidic environment, and a sequence that controls the positive activation of genes by pH, the PACC recognition site, was found in *PG*1 (Caracuel et al. 2003). Although a PACC homologue in *B. cinerea* has not been described to date, the sequence corresponding to the PACC consensus recognition site has been identified in all endo*PG* genes (Wubben et al. 1999; Manteau et al. 2003). Furthermore, the differential expression of endo*PG* by *B. cinerea* (ten Have et al. 2001) correlated well with the pH characteristics of apple and zucchini, which have low and neutral pH, respectively. In the fungus *Penicillium expansum*, PG (*PEPG*1) showed high expression at pH 4.0 and minor expression above pH 5.0. Similar results were also obtained for the endo*PG* gene, *PG*1, of *S. sclerotiorum*: transcription of the PACC homologue, *PAC1*, declined during acidification, concomitant with an increase in *pg*1 expression, and this gene was found to contain the PACC recognition site in its promoter (Rollins and Dickman 2001).

Other putative virulence factors of *B. cinerea*, including oxalic acid, laccase and protease, are also released in a pH-regulated manner at pH values in the range of 3.1–6.0. Activation of these factors at pH levels within the natural range of the host tissue is indicative of the significance of the differential expression of virulence factors in different hosts (Lumsden 1976; Movahedi and Heale 1990; Manteau et al. 2003; Vernekar et al. 1999; Ye and Ng 2002; Caracuel et al. 2003). The ability of different races of the fungus to fine tune enzyme expression in response to the ambient pH in the host further highlights the importance of the specific regulatory system that is activated under a changing environmental pH which can lead to the activation of quiescent infections (Prusky and Yakoby 2003).

The ability to modify pH may be expressed in either direction, and fungi that raise or reduce it are described as 'alkalinizing fungi' or 'acidifying fungi,' respectively.

Alkalinizing fungi

Alkalinization of the infection site by fungi is achieved by active secretion of ammonia, which is largely the product of protease activity and deamination of amino acids. The pathogenicity of *C. gloeosporioides* and expression of the virulence factor PL-B both depend on

raising the ambient pH. In the case of polyphage pathogens such as *A. alternata*, a threefold to tenfold increase in ammonia concentration, and a pH elevation of 0.2 to 2.4 pH units were detected in several hosts: tomato, pepper, melon, cherry and persimmon (Eshel et al. 2002).

Acidifying fungi

Other post-harvest pathogens, such as *P. expansum, P. digitatum, P. italicum* (Prusky and Yakoby 2003), *B. cinerea* (Manteau et al. 2003) and *S. sclerotiorum* (Bateman and Beer 1965; Ruijter et al. 1999), utilize tissue acidification to support their attack via the secretion of organic acids. *Sclerotinia sclerotiorum* and *B. cinerea* decrease the host pH by secreting large amounts of oxalic acid (Rollins and Dickman 2001; Manteau et al. 2003), whereas *Penicillium* (Prusky and Yakoby 2003; Prusky et al. 2004) (Fig. 2) and *Aspergillus* (Ruijter et al. 1999) secrete mainly gluconic and citric acids. *Penicillium expansum* isolates expressing higher activities of glucose oxidase and ability to secrete gluconic acid, showed increasing decay development in apple fruits. In contrast, reduction of glucose oxidase activity, by lowering the concentration of oxygen in the atmosphere, inhibited gluconic acid accumulation and reduced the decay development of *P. expansum* (Hadas et al. 2007). Acidic pH-specific expression of other members of the PG family was found in *S. sclerotiorum* (Lumsden 1976) and *BCPG*3 in *B. cinerea* (Wubben et al. 2000). Taken together, these results suggest that environmental acidification is important for fungal attack. It should also be noted that acidifying fungi possess the capacity to raise low pH levels to a favourable optimum (Zhang et al. 2005).

Modulators of the activation of quiescent infections and signal activation

The transition from biotrophism to a necrotrophic–saprophytic stage appears to be related to factors that are modulated at the intracellular level, and that are affected by nutrients and ambient pH. Each of the secreted compounds (organic acid or ammonia) plays a critical physiological role in the initiation of necrotrophic development. Speculation regarding the mechanisms by which secretion of organic acids enhances virulence centers on three modes of action. First, oxalate may be

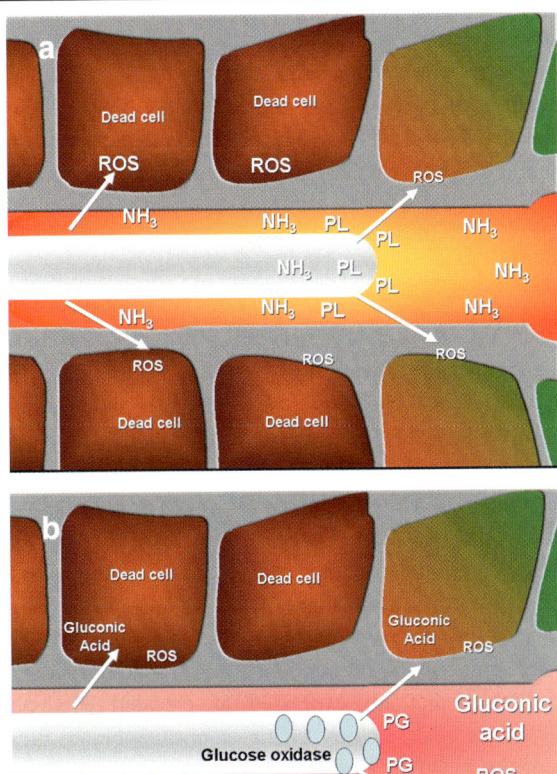

Fig. 2 Model for the activation of quiescent infections in post-harvest pathogens. In this model, two options are required for the transformation of quiescent to active infections pending on the type of pathogen (alkalinize or acidify the host tissue). The module A describes the increase in pathogenicity by the ammonification of the host tissue and ROS production by *Colletotrichum* and *Alternaria*; PL: pectate lyase. Module B describes the increase in pathogenicity by acidification of the host tissue, secretion of gluconic acid and ROS accumulation by *Pencillium expansum*; PG: polygalacturonase

directly toxic to host plants and may weaken them, thereby facilitating invasion. Second, it has been hypothesized that chelation of cell-wall Ca^{2+} by oxalic and gluconic acids loosens the plant cell wall (Bateman and Beer 1965; Hadas et al. 2007). Finally, oxalate secretion may suppress ROS generation together with the associated plant defence responses, thereby contributing to activation of the necrotrophic mode of development (Cessna et al. 2000).

Ammonification of the tissue also has significant physiological and biochemical effects on the host.

Recent results have shown that ammonification of the tissue enhances the accumulation of ROS in infected tissue (Alkan et al., unpublished data). Furthermore healthy plant cells have an electrochemical proton gradient across the plasma membrane, that is important for ion uptake, solute transport, and cell-wall growth. Several studies have demonstrated that transient extracellular alkalinization is essential for the induction of defence responses by fungal elicitors (Schaller and Oecking 1999). However, an essentially different feature of the activation of necrotrophic processes in *Colletotrichum* and *Alternaria* is the extreme elevation of pH that accompanies the accumulation of ammonia. The increased pH of the medium can affect all membrane functions, such as ion channels and plasma membrane ATPase (Gerendas and Ratcliffe 2000). Ammonium-induced alkalization leads to accumulation of weak bases and, subsequently, to elevation of the cytoplasmic and vacuolar pH (Kosegarten et al. 1997). Ammonium-toxicosis symptoms can lead to host stress responses that are expressed as elevated ethylene synthesis and various changes in membrane flux (Ingermarsson et al. 1987). Ammonium toxicosis of the host can be aggravated by the fungus, because ammonia can activate the expression of genes encoding pathogenicity factors, such as *PEL*B and PL secretion, in *C. gloeosporioides* (Kramer-Haimovich et al. 2006). This suggests that ammonification by *Colletotrichum* and *Alternaria* at the edge of the infection site may lead to deregulation of the host responses that, in turn, would facilitate the activation of fungal pathogenicity genes.

Present results suggest that one of the signals affecting secretion of organic acid and/or ammonia is the host environment pH, and this same signal may activate ammonium transporters in fungi. It is hypothesized that proteins such as the putative yeast outward transporter YaaH (Guaragnella and Butow 2003) are responsible for the elimination of excess ammonia. Other proteins, such as ammonium permeases (MEPs) of *Saccharomyces cerevisiae*, are members of a unique family of cytoplasmic membrane transporters that are specific for the ammonium ion (Pao et al. 1998). It has been suggested that members of the MEP/Amt family actively transport the charged species NH_4^+ across the cytoplasmic membrane (Howitt and Udvardi 2000), but no published reports have addressed the role of these transporters in fungal pathogenicity. Further studies of these transporters could help to explain the mechanism(s) underlying ammonium secretion by various post-harvest pathogen species.

New approaches for disease control of post-harvest pathogens

Harvested fruit and vegetables are agricultural produce that require stringent control of post-harvest handling, to avoid disease and to preserve the quality of the produce (Kobiler et al. 2001; Prusky et al. 2001a). This can be achieved by integration of safer synthetic fungicides, biological antagonists (Janisiewicz and Korsten 2002) and physical treatments (Prusky et al. 2001a). The studies that revealed host environment alkalinization by ammonia secretion during *A. alternata* and *C. gloeosporioides* colonization of fruits have opened a new approach to the modulation of disease development and control (Prusky et al. 2001a, b, Prusky and Yakoby 2003). For example, post-harvest treatment of mango fruits with HCl reduced the incidence of *Alternaria* rot after storage and shelf life (Prusky et al. 2006). These results suggest that acidic solutions may reduce the incidence of *Alternaria* rot by modulation of the pH at the infection court. In vitro and in vivo studies demonstrated that acidic solutions were also very efficient in inhibiting spore germination and germ-tube elongation.

Prusky et al. (2006) suggested that acidification of the fruit surface would provide an attractive approach for control of alkalinizing post-harvest pathogens, such as *Alternaria alternata*. This principle can apply to the use of alkalinizing agents, e.g., $NaHCO_3$, against acidifying pathogens such as *P. expansum*, *P. italicum*, and *P. digitatum* (Prusky et al. 2001b; Porat et al. 2002; Smilanick et al. 1999, 2005). It should be acknowledged that decreasing external pH to control alkalinizing fungi might also result in more favourable conditions for acidifying fungi (Zhang et al. 2005). An alternative approach for prevention of major pathogens in harvest produce could be by breading the fruit to unfavourable pH to the pathogens; however, this might lead to changes in the quality and taste of the fruits.

Summary

It has become clear in recent years that the activation of quiescent infections is not a simple process whereby a

decline in host resistance leads to activation of fungal attack. Rather, activation of quiescent biotrophic infections seems to involve a coordinated series of events (Fig. 2). One component in this complex process lies in the physiological and biochemical changes that occur in the host during ripening and senescence and that lead to a decreased host response and increased susceptibility. In parallel, activation of quiescent fungal infections involves processes that compromise host defences, directly or indirectly, by detoxification of antifungal agents (Prusky 1996). The physiological changes that accompany fruit ripening and host senescence – changes in, e.g., host pH, sugar content and cell-wall components, and oxidation of wounded tissue – trigger responses by the infecting fungus. The acidification of the tissue by organic acids (oxalic and gluconic) or its alkalinization by ammonia, and the possible modulation of the host ROS response and fungal ROS production, may contribute to rapid necrotization of the tissue. Further amplification of the decay can result from activation of gene expression and release of cell wall-degrading enzymes.

Despite intensive research, we still do not have sufficient knowledge of the range of tools utilized by post-harvest pathogens to elicit the transition from a quiescent biotrophic stage to an active necrotrophic one. In *C. gloeosporioides*, alkalinization resulted in significant changes in the expression of transcription factors, sugar transporters, and a wide range of primary metabolic genes (Miyara et al. 2005). For example, the elevated expression of glutamine synthase may reflect a requirement to handle the tremendous amount of ammonia secreted into the host tissue (Eshel et al. 2002). Further clarification of the role of putative signals (pH, nitrogen and sugar) in post-harvest pathogenesis during fruit ripening is clearly needed. Nevertheless, the current state of knowledge of fungal modulation of host pH has already opened new avenues for the control of post-harvest pathogens (Prusky et al. 2006).

References

Aguirre, J., Hansberg, W., & Navarro, R. (2006). Fungal responses to reactive oxygen species. *Medical Mycology*, *44*, 101–107.

Bailey, J. A., & Jeger, M. J. (1992). *Colletotrichum: Biology, Pathology and Control*. Wallingford: CAB International.

Bateman, D. F., & Beer, S. V. (1965). Simultaneous production and synergistic action of oxalic acid and polygalacturonase during pathogenesis by *Sclerotium rolfsii. Phytopathology*, *58*, 204–211.

Beno-Moualem, D., & Prusky, D. (2000). Early events in the development of quiescent infection of avocado fruits against *C. gloeosporioides. Phytopathology*, *90*, 553–559.

Caracuel, Z., Roncero, M. I. G., Espeso, E. A., Gonzales-Verdejo, C. I., Garcia-Maceira, F. I., & Di Pietro, A. (2003). The pH signaling transcription factor PacC controls virulence in the plant pathogen *Fusarium oxysporum. Molecular Microbiology*, *48*, 765–779.

Cessna, S. G., Searsa, V. E., Dickman, M. B., & Low, P. S. (2000). Oxalic acid, a pathogenicity factor for *Sclerotinia sclerotiorum*, suppresses the oxidative burst of the host plant. *Plant Cell*, *12*, 2191–2199.

Chen, C., & Dickman, M. B. (2005). Proline suppresses apoptosis in the fungal pathogen *Colletotrichum trifolii. Proceedings of the National Academy of Sciences of the USA*, *102*, 3459–3464.

Coates, L. M., Muirhead, I. F., Irwing, J. A. G., & Gowanlock, D. H. (1993). Initial infection processes by *Colletotrichum gloeosporioides* on avocado fruit. *Mycological Research*, *97*, 1363–1370.

Denison, S. H. (2000). pH regulation of gene expression in fungi. *Fungal Genetics and Biology*, *29*, 61–71.

Drori, N., Kramer-Haimovich, H., Rollins, J., Dinoor, A., Okon, Y., Pines, O., et al. (2003). External pH and nitrogen source affect secretion of pectate lyase by *Colletotrichum gloeosporioides. Applied and Environmental Microbiology*, *69*, 3258–3262.

Eshel, D., Miyara, I., Ailinng, T., Dinoor, A., & Prusky, D. (2002). pH regulates endoglucanase expression and virulence of *Alternaria alternata* in persimmon fruits. *Molecular Plant-Microbe Interactions*, *15*, 774–779.

Fourie, J. F., & Holz, G. (1995). Initial infection process by *Botrytis cinerea* on nectarine and plum fruits and the development of decay. *Phytopathology*, *85*, 82–87.

Gerendas, J., & Ratcliffe, R. G. (2000). Intracellular pH regulation in maize root tips exposed to ammonium at high external pH. *Journal of Experimental Botany*, *51*, 207–219.

Gil-ad, N. L., Bar-Nun, N., Noy, T., & Mayer, A. M. (2000). Enzymes in *Botrytis cinerea* capable of breaking down hydrogen peroxide. *FEMS Microbiology Letters*, *190*, 121–126.

Govrin, M., & Levin, A. (2000). The hypersensitive response facilitates plant infection by the necrotorophic pathogen *Botrytis cinerea. Current Biology*, *10*, 751–757.

Guaragnella, N., & Butow, R. A. (2003). ATO3 encoding a putative outward ammonium transporter is an RTG-independent retrograde responsive gene regulated by GCN4 and the Ssy1-Ptr3-Ssy5 amino acid sensor system. *Journal of Biological Chemisty*, *278*, 45882–45887.

Hadas, Y., Goldberg, I., Pines, O., & Prusky, D. (2007). The relationship between expression of glucose oxidase, gluconic acid accumulation, acidification of host tissue and the pathogenicity of *Penicillium expansum. Phytopathology*, *97*, 384–390.

Heath, M. C. (1997). Signalling between pathogenic rust fungi and resistant or susceptible host plants. *Annals of Botany*, *80*, 713–720.

Howitt, S. M., & Udvardi, M. K. (2000). Structure, function and regulation of ammonium transporters in plants. *Biochimica Biophysica Acta, 1465*, 152–170.

Ingermarsson, B., Oscarsson, P., Ugglas, M. A., & Larsson, C. M. (1987). Nitrogen utilization, III. Short-term effects of ammonium on nitrate uptake and nitrate reduction. *Plant Physiology, 85*, 865–867.

Janisiewicz, W. J., & Korsten, L. (2002). Biological control of postharvest disease of fruits. *Annual Reviews of Phytopathology, 40*, 411–441.

Kobiler, I., Shalom, Y., Roth, I., Akerman, M., Vinokour, Y., Fuchs, Y., et al. (2001). Effect of 2,4-dichlorophenoxyacetic acid on the incidence of side and stem-end rots in mango fruits. *Postharvest Biology and Technology, 23*, 23–32.

Kosegarten, H., Grolig, F., Wieneke, J., Wilson, G., & Hoffmann, B. (1997). Differential ammonia-elicited changes of cytosolic pH in root hair cells of rice and maize as monitored by 2′,7′-bis-(2-carboxyethyl)-5 (and -6)-carboxyfluorescein-fluorescence ratio. *Plant Physiology, 113*, 451–461.

Kramer-Haimovich, H., Servi, E., Katan, T., Rollins, J., Okon, Y., & Prusky, D. (2006). Effect of ammonia production by *Colletotrichum gloeosporioides* on *pelB* activation, pectate lyase secretion, and fruit pathogenicity. *Applied and Environmental Microbiology, 72*, 1034–1039

Latunde-Dada, A. O., O'Connell, R. J., Nash, C., Pring, R. J., Lucas, J. A., & Bailey, J. A. (1996). Injection process and identity of the hemibiotrophic anthracnose fungus (*Colletotrichum destructivum* O'Gara) from cowpea (*Vigna unguiculata* (L.) Walp.). *Mycological Research, 100*, 1133–1141.

Lumsden, R. D. (1976). Pectolytic enzymes of *Sclerotinia sclerotiorum* and their localization in infected bean. *Canadian Journal of Botany, 54*, 2630–2641.

Manteau, S., Abouna, S., Lambert, B., & Legendre, L. (2003). Differential regulation by ambient pH of putative virulence factors secretion by the phytopathogenic fungus *Botrytis cinerea. FEMS Microbiology and Ecology, 43*, 359–366.

Mellersh, D. G., Foulds, I. V., Higgins, V. J., & Heath, M. C. (2002). H_2O_2 plays different roles in determining penetration failure in three diverse plant fungal interactions. *Plant Journal, 29*, 257–268.

Mendgen, K., & Hahn, M. (2001). Plant infection and the establishment of fungal biotrophy. *Trends in Plant Science, 6*, 496498.

Miyara, I., Sherman, A., & Prusky, D. (2005). Regulation of gene expression on *Colletotrichum gloeosporioides* by alkalinization. *In*:23rd Fungal Genetics Asilomar, March, 15–20, 2005. Poster Abstracts. Poster number 324 http://www.fgsc.net/asil2005/asil2005.htm.

Movahedi, S., & Heale, J. B. (1990). Purification and characterization of an aspartic proteinase secreted by *Botrytis cinerea* Pers ex. Pers in culture and in infected carrots. *Physiological and Molecular Plant Pathology, 36*, 289–302.

O'Connell, R. J., Bailey, J. A., & Richmond, D. V. (1985). Cytology and physiology of infection of *Phaseolus vulgaris* by *Colletotrichum lindemuthianum. Physiological and Molecular Plant Pathology, 27*, 75–98.

Pao, S. S., Paulsen, I. T., & Saier Jr, M. H. (1998). Major facilitator superfamily. *Microbiology and Molecular Biology Reviews, 62*, 1–34.

Perfect, S. E., Bleddyn Hughes, H., O'Connell, R. J., & Green, J. R. (1999). *Colletotrichum*: A model genus for studies on pathology and fungal–plant interactions. *Fungal Genetics and Biology, 27*, 186–198.

Pezet, R., Viret, O., Perret, C., & Tabacchi, R. (2003). Latency of *Botrytis cinerea* Pers.:Fr. and biochemical studies during growth and ripening of two grape berry cultivars, respectively susceptible and resistant to grey mould. *Journal of Phytopathology, 151*, 208–214.

Porat, R., Daus, A., Weiss, B., Cohen, L., & Droby, S. (2002). Effects of combining hot water, sodium bicarbonate and biocontrol on postharvest decay of citrus fruit. *Journal of Horticultural Science and Biotechnology, 7*, 441–445.

Prusky, D. (1996). Pathogen quiescence in postharvest diseases. *Annual Reviews of Phytopathology, 34*, 413–434.

Prusky, D., Eshel, D., Kobiler, I., Yakoby, N., Beno-Moualem, D., Ackerman, M., et al. (2001a). Postharvest chlorine treatments for the control of the persimmon black spot disease caused by *Alternaria alternata. Postharvest Biology and Technology, 22*, 271–277.

Prusky, D., Kobiler, I., Akerman, M., & Miyara, I. (2006). Effect of acidic solutions and acid Prochloraz on the control of postharvest decays caused by *Alternaria alternata* in mango and persimmon fruit. *Postharvest Biology and Technology, 42*, 134–141.

Prusky, D., Kobiler, I., & Jacoby, B. (1988). Involvement of epicatechin in cultivar susceptibility of avocado fruits to *Colletotrichum gloeosporioides* after harvest. *Phytopathologische Zeitschrift, 123*, 140–146.

Prusky, D., McEvoy, J. L., Leverentz, B., & Conway, W. S. (2001b). Local modulation of host pH by *Colletotrichum* species as a mechanism to increase virulence. *Molecular Plant-Microbe Interactions, 14*, 1105–1113.

Prusky, D., McEvoy, J. L., Saftner, R., Conway, W. S., & Jones, R. (2004). The relationship between host acidification and virulence of *Penicillium* spp. on apple and citrus fruit. *Phytopathology, 94*, 44–51.

Prusky, D., & Yakoby, N. (2003). Pathogenic fungi: Leading or led by ambient pH? *Molecular Plant Pathology, 4*, 509–516.

Rolke, Y., Liu, S. J., Quidde, T., Williamson, B., Schouten, A., Weltring, V., et al. (2004). Functional analysis of H_2O_2-generating systems in *Botrytis cinerea*: The major Cu–Zn-superoxide dismutase (BCSOD1) contributes to virulence on French bean, whereas a glucose oxidase (BCGOD1) is dispensable. *Molecular Plant Pathology, 5*, 17–27.

Rollins, J. A., & Dickman, M. B. (2001). pH signaling in *Sclerotinia sclerotiorum*: Identification of *pac*C/RIM1 homolog. *Applied and Environmental Microbiology, 67*, 75–81.

Ruijter, G. J. G., van de Vondervoort, P. J. I., & Visser, J. (1999). Oxalic acid production by *Aspergillus niger*: An oxalate-non-producing mutant produces citric acid at pH 5 and in the presence of manganese. *Microbiology, 145*, 2569–2576.

Schaller, A., & Oecking, C. (1999). Modulation of plasma membrane H^+-ATPase activity differentially activates wound and pathogen defense responses in tomato plants. *Plant Cell, 11*, 263–272.

Shetty, N. P., Mehrabi, R., Lutken, H., Haldrup, A., Kema, G. H., Collinge, D. B., et al. (2007). Role of hydrogen peroxide during the interaction between the hembiotrophic fungal pathogen *Septoria tritici* and wheat. *New Phytologist, 11,* 637–647.

Smilanick, J. L., Mansour, M. F., Margosan, D. A., & Mlikota Gabler, F. (2005). Influence of pH and NaHCO₃ on effectiveness of imazalil to inhibit germination of *Penicillium digitatum* and to control postharvest green mold on citrus fruit. *Plant Disease, 89,* 640–648.

Smilanick, J. L., Margosan, D., Mlikota, F., Usall, J., & Michael, I. F. (1999). Control of citrus green mold by carbonate and bicarbonate salts and the influence of commercial postharvest practices on their efficacy. *Plant Disease, 83,* 139–145.

ten Have, A., Beuil, W. O., Wubben, J. P., Visser, J., & van Kan, J. A. L. (2001). *Botrytis cinerea* endopolygalacturonase genes are differentially expressed in various plant tissues. *Fungal Genetics and Biology, 33,* 97–105.

Unger, C. H., Kleta, S., Jandl, G., & Tiedemann, A. V. (2005). Suppression of the defense-related oxidative burst in bean leaf tissue and bean suspension cells by the necrotrophic pathogen *Botrytis cinerea*. *Journal of Phytopathology, 153,* 15–26.

van Baarlen, P., Staats, M., & van Kan, J. A. L. (2004). Induction of programmed cell death in lily by the fungal pathogen Botrytis elliptica. *Molecular Plant Pathology, 5,* 559–574.

Vernekar, J. V., Ghatge, M. S., & Deshpande, V. V. (1999). Alkaline protease inhibitor: A novel class of antifungal proteins against phytopathogenic fungi. *Biochemical and Biophysical Research Communications, 262,* 702–707.

Wang, X., Lichter, A., Kobiler, I., Leikin-Frenkel, A., & Prusky, D. (2004). Expression of Δ^{12} fatty acid desaturase during the induced accumulation of the antifungal diene involved in resistance to *Colletotrichum gloeosporioides* in avocado fruits. *Molecular Plant Pathology, 5,* 575–585.

Wubben, J. P., Mulder, W., ten Have, A., van Kan, J. A., & Visser, J. (1999). Cloning and partial characterization of endopolygalacturonase genes from *Botrytis cinerea*. *Applied and Environmental Microbiology, 65,* 1596–1602.

Wubben, J. P., ten Have, A., van Kan, J. A. L., & Visser, J. (2000). Regulation of endopolygacturonase gene expression in *Botrytis cinerea* by galacturonic acid, ambient pH and carbon catabolite repression. *Current Genetics, 37,* 152–157.

Yakoby, N., Beno-Moualem, D., Keen, N. T., Dinoor, A., Pines, O., & Prusky, D. (2001). *Colletotrichum gloeosporioides pel*B, is an important factor in avocado fruit infection. *Molecular Plant-Microbe Interactions, 14,* 988–995.

Yakoby, N., Kobiler, I., Dinoor, A., & Prusky, D. (2000). pH regulation of pectate lyase secretion modulates the attack of *Colletotrichum gloeosporioides* on avocado fruits. *Applied and Environmental Microbiology, 66,* 1026–1030.

Ye, X. Y., & Ng, T. B. (2002). A new peptidic protease inhibitor from *Vicia faba* seeds exhibits antifungal, HIV-1 reverse transcriptase inhibiting and mitogenic activities. *Journal of Peptide Science, 8,* 565–662.

Zhang, Z., Dvir, O., Pesis, E., Pick, U., & Lichter, A. (2005). Weak organic acids and inhibitors of pH homeostasis suppress growth of *Penicillium* infesting litchi fruits. *Journal of Phytopathology, 153,* 1–7.

Eur J Plant Pathol (2008) 121:291–302
DOI 10.1007/s10658-007-9237-2

REVIEW

What can we learn from clubroots: alterations in host roots and hormone homeostasis caused by *Plasmodiophora brassicae*

Jutta Ludwig-Müller · Astrid Schuller

Received: 13 April 2007 / Accepted: 4 October 2007
© KNPV 2007

Abstract The clubroot disease of cruciferous crops is caused by an obligate biotrophic protist, *Plasmodiophora brassicae*. The disease is characterized by the development of large root galls accompanied by changes in source-sink relations and the hormonal balance within the plant. Since the disease is difficult to control, it is of high economic interest to understand the events leading to gall formation. In this review we will give an overview on the current knowledge of changes brought about in the host root by this obligate biotrophic pathogen. Emphasis will be on the regulation of changes in plant hormone homeostasis, mainly auxins and cytokinins; the possible role of secondary metabolites, especially indole glucosinolates, in gall formation and auxin homeostasis will be discussed. Also, results from mutant analysis and microarrays using the model plant *Arabidopsis thaliana* are presented.

Keywords *Arabidopsis thaliana* · Biotrophic protist · Brassicaceae · Clubroot disease · Plant hormones

J. Ludwig-Müller (✉) · A. Schuller
Institut für Botanik, Technische Universität Dresden,
01062 Dresden, Germany
e-mail: Jutta.Ludwig-Mueller@tu-dresden.de

Introduction

Obligate biotrophic plant pathogens like *Plasmodiophora brassicae* establish an intricate interaction with their host during at least some parts of the infection process. Their dependence on host carbon sources is obvious but they might also be connected to host metabolism with respect to additional compounds like organic nitrogen, vitamins and minerals. They influence host physiology and alter host regulatory networks.

Clubroot disease development is restricted mainly to members of the mustard family and to a few other plants (Ludwig-Müller et al. 1999b), although *P. brassicae* is capable of infecting the root hairs of several non-cruciferous hosts in Gramineae, Rosaceae, Papaveraceae, Polygonaceae, Resedaceae, and Leguminosae (Webb 1949; MacFarlane 1952). In addition to crop plants, it is known that also wild crucifer species like *Capsella bursa-pastoris* (Buczacki and Ockendon 1979), *Cardamine* sp. (Tanaka et al. 1993), and *Arabidopsis thaliana* (Mithen and Magrath 1992) present on field plots can be infected with the clubroot pathogen.

This plant disease is one of the most devastating within the family of Brassicaceae, and the disease is difficult to control by either chemical or cultural means (Crisp et al. 1989). Clubroot-infected plants are usually dwarfed compared to healthy plants and when such plants are pulled out of the soil, the root system shows typical gall formation (Fig. 1). At maturity, the

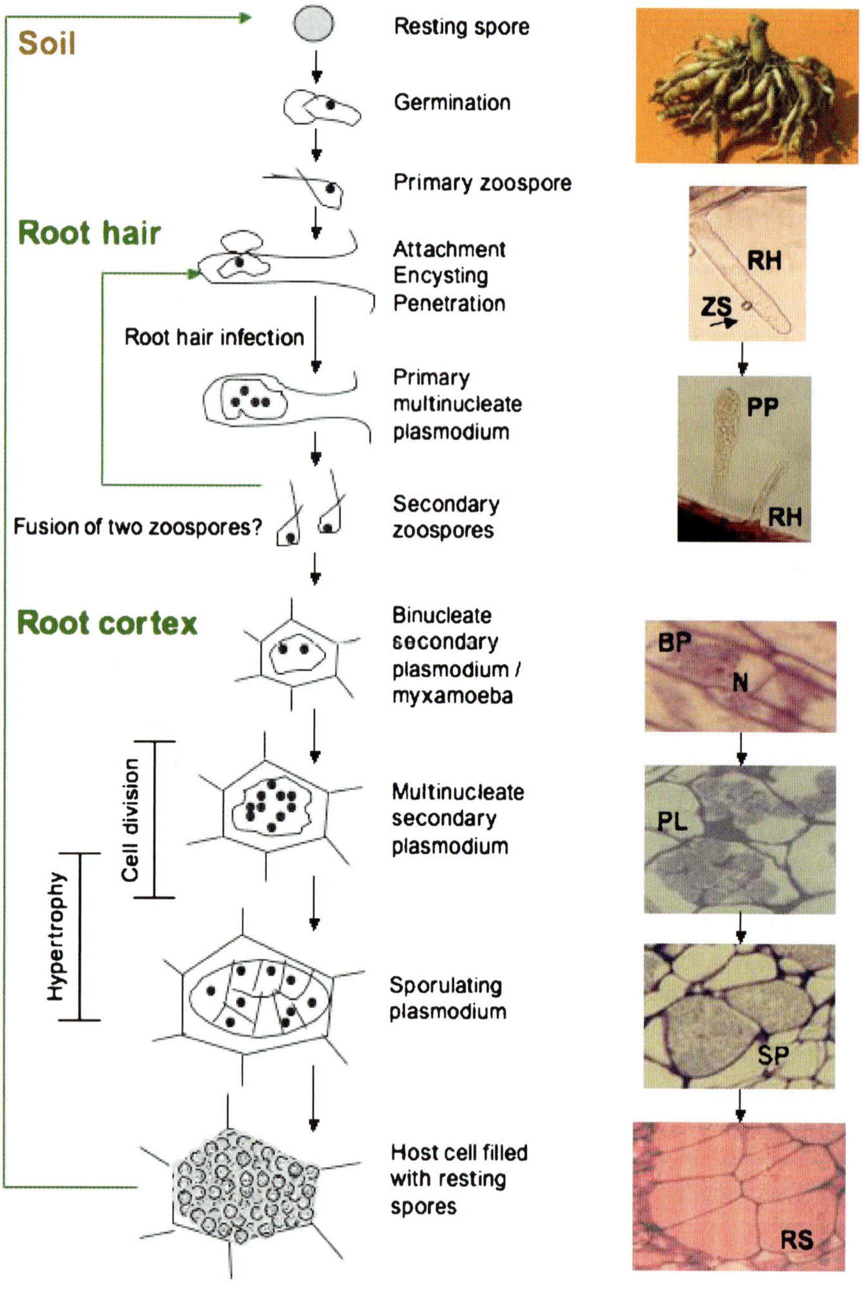

Fig. 1 The life-cycle of *Plasmodiophora brassicae* (*left panel*) illustrated by microscopic pictures of characteristic developmental stages of the pathogen (*right panel*) and a mature root gall of *Brassica* sp. The developmental stages during which cell divisions and hypertrophy occur are marked. A detailed description is given in the text. Abbreviations as they appear from the top of right panel: *ZS* zoospore (*arrow*), *RH* root hair, *PP* multinucleate primary plasmodium, *BP* binucleate secondary plasmodium (myxamoeba), *N* host nucleus, *PL* multinucleate secondary plasmodium, *SP* sporulating plasmodium, *RS* resting spores. Pictures were taken by Claudia Seidel and Jutta Ludwig-Müller

galls turn brown and a large portion of the infected roots remains under ground when the plants are harvested. Thereby the spores are liberated from the plant tissue and can remain infectious for up to 15 years (Mattusch 1977). Traditional control of clubroot includes raising the pH of the soil and rotating susceptible crops with non-cruciferous crops (Webster and Dixon 1991a; b). Growing resistant *Brassica* cultivars is another widely used alternative (Hirai 2006).

Two ecotypes of the model plant *A. thaliana* were identified, which showed a clear incompatible interaction with a distinct *P. brassicae* isolate (Fuchs and Sacristan 1996). The resistance reaction was accompanied by a hypersensitive response in roots. Infected cells were surrounded by necrotic boundaries, which restricted growth of the pathogen (Kobelt et al. 2000). These resistant ecotypes were characterized by the absence of typical clubroot symptoms, a slight reduction in number of lateral roots and the occurrence of

host cell necrosis, which were macroscopically visible as brown spots. Resistance to *P. brassicae* in these ecotypes is conferred by a dominant allele of a single nuclear gene. In contrast, genetic analysis of resistance to clubroot in *B. oleracea* has indicated that it is multigenic (Voorrips and Kanne 1997; Hirai 2006). Once *Brassica* cultivars bred for resistance are in the field for several years, the resistance may be overcome by new virulent *P. brassicae* strains (Mattusch 1994). This is only one example of differences occurring between clubroot infection in *Brassica* spp. and *Arabidopsis*. Others that concern hormone homeostasis will be discussed below.

Short introduction to the life-cycle and biology of the clubroot pathogen

The employment of molecular systematics indicated that *P. brassicae* belongs to the kingdom of Protista (Margulis et al. 1989) where it was grouped in the phylum Cercozoa (Cavalier-Smith and Chao 1997, 2003; Bulman et al. 2001). Together with other plant pathogens of the genus *Polymyxa* and *Spongospora*, *Plasmodiophora* comprises the order of the Plasmodiophorida (Cavalier-Smith and Chao 1997; Bulman et al. 2001). These have been considered as a monophyletic group based on: (1) an unusual form of nuclear division called cruciform division, (2) the presence of biflagellated, heterocont zoospores, (3) the presence of multinucleate plasmodia, (4) environmentally-resistant resting spores, and (5) obligate intracellular parasitism (Braselton 1995). This taxonomic unit was confirmed based on molecular systematic evidence using, for example, actin and ubiquitin sequences (Archibald and Keeling 2004; Bass et al. 2005) and SSU rRNA (Bulman et al. 2001).

Despite the fact that *P. brassicae* was identified as the causal agent of clubroot disease at the end of the nineteenth century (Woronin 1878), its life-cycle is still not entirely clear. The life-cycle of *P. brassicae* consists of two phases (Fig. 1): the primary phase, which is restricted to root hairs and epidermal cells of the host, and the secondary phase which occurs in the cortex and stele of roots and hypocotyl and leads to abnormal development (Ingram and Tommerup 1972).

When the haploid resting spores reach the vicinity of host roots, they start to germinate and primary zoospores are released. These zoospores become attached to a root hair opposite the point of flagellar insertion (Aist and Williams 1971) thereby starting the primary infection phase. The parasite forms a cyst which produces a tubular structure (rohr) containing a projectile-like structure (stachel), which is used to penetrate the host cell wall (reviewed in Braselton 1995). After formation of an adhesorium the parasite is injected into the root hair by the stachel where it appears as a small spherical amoeba. During further development the amoeba enlarges and several nuclear divisions occur (Williams et al. 1971) leading to the formation of a primary multinucleate plasmodium. This plasmodium develops through cleavage into zoosporangia and the development of the pathogen during the first infection cycle is completed with the release of 4–16 uninucleate zoospores to the exterior (Ayers 1944; Ingram and Tommerup 1972). Currently, the release of the secondary zoospores into the soil, and the fusion of the secondary zoospores prior to the infection of the root cortex that apparently occurs, is not resolved in detail. However, since it was possible to inoculate plants with single resting spores leading to club formation, it is assumed that for successful infection and symptom development, different mating types of *P. brassicae* are not required (Narisawa et al. 1996).

Once the pathogen has penetrated the cortex, secondary infection starts with the distribution of the pathogen in the form of a binucleate secondary plasmodium (myxamoeba) within the host tissue. However, the mode of distribution is not yet entirely clear (Buczacki 1983; Mithen and Magrath 1992; Kobelt 2000). According to Ingram and Tommerup (1972) mitotic divisions of the plasmodial nuclei occur in the binucleate secondary plasmodium and multinucleate secondary plasmodia are formed; these were observed in the central stele and were localized in the cells of the cambium and phloem parenchyma (Ludwig-Müller et al. 1999a; b; Kobelt 2000). This stage is accompanied by pronounced cell divisions and hypertrophic cells are formed. During further pathogen development cleavage of the plasmodium results in the formation of numerous resting spores. Although meiosis occurs during the cleavage of sporulating plasmodia into resting spores, it is still not known at which point in the life cycle karyogamy takes place (Braselton 1995).

Despite many efforts it has not been possible to cultivate the pathogen outside its host (Arnold et al.

1996). Several defined in vitro systems such as callus (Dekhuijzen 1975; 1981), hairy root cultures (Graveland et al. 1992; Asano et al. 2006) or suspension cultures (Asano and Kageyama 2006) have been described, but there is virtually no growth and no offspring of *P. brassicae*, so that these systems cannot replace inoculation in the soil for gall formation. Due to its obligate parasitism, information on physiological processes in *P. brassicae* is very scarce. The first sequence coding for an mRNA of *P. brassicae* induced *in planta* was reported by Ito et al. (1999), but after database searches it was difficult to assign a function to this mRNA. In contrast, the *PbTPS1* gene of *P. brassicae*, which was also expressed *in planta*, was identified as a trehalose-6-phosphate synthase (TPS) gene, based on the high homology of the predicted protein sequence to other TPS proteins (Brodmann et al. 2002). The molecular cloning of the *PbSTKL1* gene from *P. brassicae* with homology to kinase revealed another gene with a possible function (Ando et al. 2006b). Bulman et al. (2006) isolated 76 ESTs of *P. brassicae* from a library for which they found some homologies to fungal genes.

Changes in the host upon infection with *P. brassicae*

In a compatible reaction, signs of induced defence in the host root hair have been reported as deposition of callose between the host plasma membrane and the cell wall at the penetration site of the pathogen (Aist and Williams 1971). During the second infection cycle no evidence for an induced defence reaction was found. Williams and McNabola (1967) pointed out that necrotic responses were never observed and they discussed this as a reflection of the high degree of compatibility between host and parasite. Based on the observation of an enlarged host nucleolus, they noticed that the maintenance of a meristematic condition appears to be the chief response of the host cell to the parasite. Also, they discussed that the integrity of the outer membrane of the plasmodial envelope, which surrounds the plasmodium, may be under host control since its degeneration occurs at the same time as host ribosomes break down.

The most damaging aspect for the host in the infection course is the pronounced cell enlargement and cell proliferation. These symptoms, as well as the

growth of leaf-like teratomata from the roots, indicate the involvement of the plant hormones auxin and cytokinin in disease development. At the same time, the gall is established as a sink tissue for photosynthetic products. Some of the results discussed below are summarized in a model in Fig. 3.

The availability of hormone-responsive reporter genes in *Arabidopsis* enabled the localization of enhanced auxin and cytokinin responsiveness to the part of the root where gall formation occurs (Siemens et al. 2006). Experiments by Dekhuijzen (1976) indicated that infected callus tissue was not hormone autonomous. There is evidence that the pathogen is able to synthesize cytokinins (Müller and Hilgenberg 1986), but for auxins this has not yet been demonstrated (Ludwig-Müller 1999).

With the introduction of *Arabidopsis* to clubroot research it was possible to perform genome-wide transcriptome analysis to investigate host gene expression during the development of the disease on a broader basis (Siemens et al. 2006). Two time points were chosen, which were significantly different from each other. At an early time point (10 days after inoculation) small secondary plasmodia of the pathogen were visible, but only about 20% of the host tissue was colonized with limited change of host cell and root morphology. At a later time point (23 days after inoculation) different developmental stages of the pathogen were present. More than 60% of the host root cells were colonized and the root morphology was drastically altered. This experimental setup will be referred to as 'early time point' and 'late time point' throughout the text. Similarly, Devos et al. (2006) carried out a proteome analysis but during even earlier stages after inoculation (4 days). Therefore the data set is not comparable to the transcriptome data, but provides additional information about changes in the root caused by *P. brassicae*.

Auxins and indole glucosinolates

So far it has to be assumed that the increase in indole-3-acetic acid (IAA) is derived from the host plant. The biosynthesis of IAA in Brassicaceae involves several possible pathways which are interconnected and might occur simultaneously. Which pathway is operating can be dependent on the developmental stage, or stress situation of a given plant species or a certain tissue

within a plant (for a recent review on auxin biosynthesis and metabolism see Woodward and Bartel 2005). Since indole glucosinolates are connected to auxin biosynthesis in Brassicaceae, we will concentrate on the changes in auxin and indole glucosinolate metabolism connected to gall formation in two different species, *Arabidopsis* and *Brassica*. Already early work on clubroot showed that during clubroot formation, an increased synthesis and turnover of the putative host auxin precursors (see Fig. 2) indole-3-acetaldoxime (IAOx), indole-3-methylglucosinolate (indole GSL) and indole-3-acetonitrile (IAN) have been detected in infected *Brassica rapa* roots (Searle et al. 1982; Rausch et al. 1983; Butcher et al. 1984). Based on this experimental evidence it was hypothesized that if the indole GSL are key factors for the biosynthesis of IAA in clubs, then plants having little or no indole glucosinolates, and as a result less auxin, might be resistant to clubroot or should develop less severe symptoms (Butcher et al. 1974). However, conflicting results on this topic have been presented in *Brassica* species, not allowing any definite conclusion (Butcher et al. 1974; Ockendon and Buczacki 1979; Chong et al. 1981, 1984; Mullin et al. 1980; Ludwig-Müller et al. 1997; summarized in Ludwig-Müller 1999). On the other hand, the glucosinolate content of non-*Brassica* plants including Caricaceae, Resedaceae and Tropaeolaceae, which were inoculated with *P. brassicae*, also increased compared to control plants and small galls were occasionally detected in *Tropaeolum majus* (Ludwig-Müller et al. 1999b) and *Lepidium sativum* (Butcher et al. 1976). It was therefore speculated that these GSL (mainly benzyl GSL) could serve as precursors for phenylacetic acid, a naturally occurring auxin in *T. majus* (Ludwig-Müller and Cohen 2002).

With the availability of transgenic plants or mutants it was possible to further test this hypothesis. A set of mutants which were selected for changes in glucosinolate patterns (Haughn et al. 1991) was used to demonstrate that *tu8*, reduced in leaf indole GSL, showed reduced club fresh weight (Ludwig-Müller et al. 1999a), which correlated with lower IAN and lower free IAA content in the mutant roots compared to the wild-type, and the development of the pathogen seemed retarded. The *TU8* gene encodes the *Arabidopsis* heterochromatin-like protein1 (Kim et al. 2004), and therefore a direct connection to indole GSL could not be demonstrated.

A double mutant in the cytochrome P450 enzymes CYP79B2 and CYP79B3, which are involved in the first step of the biosynthesis of indole glucosinolates (Hull et al. 2000; Mikkelsen et al. 2000) and are indole glucosinolate-deficient, showed no differences in gall formation compared to wild-type plants (Siemens et al. 2007), although free IAA levels were comparable to the wild-type in the mutant galls. This demonstrates that indole glucosinolates are not the primary source of elevated levels of IAA in galls of mutant plants. One explanation might be that a block early in the pathway (Fig. 2) can compensate IAA levels possibly via a different route, whereas blocking steps later in the pathway, i.e. nitrilase (see below), does not allow for alternatives, thus resulting in smaller galls with less IAA.

The myrosinase-glucosinolate system present in crucifers is involved in several aspects of plant development and defence. Hydrolysis of GSL by myrosinase enzymes can produce substances with a remarkably wide spectrum of biological activities in addition to precursors for plant hormones (Halkier and Gershenzon 2006). One route for indole GSL metabolism is the conversion by myrosinase to IAN (Fig. 2). Myrosinase transcript was induced during late stages of clubroot in *B. rapa* (Grsic et al. 1999) and Devos et al. (2006) found an induction of a myrosinase in *Arabidopsis* by proteome analysis during very early stages of infection.

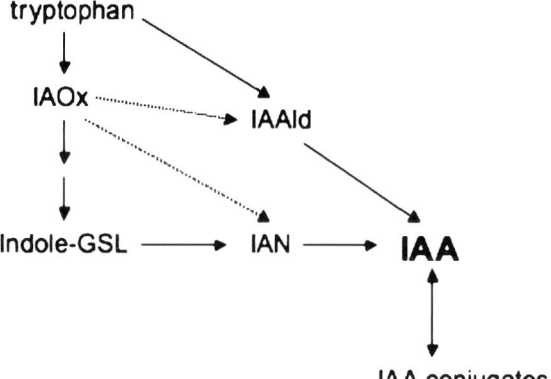

Fig. 2 The major pathways to indole glucosinolate and IAA synthesis possibly regulated during clubroot. *Solid arrows* indicate that for these pathways genes are known which encode the enzymes necessary for conversion, *dashed arrows* indicate that this pathway has been postulated or that enzymatic activities have been found, but so far no genetic evidence exists for the pathway

The next step in the IAA biosynthetic pathway would be the conversion of IAN to IAA by nitrilase (Fig. 2). Not surprisingly, nitrilase mRNA was increased during late stages of gall formation in *B. rapa* accompanied by an increase in nitrilase activity (Grsic et al. 1999). Transcript induction for nitrilase in *Arabidopsis* also correlated with the nitrilase protein localized exclusively in cells containing sporulating plasmodia (Grsic-Rausch et al. 2000; Neuhaus et al. 2000), indicating a role for this enzyme in cell enlargement (Grsic-Rausch et al. 2000). The *Arabidopsis nit1* mutant was more tolerant to clubroot, and antisense plants for the *NIT2* gene revealed retarded gall development (Grsic-Rausch et al. 2000; Neuhaus et al. 2000). In *B. rapa*, a different pathway for IAA synthesis via the intermediate indole-3-acetaldehyde (IAAld) might be also operating, because aldehyde oxidases were up-regulated during clubroot formation (Ando et al. 2006a). However, from transcriptome analysis there is no evidence that in *Arabidopsis* the corresponding genes are differentially expressed.

Since the nitrilase pathway seems to be induced only during later stages of infection, another possibility for the increase of IAA during earlier time points has to be considered. An additional pathway to control free IAA levels is the hydrolysis from inactive conjugates (Fig. 2; Woodward and Bartel 2005). Ludwig-Müller et al. (1996) have described

and characterized an auxin conjugate hydrolase activity from *B. rapa* which showed an increased enzyme activity in *P. brassicae*-infected roots compared to controls. In addition, four different amidohydrolase genes with high homology to the respective *Arabidopsis* genes were cloned from infected *B. rapa* roots (Schuller and Ludwig-Müller 2002, 2006). Using Real Time RT-PCR it was not possible to conclusively demonstrate the induction of hydrolase mRNA during infection (Schuller and Ludwig-Müller 2006). This result, together with the observation that the *B. rapa* hydrolases prefer IAA-alanine as substrate, led to the hypothesis that the activity found in root galls specific for IAA-aspartate, a conjugate most likely involved in the degradation of IAA but not in its hydrolysis (Östin et al. 1998), is not a substrate for the host hydrolase family. Therefore, maybe *P. brassicae* could be involved in IAA conjugate hydrolysis (Fig. 3), but this hypothesis has yet to be tested. The strong increase of IAA conjugates in *B. rapa* at different time points during infection (Ludwig-Müller et al. 1996; Devos et al. 2005) may also be indicative of increased IAA synthesis and subsequent conjugation. The latter may avoid a potentially toxic accumulation of IAA in the infected plant (Devos et al. 2005). Although transcriptome analysis did not show evidence for the up-regulation of auxin conjugate hydrolases in *Arabidopsis*, a proteome approach revealed IAR4, identified in a screen for

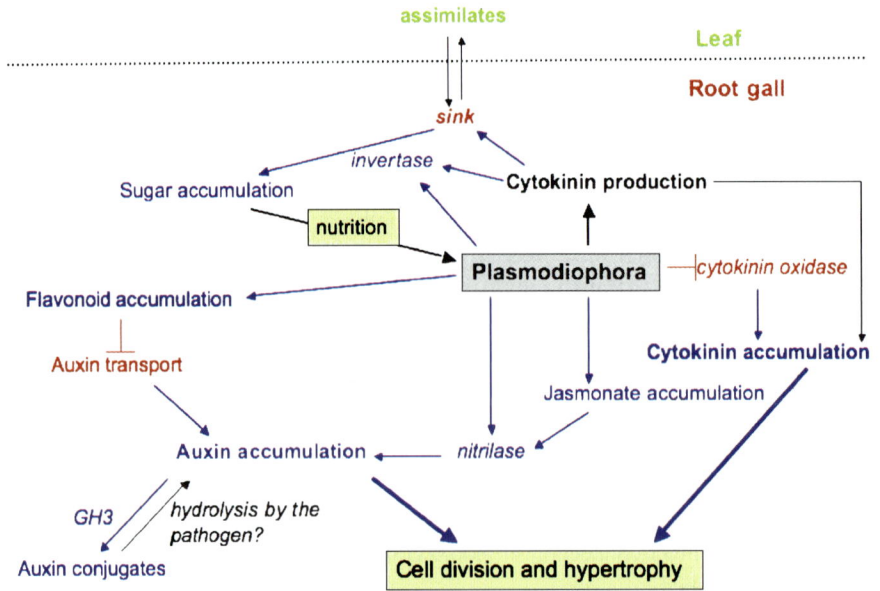

Fig. 3 A current model explaining the findings on hormone homeostasis and metabolism in clubroots of *Arabidopsis thaliana*. *Red* denotes down-regulation, *blue* up-regulation of pathways, compounds, enzymes or genes. *Black arrows* indicate that the pathway is directly influenced by *P. brassicae*. Data are summarized from direct experimental evidence or transcriptome analysis

auxin conjugate resistance, as an upregulated protein (Devos et al. 2006). The enzyme possibly acts on the conversion of indole-3-pyruvate to indole-3-acetyl-coenzyme A, which is a potential precursor of IAA amino-conjugates (LeClere et al. 2004). The reverse reaction from free IAA to auxin conjugates is catalyzed by six members of a large gene family in *Arabidopsis*, called *GH3*, which adenylate IAA and form amino acid conjugates (Staswick et al. 2005). Several *GH3* genes are auxin-inducible which makes them ideal tools to control IAA levels when excess auxin is present. The strong up-regulation of some members of the auxin-inducible *GH3* family involved in IAA amino acid conjugate formation point to the crucial role of auxin homeostasis during club development (Horn et al. 2006; Siemens et al. 2006). It is interesting to note that one conjugate formed is with aspartate (Staswick et al. 2005), linking IAA conjugate formation to the possibility that IAA amino acid conjugates with aspartate could be hydrolyzed by the pathogen (Fig. 3).

Also, several auxin-induced genes and auxin response factors were differentially regulated. The *axr3* mutant coding for the Aux/IAA protein IAA17 (for a recent review of the function of these proteins see Quint and Gray 2006), was more tolerant to clubroot (Alix et al. 2007). Other mutants connected to auxin metabolism or signalling were not affected (e.g. *axr1*, *axr2*, *ilr1*; Siemens et al. 2002). Furthermore, several putative auxin transport proteins were up-regulated according to transcriptome analysis, although a mutant of the auxin influx carrier AUX1 (Yang et al. 2006) did not show any differences in gall formation compared to the wild-type (Siemens et al. 2002). Differences in the observation that some mutants did not show tolerance or resistance even though their transcripts were increased, might be due to their position and importance in the pathway, whether the mutation can be compensated by others and whether parallel pathways are present. Since the IAA distribution seems to be important for gall development (Devos et al. 2006) it is tempting to speculate that fine tuning of the auxin transport pathway is also involved in regulating the auxin response (Fig. 3). Flavonoids are discussed as modulators of auxin efflux carriers of the PIN family (Besseau et al. 2007; Peer et al. 2004) and the accumulation of these compounds has been observed in root galls (Ludwig-Müller et al. 2006).

Cytokinins

It was shown that the amount of free and bound cytokinins was two to three times higher in clubs than in control roots (Dekhuijzen 1980). Dekhuijzen (1981) also showed that the contents of bound and free cytokinins were different in the host cytoplasm and plasmodia of the pathogen. Plasmodia contained zeatin riboside and its glucose derivative, whereas the host cytoplasm contained zeatin riboside and small amounts of the glucose-6-phosphate derivatives of zeatin and zeatin riboside (Dekhuijzen, 1981). From these results, it was concluded that plasmodia synthesize cytokinins which are released into the host cytoplasm inducing host cell division. Müller and Hilgenberg (1986) isolated young secondary plasmodia from clubs 23 days after inoculation by a two-step percoll gradient and showed that isolated secondary plasmodia of *P. brassicae* were not only able to take up ^{14}C-adenine in vitro, but also to incorporate adenine into *trans*-zeatin. It can thus be assumed that the increase in cytokinins is at least partly derived from active synthesis of zeatin by plasmodia of *P. brassicae*. The authors speculated that only a very small amount of *trans*-zeatin with its high biological activity would be sufficient to induce a sink in clubbed roots (Müller and Hilgenberg 1986). Transcriptome data showed that putative cytokinin biosynthesis genes as well as root-specific cytokinin oxidases/dehydrogenases, which are involved in the degradation of cytokinins (Werner et al. 2003), were down-regulated in *Arabidopsis* root galls. The down-regulation of the cytokinin-degradation capacity within colonized cells in combination with cytokinin production inside these cells by the pathogen can turn these cells into strongly dividing tissue by small amounts of pathogen-derived cytokinin (Siemens et al. 2006). This is strongly corroborated by the fact that lowering cytokinin levels by overexpressing a cytokinin oxidase gene in *Arabidopsis* leads to strong tolerance against *P. brassicae* infection (Siemens et al. 2006). However, in *B. rapa* in contrast to the situation in *Arabidopsis*, an increased transcription of isopentenyl transferase genes, involved in cytokinin biosynthesis, was described in root galls (Ando et al. 2005). These differences might be explained by differences in the host life-cycle because for *Arabidopsis*, only whole root systems were analyzed, whereas individual parts of the root were investigated in the study with *B. rapa*. The authors have separated gall tissue from tissue

without symptoms and also looked at younger or older parts of the gall which could influence the outcome of such an analysis (Ando et al. 2005).

Cytokinins are also known to be involved in the attraction of nutrients. The induction of a strong metabolic sink by *P. brassica* has been described decades ago (Keen and Williams 1969a, b). Reallocation of assimilates from leaves to infected hypocotyls has been detected in this phase of infection (Keen and Williams 1969b). Evans and Scholes (1995) showed the redirection of soluble sugars and the altering of carbon partitioning in infected *Arabidopsis* plants. Starch accumulation has been observed (Brodmann et al. 2002), but starch synthesis mutants (*adg1, adg2*) of *Arabidopsis* did not showed any degree of tolerance to clubroot infection (Siemens et al. 2002). Most of the genes involved in sugar or starch metabolism were up-regulated either at the first or second time point, including sucrose synthase, invertase, starch synthase, and ß-amylase (Siemens et al. 2006). In this context it is interesting to note that invertases can be induced by cytokinins (Ehneß and Roitsch 1997). In addition, several sugar transporters were differentially regulated, which indicates a strong change in trafficking of metabolites between different cells and compartments. Interestingly, a trehalase gene was also up-regulated at the late time point, which might be involved in the control of trehalose levels in the plant, because *P. brassicae* produced trehalose in its resting spores (Brodmann et al. 2002). The disaccharide trehalose is a storage form of carbon but also a protectant against various stresses such as desiccation and heat stress in bacteria and yeast (Müller et al. 1999). In an analogy, it could be seen as storage in clubroot-diseased *Arabidopsis* plants because trehalose was found only in infected tisues (Brodmann et al. 2002), but it is not clear whether the trehalose produced in plant tissue is derived from the pathogen. However, it could also protect the resting spores of *P. brassicae* from desiccation.

Hormones involved in biotic and abiotic stress signalling

Abscisic acid (ABA), ethylene, jasmonic acid (JA), and salicylic acid (SA) are plant hormones that play important roles during abiotic and biotic stress signalling. With respect to biotic stress, there is evidence that several signalling pathways involving plant hormones exist, but not all compounds are involved at the same time. It is a generally accepted view that, for example, ethylene acts synergistically with JA in the activation of defences against necrotrophic pathogens and as an antagonist in the SA-dependent resistance against biotrophic pathogens. However, SA and ethylene can also act together in the control of resistance to selected biotrophs. Consequently, mutants with constitutive ethylene, jasmonate or salicylate response are more tolerant to biotrophic pathogens (for a recent review on the complex networks see e.g. Adie et al. 2007). While there is increasing knowledge on signalling pathways during infection of *Arabidopsis* with leaf pathogens, no such evidence has been detailed for the root pathogen *P. brassicae*.

The regulatory function of abscisic acid has been extensively studied in relation to plant abiotic stress responses, such as drought, salt, and cold (Davies et al. 2005; Gusta et al. 2005), but it has recently been acknowledged to regulate responses to biotic stress as well in an antagonistic manner with ethylene and JA (Adie et al. 2007). First indications that ABA plays a role during clubroot have been obtained from *B. rapa* roots (Devos et al. 2005). These authors found an increase in ABA when gall development occurred. Analyzing the transcriptome data of *Arabidopsis* concerning the ABA influence during clubroot disease indicates that genes coding for ABA responsive/ induced proteins as well as genes that are involved in ABA signalling were strongly increased during the late time point. In addition, several drought-responsive genes were also up-regulated (the total dataset of the microarray can be found at Array Express of the European Bioinformatics Institute, experiment no. E-MEXP-254).

For genes involved in ethylene synthesis or genes annotated to be ethylene-responsive, no consistent pattern was obtained, although several genes encoding biosynthetic enzymes and ethylene response factors were differentially regulated according to the microarray data. Mutant analysis showed that ethylene response mutants were not tolerant to clubroot (Siemens et al. 2002; Alix et al. 2007). However, the *alh1* mutant, defective in the cross-talk between ethylene and auxins (Vandenbussche et al. 2003), probably at the site of auxin transport, shows a resistant phenotype upon *P. brassicae* infection (Devos et al. 2006).

 Springer

JA and its methyl ester have been found in a large number of plant species (Parthier 1991). They are known to regulate a number of different physiological processes in plants including the induction of senescence, vegetative storage proteins, and proteinase inhibitors (Koda 1992). They also have a role in interplant signalling (Pena-Cortes et al. 2004) and in signal transduction in relation to defence gene induction (Pauw and Memelink 2004; Pozo et al. 2004). Endogenous JA concentrations increased between 21 and 35 days after inoculation in infected *B. rapa* roots (Grsic et al. 1999) indicating a role for JA as a signal during clubroot development. Analysis of the transcription of genes involved in the jasmonate biosynthesis pathway showed a strong increase in expression for one lipoxygenase gene at the late time point, whereas genes encoding the following proteins in the biosynthesis pathway were either not regulated or they had rather a tendency to show down-regulation. The gene encoding JAR1, the enzyme responsible for adenylation of JA and subsequent synthesis of JA amino acid conjugates (Staswick et al. 2002; Staswick and Tiryaki 2004) was down-regulated and consequently the *jar1* mutant was more sensitive to clubroot (Siemens et al. 2002). It was also shown that indole GSL is selectively induced by jasmonic acid (JA) or methyl-JA treatment in *B. rapa* (Ludwig-Müller et al. 1997) and that nitrilase and myrosinase were inducible by JA (Grsic et al. 1999). These findings indicate possible links between auxin and JA response during clubroot development.

Conclusion

A general problem researchers have to deal with during biochemical and molecular investigations on clubroot is the intimate contact between pathogen and host. The data available so far have mostly been obtained with total root material so that very subtle changes of metabolism or gradients in hormone concentrations could not easily be detected when infected and healthy tissues were compared. One has to take into account that the pathogen may be able to induce local changes, which have to be resolved on a cytochemical level. Therefore, it will be necessary in the future to obtain data using either immunocytological studies, as has been done with antibodies against nitrilase (Grsic-

Rausch et al. 2000), or with promoter-reporter lines which will allow at least detection of responsiveness to hormones on a more cellular level (Siemens et al. 2006). This approach will of course also allow the monitoring of other transcriptional activities in more detail. Also, the use of reverse genetics or tissue-specific expression of transcripts to change gene expression in a pathogen-inverse manner will allow us to investigate the function of individual genes in the context of clubroot disease.

Acknowledgement Work in the author's laboratory was funded by the Deutsche Forschungsgemeinschaft (DFG) and the State of Saxony (Sächsisches Landesamt für Umwelt und Geologie).

References

Adie, B., Chico, J. M., Rubio-Somoza, I., & Solano, R. (2007). Modulation of plant defenses by ethylene. *Journal of Plant Growth Regulation, 26*, 160–177.

Aist, J. R., & Williams, P. H. (1971). The cytology and kinetics of cabbage root hair penetration by *Plasmodiophora brassicae. Canadian Journal of Botany, 49*, 2023–2034.

Alix, K., Lariagon, C., Delourme, R., & Manzanares-Dauleux, M. J. (2007). Exploiting natural genetic diversity and mutant resources of *Arabidopsis thaliana* to study the *A. thaliana-Plasmodiophora brassicae* interaction. *Plant Breeding, 126*, 218–221.

Ando, S., Asano, T., Tsushima, S., Kamachi, S., Hagio, T., & Tabei, Y. (2005). Changes in gene expression of putative isopentenyltransferase during clubroot development of Chinese cabbage (*Brassica rapa* L.). *Physiological and Molecular Plant Pathology, 67*, 59–67.

Ando, S., Tsushima, S., Tagiri, A., Kamachi, S., Konagaya, K.-I., & Hagio, T., et al. (2006a). Increase in *BrAO1* gene expression and aldehyde oxidase activity during clubroot development in Chinese cabbage (*Brassica rapa* L.). *Molecular Plant Pathology, 7*, 223–234.

Ando, S., Yamada, T., Asano, T., Kamachi, S., Tsushima, S., & Hagio, T., et al. (2006b). Molecular cloning of PbSTKL1 gene from *Plasmodiophora brassicae* expressed during clubroot development. *Journal of Phytopathology, 154*, 185–189.

Archibald, J. M., & Keeling, P. J. (2004). Actin and ubiquitin protein sequences support a Cercozoan/Foraminiferan ancestry for the Plasmodiophorid plant pathogens. *Journal of Eukaryotic Microbiology, 51*, 113–118.

Arnold, D. L., Blakesley, D., & Clarkson, J. M. (1996). Evidence for the growth of *Plasmodiophora brassicae* in vitro. *Mycological Research, 100*, 535–540.

Asano, T., & Kageyama, K. (2006). Growth and movement of secondary plasmodia of *Plasmodiophora brassicae* in turnip suspension-culture cells. *Plant Pathology, 5*, 145–151.

Asano, T., Kodama, A., & Kageyama, K. (2006). Susceptibility of hairy root lines of *Brassica* species to *Plasmodiophora brassicae* and in an in vitro subculture system. *Journal of General Plant Patholology*, *72*, 85–91.

Ayers, G. W. (1944). Studies on the life history of the club root organism *Plasmodiophora brassicae*. *Canadian Journal of Research*, *22*, 143–149.

Bass, D., Moreira, D., Lopez-Garcia, P., Polet, S., Chao, E. E., & von der Heyden, S., et al. (2005). Polyubiquitin insertions and the phylogeny of Cercozoa and Rhizaria. *Protist*, *156*, 149–161.

Besseau, S., Hoffmann, L., Geoffroy, P., Lapierr, C., Pollet, B., & Legrand, M. (2007). Flavonoid accumulation in Arabidopsis repressed in lignin synthesis affect auxin transport and plant growth. *The Plant Cell*, *19*, 148–162.

Braselton, J. P. (1995). Current status of the Plasmodiophorids. *Critical Reviews of Microbiology*, *21*, 263–275.

Brodmann, D., Schuller, A., Ludwig-Müller, J., Aeschbacher, R. A., Wiemken, A., & Boller, T., et al. (2002). Induction of trehalase in *Arabidopsis* plants infected with the trehalose-producing pathogen *Plasmodiophora brassicae*. *Molecular Plant-Microbe Interaction*, *15*, 693–700.

Buczacki, S. T. (1983). Plasmodiophora. An inter-relationship between biological and practical problems. In S. T. Buczacki (Ed.) *Zoosporic plant pathogens* pp. 161–191. Academic: London.

Buczacki, S. T., & Ockendon, J. G. (1979). Preliminary observations on variation in susceptibility to clubroot among collections of some wild crucifers. *Annals of applied Biology*, *92*, 113–118.

Bulman, R. S., Kühn, S. F., Marschall, J. W., & Schnepf, E. (2001). A phylogenetic analysis of the SSR rRNA from members of the Plasmodiophorida and Phagomyxida. *Protist*, *152*, 43–51.

Bulman, S., Siemens, J., Ridgeway, H., Eady, C., & Conner, A. (2006). Identification of genes from the obligate intracellular plant pathogen, *Plasmodiophora brassicae*. *FEMS Microbiological Letters*, *264*, 198–204.

Butcher, D. N., El-Tigani, S., & Ingram, D. S. (1974). The role of indole glucosinolates in the clubroot disease of the Cruciferae. *Physiological Plant Pathology*, *4*, 127–141.

Butcher, D. N., Searle, L. M., & Mousdale, D. M. A. (1976). The role of glucosinolates in the club root disease of the cruciferae. *Mededeligen Faculteit Landbouwwetenschappen Rijksuniersiteit Gent*, *41/2*, 525–532.

Butcher, D. N., Chamberlain, K., Rausch, T., & Searle, L. M. (1984). Changes in indole metabolism during the development of clubroot symptoms in Brassicas. In: Biochemical Aspects of Synthetic and Naturally Occurring Plant Growth Regulators. *British Plant Growth Regulator Group, Monograph*, *11*, 91–101.

Cavalier-Smith, T., & Chao, E. E. (1997). Sarcomonad ribosomal RNA sequences, rhizopod phylogeny, and the origin of euglyphid amoebae. *Archiv für Protistenkunde*, *147*, 227–236.

Cavalier-Smith, T., & Chao, E. E. (2003). Phylogeny and classification of phylum Cercozoa (Protozoa). *Protist*, *154*, 341–358.

Chong, C., Chiang, M. S., & Crete, R. (1981). Thiocyanate ion content in relation to clubroot disease severity in cabbages. *Horticultural Science*, *16*, 663–664.

Chong, C., Chiang, M. S., & Crete, R. (1984). Studies in glucosinolates in clubroot resistant selections and susceptible commercial cultivars of cabbages. *Euphytica*, *34*, 65–73.

Crisp, P., Crute, I. R., Sutherland, R. A., Angell, S. M., Bloor, K., & Burgess, H., et al. (1989). The exploitation of genetic ressources of *Brassica oleracea* in breeding for resistance to clubroot *Plasmodiophora brassicae*. *Euphytica*, *42*, 215–226.

Davies, W. J., Kudoyarova, G., & Hartung, W. (2005). Long-distance ABA signaling and its relation to other signaling pathways in the detection of soil drying and the mediation of the plant's response to drought. *Journal of Plant Growth Regulation*, *24*, 285–295.

Dekhuijzen, H. M. (1975). The enzymatic isolation of secondary vegetative plasmodia of *Plasmodiophora brassicae* from callus tissue of *Brassica campestris*. *Physiological Plant Pathology*, *6*, 187–192.

Dekhuijzen, H. M. (1976). The role of growth hormones in club root formation. *Mededeligen Faculteit Landbouwwetenschappen Rijksuniersiteit Gent*, *41*, 517–523.

Dekhuijzen, H. M. (1980). The occurrence of free and bound cytokinins in clubroots and *Plasmodiophora brassicae* infected turnip tissue cultures. *Physiologia Plantarum*, *49*, 169–176.

Dekhuijzen, H. M. (1981). The occurrence of free and bound cytokinins in plasmodia of *Plasmodiophora brassicae* isolated from tissue cultures of clubroots. *Plant Cell Reports*, *1*, 18–20.

Devos, S., Vissenberg, K., Verbelen, J.-P., & Prinsen, E. (2005). Infection of Chinese cabbage by *Plasmodiophora brassicae* leads to a stimulation of plant growth: impacts on cell wall metabolism and hormonal balance. *New Phytologist*, *166*, 241–250.

Devos, S., Laukens, K., Deckers, P., Van Der Straeten, D., Beeckman, T., & Inze, D., et al. (2006). A hormone and proteome approach to picturing the initial metabolic events during *Plasmodiophora brassicae* infection on *Arabidopsis*. *Molecular Plant-Microbe Interaction*, *19*, 1431–1433.

Ehneß, R., & Roitsch, T. (1997). Coordinated induction of extracellular invertase and glucose transporters in *Chenopodium rubrum* by cytokinins. *The Plant Journal*, *11*, 539–548.

Evans, J. L., & Scholes, J. D. (1995). How does clubroot alter the regulation of carbon metabolism in its host. *Aspects of Applied Biology*, *42*, 125–132.

Fuchs, H., & Sacristan, M. D. (1996). Identification of a gene in *Arabidopsis thaliana* controlling resistance to clubroot (*Plasmodiophora brassicae*) and characterization of the resistance response. *Molecular Plant-Microbe Interaction*, *9*, 91–97.

Graveland, R., Dale, P., & Mithen, R. (1992). Gall development in hairy root cultures infected with *Plasmodiophora brassicae*. *Mycological Research*, *96*, 225–228.

Grsic, S., Kirchheim, B., Pieper, K., Fritsch, M., Hilgenberg, W., & Ludwig-Müller, J. (1999). Induction of auxin biosynthetic enzymes by jasmonic acid and in clubroot diseased Chinese cabbage plants. *Physiologia Plantarum*, *105*, 521–531.

Grsic-Rausch, S., Kobelt, P., Siemens, J., Bischoff, M., & Ludwig-Müller, J. (2000). Expression and localization of nitrilase during symptom development of the clubroot

disease in *Arabidopsis thaliana*. *Plant Physiology*, *122*, 369–378.

Gusta, L. V., Trischuk, R., & Weiser, C. J. (2005). Plant cold acclimation: The role of abscisic acid. *Journal of Plant Growth Regulation*, *24*, 308–318.

Halkier, B. A., & Gershenzon, J. (2006). Biology and biochemistry of glucosinolates. *Annual Review of Plant Biology*, *57*, 303–333.

Haughn, G. W., Davin, L., Giblin, M., & Underhill, E. W. (1991). Biochemical genetics of plant secondary metabolites in *Arabidopsis thaliana*. The glucosinolates. *Plant Physiology*, *97*, 217–226.

Hirai, M. (2006). Genetic analysis of clubroot resistance in *Brassica* crops. *Breeding Science*, *56*, 223–229.

Horn, C., Siemens, J., & Ludwig-Müller, J. (2006). The GH3-gene family of *Arabidopsis thaliana* and the obligate pathogen *Plasmodiophora brassicae*. (Paper presented at the 15th Crucifer Genetics Workshop: Brassic 2006, Wageningen, The Netherlands).

Hull, A. K., Vij, R., & Celenza, J. L. (2000). Arabidopsis cytochrome P450s that catalyze the first step of tryptophan-dependent indole 3-acetic acid biosynthesis. *Proceedings of the National Academy of Sciences of the United States of America*, *97*, 2379–2384.

Ingram, D. S., & Tommerup, I. C. (1972). The life history of *Plasmodiophora brassicae* Woron. *Proceedings of the Royal Society B*, *180*, 103–112.

Ito, S., Ichinose, H., Yanagi, C., Tanaka, S., Kameya-Iwaki, M., & Kishi, F. (1999). Identification of an in planta-induced mRNA of *Plasmodiophora brassicae*. *Journal of Phytopathology*, *147*, 79–82.

Keen, N. T., & Williams, P. H. (1969a). Synthesis and degradation of starch and lipids following infection of cabbage by *Plasmodiophora brassicae*. *Phytopathology*, *59*, 778–785.

Keen, N. T., & Williams, P. H. (1969b). Translocation of sugars into infected cabbage tissues during club root development. *Plant Physiology*, *44*, 748–754.

Kim, J. H., Durrett, T. P., Last, R. L., & Jander, G. (2004). Characterization of the *Arabidopsis* TU8 glucosinolate mutation, an allele of *TERMINAL FLOWER2*. *Plant Molecular Biology*, *54*, 671–682.

Kobelt, P. (2000). Die Verbreitung von sekundären Plasmodien von *Plasmodiophora brassicae* (Wor.) im Wurzelgewebe von *Arabidopsis thaliana* nach immunhistologischer Markierung des plasmodialen Zytoskeletts. Dissertation, Institut für Angewandte Genetik, Freie Universität Berlin, Germany.

Kobelt, P., Siemens, J., & Sacristan, M. D. (2000). Histological characterisation of the incompatible interaction between *Arabidopsis thaliana* and the obligate biotrophic pathogen *Plasmodiophora brassicae*. *Mycological Research*, *104*, 220–225.

Koda, Y. (1992). The role of jasmonic acid and related compounds in the regulation of plant development. *International Review of Cytology*, *135*, 155–199.

LeClere, S., Rampey, R. A., & Bartel, B. (2004). *IAR4*, a gene required for auxin conjugate sensitivity in *Arabidopsis*, encodes a pyruvate dehydrogenase E1α homolog. *Plant Physiology*, *135*, 989–999.

Ludwig-Müller, J. (1999). *Plasmodiophora brassicae*, the causal agent of clubroot disease: A review on molecular and biochemical events in pathogenesis. *Journal of Plant Disease and Plant Protection*, *106*, 109–127.

Ludwig-Müller, J., & Cohen, J. D. (2002). Identification and quantification of three active auxins in different tissues of *Tropaeolum majus*. *Physiologia Plantarum*, *115*, 320–329.

Ludwig-Müller, J., Epstein, E., & Hilgenberg, W. (1996). Auxin-conjugate hydrolysis in Chinese cabbage: Characterization of an amidohydrolase and its role during the clubroot disease. *Physiologia Planarum*, *97*, 627–634.

Ludwig-Müller, J., Schubert, B., Pieper, K., Ihmig, S., & Hilgenberg, W. (1997). Glucosinolate content in susceptible and tolerant Chinese cabbage varieties during the development of the clubroot disease. *Phytochemistry*, *44*, 407–414.

Ludwig-Müller, J., Pieper, K., Ruppel, M., Cohen, J. D., Epstein, E., & Kiddle, G., et al. (1999a). Indole glucosinolate and auxin biosynthesis in *Arabidopsis thaliana* L. glucosinolate mutants and the development of the clubroot disease. *Planta*, *208*, 409–419.

Ludwig-Müller, J., Bennett, R. N., Kiddle, G., Ihmig, S., Ruppel, M., & Hilgenberg, W. (1999b). The host range of *Plasmodiophora brassicae* and its relationship to endogenous glucosinolate content. *New Phytologist*, *144*, 443–458.

Ludwig-Müller, J., Siemens, J., Horn, C., & Päsold, S. (2006). Metabolic and hormonal changes during root gall development after infection of Arabidopsis with *Plasmodiophora brassica*. (Paper presented at the Plant Genetics Conference, Kiel, Germany).

MacFarlane, I. (1952). Factors affecting the survival of *Plasmodiophora brassicae* Wor. in the soil and its assessment by a host test. *Annals of Applied Biology*, *39*, 239–256.

Margulis, L., Corliss, J. O., Melkonian, M., & Chapman, D. J. (1989). *Handbook of Protoctista*. Jones and Partlett Publishers: Boston.

Mattusch, P. (1977). Epidemiology of crucifers caused by *Plasmodiophora brassicae*. In S. T. Buczacki, & P. H. Williams (Eds.) *Woronin + 100 international conference on clubroot* (pp. 24–28). Madison, WI: University of Wisconsin.

Mattusch, P. (1994). Kohlhernieanfälligkeit eines Chinakohlsortiments. *Gemüse*, *30*, 357–359.

Mikkelsen, M. D., Hansen, C. H., Wittstock, U., & Halkier, B. A. (2000). Cytochrome P450 CYP79B2 from Arabidopsis catalyzes the conversion of tryptophan to indole-3-acetaldoxime, a precursor of indole glucosinolates and indole-3-acetic acid. *Journal of Biological Chemistry*, *275*, 33712–33717.

Mithen, R., & Magrath, R. (1992). A contribution to the life history of *Plasmodiophora brassicae*: secondary plasmodia development in root galls of *Arabidopsis thaliana*. *Mycological Research*, *96*, 877–885.

Müller, J., Wiemken, A., & Aeschbacher, R. (1999). Trehalose metabolism in sugar sensing and plant development. *Plant Science*, *147*, 37–47.

Müller, P., & Hilgenberg, W. (1986). Isomers of zeatin and zeatin riboside in clubroot tissue: Evidence for trans-zeatin biosynthesis by *Plasmodiophora brassicae*. *Physiologia Plantarum*, *66*, 245–250.

Mullin, W. J., Proudfoot, K. G., & Collins, M. J. (1980). Glucosinolate content and clubroot of rutabaga and turnip. *Canadian Journal of Plant Science*, *60*, 605–612.

Narisawa, K., Kageyama, K., & Hashiba, T. (1996). Efficient root infection with single resting spores of *Plasmodiophora brassicae. Mycological Research, 100*, 855–858.

Neuhaus, K., Grsic-Rausch, S., Sauerteig, S., & Ludwig-Müller, J. (2000). *Arabidopsis* plants transformed with nitrilase 1 or 2 in antisense direction are delayed in clubroot development. *Journal of Plant Physiology, 156*, 756–761.

Ockendon, J. G., & Buczacki, S. T. (1979). Indole glucosinolate incidence and clubroot susceptibility of three cruciferous weeds. *Transactions of the British mycological Society, 72*, 156–157.

Östin, A., Kowalczyk, M., Bhalerao, R. P., & Sandberg, G. (1998). Metabolism of indole-3-acetic acid in Arabidopsis. *Plant Physiology, 118*, 285–296.

Parthier, B. (1991). Jasmonates, new regulators of plant growth and development: Many facts and few hypotheses on their actions. *Botanica Acta, 104*, 446–454.

Pauw, B., & Memelink, J. (2004). Jasmonate-responsive gene expression. *Journal of Plant Growth Regulation, 23*, 200–210.

Peer, W. A., Bandyopadhyay, A., Blakeslee, J. J., Makam, S. N., Chen, R. J., & Masson, P. H., et al. (2004). Variation in expression and protein localization of the PIN family of auxin efflux facilitator proteins in flavonoid mutants with altered auxin transport in *Arabidopsis thaliana. The Plant Cell, 16*, 1898–1911.

Peña-Cortés, H., Barrios, P., Dorta, F., Polanco, V., Sánchez, C., & Sánchez, E., et al. (2004). Involvement of jasmonic acid and derivatives in plant response to pathogen and insects and in fruit ripening. *Journal of Plant Growth Regulation, 23*, 246–260.

Pozo, M. J., Van Loon, L. C., & Pieterse, C. M. J. (2004). Jasmonates – Signals in plant-microbe interactions. *Journal of Plant Growth Regulation, 23*, 211–222.

Quint, M., & Gray, W. M. (2006). Auxin signaling. *Current Opinion in Plant Biology, 9*, 448–453.

Rausch, T., Butcher, D. N., & Hilgenberg, W. (1983). Indole-3-methylglucosinolate biosynthesis and metabolism in clubroot diseased plants. *Physiologia Plantarum, 58*, 93–100.

Schuller, A., & Ludwig-Müller, J. (2002). Isolation of differentially expressed genes involved in clubroot disease. *Plant Protection Science, 38*, 483–486.

Schuller, A., & Ludwig-Müller, J. (2006). A family of auxin conjugate hydrolases from *Brassica rapa*: Characterization and expression during clubroot disease. *New Phytologist, 171*, 145–158.

Searle, L. M., Chamberlain, K., Rausch, T., & Butcher, D. N. (1982). The conversion of 3-indolemethylglucosinolate to 3-indoleacetonitrile by myrosinase and its relevance to the clubroot disease of the cruciferae. *Journal of Experimental Botany, 33*, 935–942.

Siemens, J., Nagel, M., Ludwig-Müller, J., & Sacristán, M. D. (2002). The interaction of *Plasmodiophora brassicae* and *Arabidopsis thaliana*: Parameters for disease quantification and screening of mutant lines. *Journal of Phytopathology, 150*, 592–605.

Siemens, J., Keller, I., Sarx, J., Kunz, S., Schuller, A., & Nagel, W., et al. (2006). Transcriptome analysis of *Arabidopsis* clubroots indicate a key role for cytokinins in disease development. *Molecular Plant-Microbe Interaction, 19*, 480–494.

Siemens, J., Glawischnig, E., & Ludwig-Müller, J. (2007). Indole glucosinolates and camalexin do not influence the development of the clubroot disease in *Arabidopsis thaliana. Journal of Phytopathology* (in press).

Staswick, P. E., Tiryaki, I., & Rowe, M. L. (2002). Jasmonate response locus JAR1 and several related Arabidopsis genes encode enzymes of the firefly luciferase superfamily that show activity on jasmonic, salicylic and indole-3-acetic acids in an assay for adenylation. *The Plant Cell, 14*, 1405–1415.

Staswick, P. E., & Tiryaki, I. (2004). The oxylipin signal jasmonic acid is activated by an enzyme that conjugates it to isoleucine in Arabidopsis. *The Plant Cell, 16*, 2117–2127.

Staswick, P. E., Serban, B., Rowe, M., Tiryaki, I., Maldonado, M. T., & Maldonado, M. C., et al. (2005). Characterization of an Arabidopsis enzyme family that conjugates amino acids to indole-3-acetic acid. *The Plant Cell, 17*, 616–627.

Tanaka, S., Ito, S., Kameya-Iwaki, M., Katumoto, K., & Nishi, Y. (1993). Occurrence and distribution of clubroot disease on two cruciferous weeds, *Cardamine flexuosa* and *C. scutata*, in Japan. *Transactions of the mycolgical Society Japan, 34*, 381–388.

Vandenbussche, F., Smalle, J., Le, J., Saibo, N. J. M., De Paepe, A., & Chaerle, L., et al. (2003). The *Arabidopsis* mutant *alh*1 illustrates a cross talk between ethylene and auxin. *Plant Physiology, 131*, 1228–1238.

Voorrips, R. E., & Kanne, H. J. (1997). Genetic analysis of resistance to clubroot (*Plasmodiophora brassicae*) in *Brassica oleracea*. I. Analysis of symptom grades. *Euphytica, 93*, 31–39.

Webb, P. C. R. (1949). Zoosporangia, believed to be those of *Plasmodiophora brassicae*, in the root hairs of non-cruciferous plants. *Nature, 163*, 608.

Webster, M. A., & Dixon, G. R. (1991a). Calcium, pH and inoculum concentration influencing colonization by *Plasmodiophora brassicae. Mycological Research, 95*, 64–73.

Webster, M. A., & Dixon, G. R. (1991b). Boron, pH and inoculum concentration influencing colonization by *Plasmodiophora brassicae. Mycological Research, 95*, 74–79.

Werner, T., Motyka, V., Laucou, V., Smets, R., van Onckelen, H., & Schmülling, T. (2003). Cytokinin-deficient transgenic *Arabidopsis* plants show multiple developmental alterations indicating opposite functions of cytokinins in the regulation of shoot and root meristem activity. *The Plant Cell, 15*, 2532–2550.

Williams, P. H., & McNabola, S. S. (1967). Fine structure of *Plasmodiophora brassicae* in sporogenesis. *Canadian Journal of Botany, 45*, 1665–1669.

Williams, P. H., Aist, S. J., & Aist, J. R. (1971). Response of cabbage root hairs to infection by *Plasmodiophora brassicae. Canadian Journal of Botany, 49*, 41–47.

Woodward, A. W., & Bartel, B. (2005). Auxin: regulation, action and interaction. *Annals of Botany, 95*, 707–735.

Woronin, M. (1878). *Plasmodiophora brassicae*, Urheber der Kohlpflanzen-Hernie. *Jahrbuch der Wissenschaften in Botanik, 11*, 548–574.

Yang, Y., Hammes, U. Z., Taylor, C. G., Schachtmann, D. P., & Nielsen, E. (2006). High-affinity auxin transport by the AUX1 influx carrier protein. *Current Biology, 16*, 1123–1127.

Eur J Plant Pathol (2008) 121:303–312
DOI 10.1007/s10658-007-9259-9

Problems with disseminating information on disease control in wheat and barley to farmers

Lise Nistrup Jørgensen · Egon Noe ·
Ghita C. Nielsen · Jens Erik Jensen ·
Jens Erik Ørum · Hans O. Pinnschmidt

Received: 25 May 2007 / Accepted: 22 November 2007 / Published online: 11 December 2007
© KNPV 2007

Abstract Plant pathologists have traditionally worked in the area of clarifying and understanding the disease cycles of specific diseases, factors influencing epidemiology, yield loss potential and host-pathogen interactions in order to be able to minimise the disease risk, build warning systems or recommend specific control thresholds in relation to the application of fungicides. The decision support system Crop Protection Online (CPO) is an example of a threshold-based system that determines economically viable fungicide strategies. The system is based on using appropriate doses aimed at minimising the overall pesticide input. CPO is used widely by advisors and many of the thresholds are generally accepted and disseminated through newsletters. The national figures for the use of fungicides in cereals have shown a major reduction during the last 20 years and their use today is much in line with the level that can be achieved from using CPO as indicated from validation trials. The number of end-users among farmers has been stable at around 3% during the last 10 years (800–1,000 farmers). Major hurdles in increasing the number of users are believed to be: (1) the requirements for carrying out assessments in the field, (2) farm sizes getting larger, leaving less time for decision making for individual fields, (3) lack of economic incentives to change from standard treatments, (4) the failure of decision support systems to interact with other computer-based programmes on the farm, (5) the lack of compatibility of decision support systems with farmers' ways of making decisions on crop protection in general, (6) the need for direct interactions with advisors. A sociological investigation into the farmers' way of making decisions in the area of crop protection has shown that arable farmers can be divided into three major groups: (a) systems-orientated farmers, (b) experienced-based farmers and (c) advisory-orientated farmers. The information required by these three groups is different and has to be looked at individually from the end-user's perspective rather than from the scientist's perspective. New ways of entering the decision support system where specific

L. N. Jørgensen (✉) · H. O. Pinnschmidt
Faculty of Agricultural Sciences, Department of Integrated
Pest Management, University of Aarhus,
4200 Slagelse, Denmark
e-mail: Lisen.jorgensen@agrsci.dk

E. Noe
Faculty of Agricultural Sciences,
Department of Agroecology and Environment,
University of Aarhus,
8830 Tjele, Denmark

G. C. Nielsen · J. E. Jensen
The Danish Advisory Services,
Udkærsvej 15,
8200 Skejby, Denmark

J. E. Ørum
Faculty of Life Sciences, Food and Resource Economics
Institute, University of Copenhagen,
Rolighedsvej,
Copenhagen, Denmark

field inspections are omitted and where regional disease data are relied on, have been investigated and tested in field trials. The results show possibilities for further developments in that direction, which might be one way of gaining more end-users.

Keywords Decision support system · Disease thresholds · Eco-sociological barriers · Farmer types · Winter wheat

Introduction

Legislation and limitations have been imposed on the use of pesticides in many countries in northern Europe for the last 20–25 years. Based on these national action plans, it has become important to encourage a low pesticide input strategy. Similar proposals to eliminate unnecessary use are also stated in the EU's Thematic Strategy on the Sustainable Use of Pesticides (Anonymous 2006b). In order to eliminate unnecessary use of fungicides, precise knowledge of the disease risk of an epidemic at field level is essential. In order to achieve this, it is very important to provide farmers with simple and robust methodologies to assess disease occurrence or risk in order to judge whether fungicides should be applied (Hansen et al. 1994; Jørgensen et al. 1996; Verret et al. 2000).

The development of thresholds, models and decision support systems (DSS) to farmers and advisors covering this need started in the late 1980s. EPIPRE was one of the first systems to be developed (Zadoks 1983). Several other systems have been developed since and new systems are still entering the field today (Röhrig 2006). Among the systems that have been on the market for the longest time are Pro_Plant and Crop Protection Online (CPO); both systems are today Internet-based (Rydahl et al. 2003; Volk et al. 2003).

More specifically, the Danish DSS CPO is a threshold-based system that determines economically viable strategies for control of diseases, pests and weeds using low pesticide input (Secher et al. 1995; Rydahl 2003). The system is designed to support advisors and farmers when they make decisions about crop protection in arable crops. CPO consists of three parts: (1) encyclopaedic information on pests, diseases and weeds, (2) recommendations for control of pests and diseases of cereals and (3) recommendations

for control of weeds in the major crops grown in Denmark. The models focus on optimising the input and minimising the cost of controlling a given problem seen in the field. Calculation of the optimal pesticide dose is therefore an important output. Focus on control strategies on diseases has been not only to look for the best disease control but also to optimise net yield rather than gross yield (Jørgensen et al. 2000). CPO is part of Planteinfo (http://www.Planteinfo.dk), which, among other things, features access to climate data, a pesticide database and a cultivar database covering the major crops grown in Denmark.

Based on field trial studies, CPO has shown a significant potential for reducing the use of herbicides in particular compared to the common spraying practice. Due to the proven reduction potential, there has been a significant interest among politicians to ensure a more widespread use of the system. This is in line with the pesticide reduction schemes implemented in Denmark for more than 20 years (Jørgensen and Kudsk 2006).

CPO is used by nearly all Danish advisors (approximately 420) in the field of arable production, and approximately 3% of Danish farmers subscribe to the system. These user numbers have been fairly stable for approximately 10 years. It has often been discussed whether it would be possible to increase the number of farmers using the system. However, and unfortunately, this has proven to be very difficult. The main obstacle is the time-consuming field registrations, which are necessary for the models to calculate field-specific recommendations. Historically, it has proven difficult to stimulate farmers to use DSS (McCown 2002). It has been shown previously that farmers have different expectations of decision-making tools depending on their styles of farming (Noe and Halberg 2002).

The major objective of this paper is to summarise the state of the art in Denmark with respect to practical disease control in winter wheat. The paper gives examples of thresholds used in CPO and results from applying and validating the existing system. A sociological investigation with focus on the potential use and barriers to applying an existing decision support system like CPO is reported. Results from trials investigating possibilities of avoiding time-demanding field registrations are also presented. The results are discussed from both an agronomic and a socio-economic angle. The results are based on a project covering both weed and disease control in cereals in order to validate the

potential from the DSS and other means of disseminating information. This paper focuses on elements concerned with disease control in cereals.

Methods

Agronomic evaluation of the system

Since the early development of CPO, control thresholds have been developed empirically and validated in field trials in order to investigate whether the recommendations given by the system are competitive with other fungicide recommendations. A total of 45 validation trials were carried out in winter wheat and 38 in spring barley distributed at different sites in Denmark which represent all relevant soil and climate conditions. The trials were conducted as standard field trials with four replicates and a randomised block design. The treatments applied in the CPO plots were based on weekly disease assessments in order to determine whether and when the disease thresholds were exceeded. Following the recommendations of different control strategies, disease assessments were carried out as per cent leaf area covered by each disease. Assessments were made at several growth stages, but the focus in this paper is on assessments carried out at approximately growth stage (GS) 75. The plots were harvested with a plot combine harvester. The dry matter content was measured, and the grain yield was corrected to 15% moisture content. The benefit from fungicide treatment is represented in terms of yield, with deduction made to cover the cost of fungicide and its application costs. A grain price of 750 DKK t^{-1} was used for the calculations.

Investigation of new approaches to CPO

During two seasons, a total of ten trials were carried out by the Danish Agricultural Advisory Service (DAAS) to compare the existing version of CPO with a new prototype, which, instead of using individual field information, relies on information from the national monitoring system using data from the nearest regional locality as a basis for assessing the need for control. The trials were carried out at ten sites in Denmark representing different soil types, climates and disease pressures. Disease assessments were all made as per cent leaf area covered by each disease. The plots were

harvested with a plot combine harvester. The dry matter content was measured, and the grain yield was corrected to 15% moisture content.

Both in validation trials and trials investigating the new approach to CPO, spraying was carried out with a plot sprayer at low pressure (2–3 bars) using flat fan nozzles and a water volume of 200–300 l ha^{-1}. Statistical analyses on the data were carried out using ANOVA, from which LSDs were calculated at the 95% confidence level.

Sociological investigation on farmers' decision process on crop protection

The sociological investigation was based on a questionnaire with 746 returns and four focus group interviews in order to understand how farmers make their decisions with respect to using pesticides, and to explore the barriers to a more extensive use of the present CPO.

The method used to identify and analyse for ideal types was previously described by Weber (1922). The statistical technique used for analysing the questionnaire data was the categorical principal component analysis. Selected variables were used to construct a system of co-ordinates based on their inter-relationships. In this study, the procedure indicated three different types of decision-making strategies for arable farmers. Following this grouping, focus group interviews were carried out with farmers who had responded to the questionnaire in order to obtain a better understanding of the three types with respect to decision-making, actions and motivational logic. The exact classification of each decision-making strategy was made based on the characteristic ways in which the participants from each group made decisions. The method and the analyses of the data are described in further detail by Langvad and Noe (2006) and Jørgensen et al. (2007a).

Results and discussion

Thresholds and cultivar ranking in wheat

Examples of the thresholds used in CPO for control of wind and splash-borne diseases in wheat are shown in Table 1. The thresholds are based on many years of

Table 1 List of some of the thresholds used for control of diseases in winter wheat in the decision support system Crop Protection Online (CPO)

Winter wheat	
Eyespot	
>35% plants attacked at growth stage (GS) 30–32. Only main shoots are assessed. The attack must spread to the next-to-the-outermost leaf sheath to be included	
Powdery mildew	**Brown rust**
Susceptible cultivars: GS 29–31: >10% plants attacked, GS 32–40: >25% plants attacked	Susceptible cultivars: GS 30–31: >25% plants attacked, GS 32–50: >10% plants attacked, GS 51–71: >25% plants attacked
Non-susceptible and partly susceptible cultivars: GS 29–31: >25% plants attacked, GS 32–40: >50% plants attacked	
Yellow rust	
Susceptible cultivars: GS 29–60: >1% plants attacked, GS 61–71: >10% plants attacked	Partly susceptible cultivars: GS 29–60: >1% plants attacked, GS 61–71: >50% plants attacked
Non-susceptible cultivars: GS 29–60: >10% plants attacked, GS 61–71: >75% plants attacked	
Septoria	
Susceptible cultivars: At least 4 days with precipitation (>1 mm) from GS 32. If the crop was sprayed before GS 52, the counting of days with precipitation begins after 10 days. If the crop is sprayed from GS 52 onwards, the counting of days with precipitation begins after 20 days. A maximum of 30 days are counted back in time. A spray is also triggered at GS 45–59 if >10% of the plants show attack on the third leaf from the top. Control of septoria, if any, can be considered until GS 71	Partly susceptible cultivars: At least 5 days with precipitation (more than 1 mm) from GS 37. Control at GS 39 at the earliest. Otherwise as in susceptible cultivars
Tan spot	
Only with wheat as previous crop and with reduced tillage between crops	Only with wheat as previous crop and with reduced tillage between crops
Susceptible cultivars: GS 31–32: >75% plants attacked, GS 33–60: >25% plants attacked, GS 61–71: >50% plants attacked	Less susceptible cultivars: GS 37–60: >50% plants attacked, GS 61–71: >75% plants attacked

empirical work (Jørgensen et al. 1996) and are occasionally revised if new information becomes available. Examples of recent changes include incorporation of thresholds for new diseases such as tan spot (*Drechslera tritici-repentis*; Jørgensen and Olsen 2007), adjustments of the use of thresholds for control of powdery mildew (*Blumeria graminis*; Jørgensen and Pinnschmidt 2004) and adjustments of recommended doses and timing for the control of septoria leaf blotch (*Septoria tritici*; Henriksen et al. 2000). In relation to future climate changes, the programme will probably need further adjustments if, for instance, brown rust (*Puccinia triticina*) becomes more common as an early epidemic disease.

Specifically, the lists of cultivars are updated every year including the latest ranking of the cultivars' susceptibility to different diseases. Growing cultivars with good disease resistance is the most important

tool for trying to reduce the potential loss from disease in an integrated control strategy (Jørgensen et al. 2003a). Major differences in resistance and response to fungicide application can be found between cultivars and therefore cultivar resistance is taken into consideration when the disease models in CPO are used. For each disease in wheat, all cultivars are divided into three groups according to their resistance level. This grouping influences the threshold and the doses recommended. The most commonly grown winter wheat cultivars in Denmark vary significantly in their resistance ranking for leaf diseases (Table 2). The disease ranking for septoria leaf blotch, powdery mildew, yellow rust (*Puccinia striiformis*) and brown rust is based on national cultivar trials carried out at approximately 15 locations annually or specific small plot trials using artificial inoculation, as it is the case for yellow rust,

Table 2 Commonly grown cultivars in Denmark 2007 and their disease resistance ranking to the most important diseases

Cultivar	Yield level dt ha^{-1}	Percent area grown 2007	Ranking of cultivars to disease resistance (1–3) 1=high resistance					
			Septoria leaf blotch	Powdery mildew	Yellow rust	Brown rust	Tanspot	Fusarium head blight
Smuggler	89.8 (101)[b]	34	2	1	1	1	2	2
Skalmeje	88.3 (99)	11	2	1	1	2	2	1
Cultivar mixture[a]	89.1 (100)	11	–	–	–	–	–	–
Samyl	88.2 (99)	7	3	3	1	2	2	2
Opus	87.3 (98)	7	2	2	1	1	2	2
Robigus	85.4 (96)	7	2	0	1	1	2	2
Abika	88.4 (99)	3	2	0	1	1	2	2
Ambition	94.2 (106)	2	1	0	0	2	2	2

[a] The mixture contained Ambition, Smuggler and Skalmeje or Ambition, Abika and Robigus

[b] Yield relative to the cultivar mixture

tan spot and fusarium head blight (*Fusarium* spp.). The ranking of cultivars is updated annually by a standing committee to ensure that any changes in susceptibility are included in the database for cultivars as well as in CPO. The cultivar database (www. sortinfo.dk) is widely used by advisors and farmers as the system combines all relevant information on current and prospective cultivars including ranking for disease resistance and estimating the likely cost for fungicide treatments in an average year and cropping situation (Detlefsen and Jensen 2001).

Agronomic evaluation of the system

Results from 38 spring barley trials and 45 winter wheat validation trials carried out in Denmark between 1998 and 2003 are summarised in Table 3.

CPO was validated and compared to various standard treatments, which represent common practices using appropriate doses. In most cases, the standard treatments in wheat cover two to three treatments with a dose ranging from 25–50% of normal field rate per treatment. Half of the trials were carried out using 2× 33% of the standard label rate. In spring barley, a single or split treatment with effective fungicides was used for comparison. In split treatments, 2×25% of the normal label rate was used, and, when comparing with a single treatment 25–50% of the normal field rate was used. In both wheat and barley, strobilurins and triazoles were used alone or in mixtures. This was the case for both standard treatments and the treatments recommended by CPO. The results showed that the system gave similar or slightly better disease control and yield responses compared to standard

Table 3 Results on disease control, yields, cost of control and Treatment Frequency Index (TFI) from validation of Crop Protection Online (CPO) in field trials in spring barley and winter wheat

Crop	Treatment	No. of trials	Percent brown rust	Percent septoria leaf blotch	Percent net blotch	Percent powdery mildew	Yield increase (dt ha^{-1})	Net yield (dt ha^{-1})	Costs of fungicide + application (DKK ha^{-1})	TFI
Spring barley	Reference [a]	38	0.1		2		4.3	1.0	180	0.5
38 trials	CPO	38	0.1		2		5.4	1.9	200	0.41
LSD$_{95}$[c]							ns			
Winter wheat	Reference [a]	45		12		2	12.3	6.6	362	0.79
45 trials	CPO	45		10		1	13.4	7.7	360	0.70
LSD$_{95}$[b]							NS			

[a] Reference treatments have been chosen based on present standards and varies across the trials

[b] LSD$_{95}$ values refer to comparison between actual version of CPO and actual reference treatments

Table 4 Reduction potential in Treatment Frequency Index (TFI) from using Crop Protection Online (CPO) in winter wheat and spring barley assessed in relation to official pesticide statistics for fungicides and the target figures from the pesticide action plan 2004–2009

	TFI fungicides	
	Winter wheat	Spring barley
Official statistics[a]	0.74	0.34
Target for 2009	0.65	0.35
CPO testing[b]	0.70	0.41
No. of trials	45	38

[a] Average of 3 years of statistics (2003–2005 sales figures)

[b] Average TFI fungicide input from 8 years' trial testing of CPO in field trials

treatments although differences were not statistically significant. Overall, the fungicide input was slightly lower compared with the standard treatments.

The optimal input with fungicides depends on the disease pressure and the climate in the individual season, but the susceptibility of the cultivar in particular plays a major role for the optimal input. The differences in margin between resistant and susceptible cultivars are largest in seasons with severe attacks of septoria leaf blotch and yellow rust (Jørgensen et al. 2003a). The dose-response curves for ear treatments targeting control of septoria leaf blotch and yield responses have generally been very flat. The economic optimum in resistant cultivars was a TFI (treatment frequency index = number of standard dosages used) between 0.25 and 0.5. Most typically, only one ear application was profitable. On

more susceptible cultivars, the optimum TFI was 0.5–0.9, and a two-spray programme gave more flexibility with regard to the dose needed at the second application (Jørgensen and Nielsen 2003b).

Validation trials from the past eight years with fungicides in cereals indicated only a very limited and uncertain reduction potential in fungicide usage (Table 4). The system offers a good opportunity to control disease only if a need can be identified. In general, the CPO recommendations for fungicide input are in agreement with the target figures envisaged by the pesticide action plan 2004–2009. During the past 20 years, a widespread use of reduced doses has led to a significant reduction in fungicide input in winter wheat (Fig. 1). On average, the number of fungicide treatments per season in wheat is approximately two, and the average TFI of fungicides is 0.35 per treatment (Kleffmann–Farmstat 2005; Anon. 2006b).

Investigation of new introductions to CPO

The results from the ten trials showed quite similar results from using either recommendations from the present version of CPO or the prototype using regional disease data rather than individual field data (Table 5). In one case, the level of powdery mildew was much higher in the monitoring system compared to the level in the actual field where the trial was carried out. This led to an unnecessarily high input at that site. In the other nine trial sites, the input from the two models was very similar. Generally, the risk of not achieving the best result increases if individual

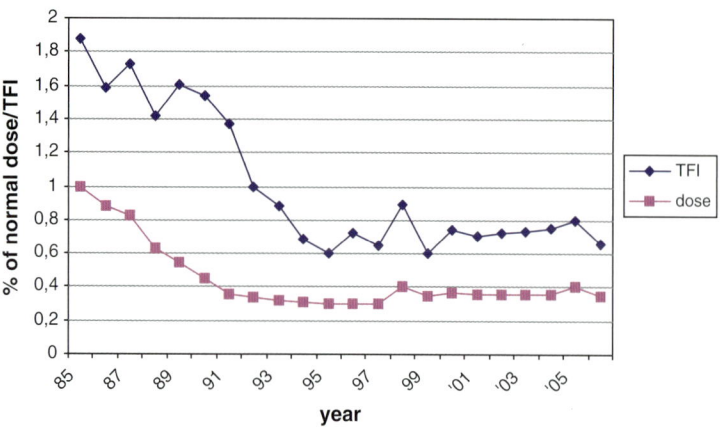

Fig. 1 Changes in Treatment Frequency Index (TFI) and dose used per application in wheat from 1985 to 2004. (Source: Kleffmann-Farmstat (2005) and pesticide statistics)

Table 5 Results from 2 years' trials comparing the present version of Crop Protection Online (CPO) with a prototype that does not require specific field assessments

Treatment	Fungicide input, TFI[a]	Percent powdery mildew	Percent septoria leaf blotch	Yield and yield increase (dt ha^{-1})	Net yield increase (dt ha^{-1})
Untreated		1.3	20.2	74.9	
CPO	0.58	0.5	9.2	8.6	3.8
CPO prototype	0.68	0.5	8.3	8.3	2.9
LSD$_{95}$ (ex. untr.)	NS	NS	NS		NS

Average of 10 trials. Net yield has been calculated after costs of fungicides and application have been deducted

[a] Treatment Frequency Index

field registrations are not carried out. However, the risk seems to be relatively small based on the ten trials carried out during two seasons compared with spraying routinely.

Based on the experiences from these trials and a wish to increase the farmers' direct use of the CPO, the plan is to develop the system in a way that integrates information from the national field database run by the advisory service with regional monitoring data and climatic information. The development of this system occurs initially in a closed group with advisors and farmers in order to investigate if the farmers will find it useful when making decisions on disease control. Further testing of the prototype should preferably be carried out over more seasons and under more variable disease pressure in order to verify the consistency of the system when direct field assessments are not made.

The sociological investigation

The sociological investigation of the farmers' way of making decisions in the area of crop protection identified three types of farmers: (a) system-orientated farmers, (b) experienced-based farmers and (c) advisory-orientated farmers. Several factors were found to illustrate some of the characteristics of the three types (Table 6). None of the three groups of farmers were prepared to carry out very detailed field investigations in order to adjust the input according to need in specific fields (Langvad and Noe 2006; Jørgensen et al. 2007b).

System-orientated strategists generally do not rely on exact field assessments. They tend to manage bigger farms with specialised arable production and demand high levels of disease control. They often carry out treatments following a partially fixed strategic plan, which they have prepared in the winter. This plan is also used as a list for planning purchases of chemicals. Experienced-based strategists base their production on a close, personal knowledge of each individual field, substituting detailed disease assessments with continual more general observations of the fields. The farm size for this group is medium to low. Strategists, who contract their decision-making out to an advisor, i.e. the advisory-contracting decision makers, give higher priority to operations other than growing crops; often they have animal production as their major activity. Typically, they evaluate disease levels in study groups or together with the advisor, but they are not likely to operate a system like CPO alone.

Table 6 Main characteristics of three farmer types – based on their way of making decisions in relation to crop protection

System-orientated farmers	Experienced-based farmers	Advisory-based farmers
Bigger farms	Medium-sized farms	No specific farm size
Systematic approach to planning	Very aware of problems in specific fields	Often main income from animal production or other special production
Specialised arable farmers prepare a spraying plan in winter, which will mostly be followed during the season	Have no specific spray plan prepared before the season. Has a 'Learning by doing' approach	Less specific interest in arable farming. Rely on recommendations from advisors
Relative high pesticide input	Normal to slightly high pesticide input	Relatively low pesticide input

The present CPO prioritises a specific economic value rationale of cost savings by reducing input of pesticides. This strategy and value rationale is different from that of all the three identified groups of farmers. CPO has to be viewed to a greater extent from the end-user's perspective rather than from the perspective of the scientist who developed the systems. An interesting phenomenon was found in the way most farmers view the risks of disease. In practice, this proved to be more closely linked with the risk of possible disruption of their management system, whether in the form of extra work, aesthetically negative consequences (e.g. from lodging or early senescence) or incidents of unacceptably low yields.

Although there is much confidence in the technical content of CPO, the sociological study showed that CPO does not fit in with many farmers' decision strategies. For most farms, crop protection is only a small part of the continuous flow of operations and decision-making made on farms. Decisions on crop protection are not based on isolated economic considerations, but are coupled with a number of other considerations and circumstances including the farm: machinery, manpower, task planning, etc.

The farmer study also provided specific suggestions for improving the system, for example connecting CPO to other computer systems used by the farmers, using CPO to a larger extent in the planning phase and independently of the time of season and giving a larger degree of varied answers that among other things include crop rotations.

Potential use

An increased distribution of CPO depends on the user interface and on the functionality of CPO being targeted to a larger extent at the individual decision strategies. The investigation suggests that DSS have a potential use by approximately 30% of target users for this kind of tool. This would, however, require that the end-users should be very actively involved in future development processes. The main obstacle to more frequent use is the time-consuming field registrations, which are necessary for the models to calculate field-specific recommendations.

CPO is used among wholesalers and distributors to a limited extent. This group has, however, generally a relatively low impact on the recommendations used by the farmers. Farmers have a long tradition for using information and recommendations given by their own independent DAAS. CPO including both the weed and disease system is broadly supported by the chemical industry. The industry partly supports the research behind the system and typically sees the system as a means of applying and supporting IPM strategies.

The models in CPO have been tested, adjusted and exported to other countries (e.g. Latvia, Lithuania and Poland). This work has underlined the importance of good cooperation between researchers and local advisory services, and the importance of models that are well documented and not too complex. The work has also illustrated that models cannot be transferred directly to new agro-ecological regions but need to be tested and adapted to local regional cropping situations (Bankina and Priekule 2003; Czembor et al. 2003; Semaskiene et al. 2003).

Conclusion

Scientists generate highly complex knowledge as a basis for developing models in the area of crop protection and pesticide use. However, the information does not always reach farmers as it is presented in a way that does not correspond with the way things are done in practice. Economic analysis of historical trial data shows that CPO models give acceptable recommendations. DSS have proven to be useful tools when pesticide use has to be optimised. This includes spraying only when economic thresholds have been exceeded. Generally, the DSS CPO is valued and trusted for its information, and is widely used by advisors and as a learning tool for students. Often, information given in newsletters to farmers is based on output from the system and as such the indirect use of the system is high. Direct use of DSS by farmers is, however, so far relatively low because the system is not in line with the way most farmers manage their crop protection.

The potential for further reductions in fungicide use in cereals is relatively small in Denmark as a major reduction took place in the 1990s. A major barrier is therefore the lack of economic incentives to change from today's standard treatments. Another major obstacle is the requirement to carry out field registration as part of using the system. Many farmers

do not find this practicable. This is at least in part linked to the fact that farm sizes are getting larger, leaving less time for decision making at the individual field level. Despite the relatively low use of fungicides in cereals in Denmark, there is still a tendency for farmers to overestimate the risk of disease and therefore to use a dose that is too high or to apply an extra spray. Farmers are generally risk-adverse and concerned about unintended disruption of their management system. They tend to partly overreact, as they are not prepared to risk crop failure or large economic loss due to diseases.

In future developments of CPO, it is recommended that farmers' decision-making practice and their reasoning logic should be addressed. Farmers cannot be considered to be a homogenous group and therefore it is naive to believe that one system will fit all farmers. It is therefore important to acknowledge the major differences and develop different approaches for supporting farmers' crop protection activities. On the basis of our studies it has become evident that a large proportion of arable farmers cannot be reached via CPO directly. To target these other groups, appropriate recommendations based on seasonal variation in disease pressure have proven to be reliable with respect to both disease control and yield. These recommendations are typically given through the DAAS.

References

Anonymous (2006a). *Bekæmpelsesmiddelstatistikken*. Denmark: Miljøstyrelsen.

Anonymous. (2006b). A thematical strategy on the sustainable use of pesticides [(COM 2006 372].

Bankina, B., & Priekule, I. (2003). *Experience of using reduced dosages of fungicides for cereal disease control in Latvia*. DIAS report no. 96. Proceedings of the Crop Protection Conference for the Baltic Sea Region, April 2003, Poznan, Poland, pp. 130–140.

Brooks, D. H. (1998). *Decision Support System for Arable Crops (DESSAC): An integrated approach to decision support*. The 1998 Brighton Conference, Pest & Diseases. British Crop Protection Council, pp. 239–246.

Czembor, J. H., Horsozkiewicz-Janka, J., & Nierobca, A. (2003). *Testing of Danish decision support system in protection of winter wheat in Poland during 2001–2003*. DIAS report no. 96. Proceedings of the Crop Protection Conference for the Baltic Sea Region, April 2003, Poznan, Poland, 165.

Detlefsen, N., & Jensen, A. L. (2001). Variety selection for winter wheat, sortsvalg. In J. Steffe (Ed.), *Proceedings from the third European conference of the European federation for information technology in agriculture, food and the environment*, June 18–20, Montpellier, France. Vol. 1, 1–5.

Hagelskjær, L., & Jørgensen, L. N. (2003). A web-based decision support system for integrated management of diseases and pest in cereals. *EPPO Bulletin, 33*, 467–471.

Hansen, J. G. H., Secher, B. J. M., Jørgensen, L. N., & Welling, B. (1994). Thresholds for control of *Septoria* spp. in winter wheat. *Plant Pathology, 43*, 183–189.

Henriksen, K. E., Jørgensen, L. N., & Nielsen, G. C. (2000). PC-plant protection – a tool to reduce fungicide input in winter wheat, winter barley and spring barley in Denmark. Brighton Crop Protection Conference. *Pest and Diseases*, 835–840.

Jørgensen, L. N., Hagelskjær, L., & Nielsen, G. C. (2003a). *Adjusting the fungicide input in winter wheat depending on variety resistance*. BCPC conference on Crop Science and Technology, Glasgow, 1115–1120.

Jørgensen, L. N., & Kudsk, P. (2006). *Twenty years' experience with reduced agrochemical inputs*. HGCA R&D conference, Lincolnshire, UK. Arable crop protection in the balance profit and the environment, 25–26 Jan. 2006, 16.1–16.10.

Jørgensen, L. N., & Nielsen, G. C. (2003b). Septoria control using threshold-based systems and fungicides. Global insights into the Septoria and Stagonospora diseases in cereals. In G. H. J. Kema, M. van Ginkel, & M. Harrabi (Eds.), *Proceedings of the 6th international symposium on septoria and stagonospora diseases of cereals* (pp. 49–61), 8–12 Dec. 2003, Tunis.

Jørgensen, L. N., Nielsen, G. C., & Henriksen, K. E. (2000). Margin over cost in disease management in winter wheat and spring barley in Denmark. Proceedings of the Brighton Crop Protection Conference. *Pest and Diseases*, 655–662.

Jørgensen, L. N., Noe, E., Langvad, A. M., Jensen, J. E., Ørum, J. E., & Rydahl, P. (2007b). Decision support systems: barriers and farmers need of support. *EPPO Bulletin, 37*, 374–377.

Jørgensen, L. N, Noe, E., Langvad, A. M., Rydahl, P., Jensen, J. E., Ørum, J. E., et al. (2007a). *Vurdering af Planteværn Onlines økonomiske og miljømæssige effekt*. Bekæmpelsesmiddelforskning fra Miljøstyrelsen nr. 115 (English summary). http://www2.mst.dk/common/Udgivramme/Frame.asp?pg=http://www2.mst.dk/Udgiv/publikationer/2007/978-87-7052-590-9/html/default.htm.

Jørgensen, L. N., & Olsen, L. V. (2007). Control of tan spot (*Drechslera tritici-repentis*) using host resistance. Tillage methods and fungicides. *Crop Protection, 26*, 1606–1616.

Jørgensen, L. N., & Pinnschmidt, H. (2004). *Yield effects and control of powdery mildew in winter wheat in the presence of septoria*. Proceedings of the 11th International Cereal Rusts and Powdery Mildews Conference, Norwich, England.

Jørgensen, L. N., Secher, B. J., & Nielsen, G. C. (1996). Monitoring diseases of winter wheat on both a field and a national level. *Crop Protection, 13*, 383–390.

Kleffmann-Farmstat (2005). Statistical data on pesticide use in DK (Company data).

Langvad, A. M., & Noe, E. (2006). (Re-)innovating tools for decision-support in the light of farmers' various strategies. In H. Langeveld & N. Röling (Eds.), *Changing European*

farming systems for a better future – new visions for rural areas (pp. 335–339). The Netherlands: Wageningen.

McCown, R. L. (2002). Probing the enigma of the decision support system for farmers: Learning from experience and from theory. *Agricultural Systems, 74*, 1–10.

Noe, E., & Halberg, N. (2002). Research experience with tools to involve farmers and local institutions in developing more environmentally friendly practices. In K. Hagedorn (Ed.), *Environmental co-operation and institutional change*. Cheltenham, England: Edward Elgar.

Röhrig, M. (2006). http://www.isip.de online *plant protection information in Germany*. Abstract from EPPO conference on computer aids for plant protection. Wageningen, 17–19 Oct. 2006.

Rydahl, P. (2003). A web-based decision support system for integrated management of weeds in cereals and sugarbeet. *EPPO Bulletin, 33*, 455–460.

Rydahl, P., Hagelskjær, L., Pedersen, L., & Bøjer, O. Q. (2003). User interfaces and system architecture of a web-based decision support system for integrated pest management. *EPPO Bulletin, 33*, 473–482.

Secher, B. J. M., Jørgensen, L. N., Murali, N. S., & Boll, P. S. (1995). Field validation of a decision support system for the control of pests and diseases in cereals in Denmark. *Pesticide Science, 45*, 195–199.

Semaskiene, R., Tamosiunas, K., & Dabkevicius, Z. (2003). *Experience of using reduced dosages of fungicide for winter wheat in Lithuania*. DIAS report no. 96. Proceedings of the Crop Protection Conference for the Baltic Sea Region, April 2003, Poznan, Poland. 123–129.

Verret, J. A., Klink, H., & Hoffmann, G. M. (2000). Regional monitoring for disease prediction and optimisation of plant protection measures: The IPM Wheat Model. *Plant Disease, 84*, 816–826.

Volk, T., Johnen, A., Newe, M., & Meier, H. (2003). *ProPlant expert.com – The online consultation system on crop protection in cereals, rapeseed, potatoes and sugar beet: Experiences with cereal disease control in the region and possibilities for regional adaptations*. Crop Protection Conference for the Baltic Sea Region, April 2003, Poznan, Poland, pp. 103–113.

Weber, M. (1922). *The Protestant ethic and the spirit of capitalism*. Los Angeles: Roxbury (US), 1998.

Zadoks, J. C. (1983). An integrated disease and pest-management scheme, EPIPRE, for wheat. *Ciba Foundation Symposia, 97*, 116–129.

Eur J Plant Pathol (2008) 121:313–322
DOI 10.1007/s10658-007-9232-7

REVIEW

Control of plant diseases by natural products: Allicin from garlic as a case study

Alan J. Slusarenko · Anant Patel · Daniela Portz

Received: 22 June 2007 / Accepted: 27 September 2007
© KNPV 2007

Abstract This review aims to increase awareness of the potential for developing plant protection strategies based on natural products. Selected examples of commercial successes are given and recent data from our own laboratory using allicin from garlic are presented. The volatile antimicrobial substance allicin (diallylthiosulphinate) is produced in garlic when the tissues are damaged and the substrate alliin (S-allyl-L-cysteine sulphoxide) mixes with the enzyme alliin-lyase (E.C.4.4.1.4). Allicin is readily membrane-permeable and undergoes thiol-disulphide exchange reactions with free thiol groups in proteins. It is thought that these properties are the basis of its antimicrobial action. We tested the effectiveness of garlic juice against a range of plant pathogenic bacteria, fungi and oomycetes *in vitro*. Allicin effectively controlled seed-borne *Alternaria* spp. in carrot, *Phytophthora* leaf blight of tomato

and tuber blight of potato as well as *Magnaporthe* on rice and downy mildew of *Arabidopsis*. In *Arabidopsis* the reduction in disease was apparently due to a direct action against the pathogen since no accumulation of salicylic acid (a marker for systemic acquired resistance, SAR) was observed after treatment with garlic extract. We see a potential for developing preparations from garlic for use in organic farming, e.g. for reducing the pathogen inoculum potential in planting material such as seeds and tubers. We have tested various encapsulation formulations in comparison to direct treatment.

Keywords Alginate · Encapsulation · Formulation · Plant protection

Introduction

In the course of evolution, plants have developed chemical defence mechanisms against potential pathogens and pests. Society's dependence on intensive agriculture and horticulture for food production has accentuated the need to reduce crop losses. Environmental considerations have highlighted the requirement for sustainable solutions in agriculture and consumer pressures for green alternatives have accompanied a boom in the organic farming sector. This has awakened new interest in natural products as a source for novel industrial plant protection strategies. The purpose of this review is to give an overview of the use of some natural products in plant protection

Note added in proof: The complete chemical synthesis of azadirachtin has now been achieved. Nature (2007) 448: 630–631

A. J. Slusarenko (✉) · D. Portz
Department of Plant Physiology,
RWTH Aachen University,
52056 Aachen, Germany
e-mail: Alan.slusarenko@bio3.rwth-aachen.de

A. Patel
Institute of Technology und Biosystems Engineering,
Federal Agricultural Research Centre (FAL),
38116 Braunschweig, Germany

and to highlight the potential of allicin from garlic, which we believe holds much promise for future development in at least some specialised areas of agriculture and horticulture.

What is a natural substance?

A scientist and a lay-person perhaps have a different understanding of what the term 'natural' conveys. Similarly, the motivations for investigating the use of natural products in plant protection might also be different. By definition, a natural or biogenic substance is either synthetized directly by a living organism or is derived from substances of biogenic origin by chemical reactions occurring without human intervention; for example by decomposition of biological materials. Thus, humus, or in the wider sense coal, oil and limestone, are examples of natural or biogenic substances. In the context relevant to this review, a natural product is viewed as a physiologically active chemical which is synthetized by plants, animals or microbes. In contrast, a synthetic chemical is one which does not occur naturally and must be synthetized from other substances by human intervention. Of course, many naturally occurring substances can also be synthetized in the laboratory, and indeed the use of a pure, chemically synthetized molecule in laboratory tests is usually a pre-requisite for the acceptance of biological activity attributed to a particular substance in a complex mixture from a natural source.

Why consider natural substances for plant protection?

In the public perception 'natural' is often equated directly with 'benign' and 'environmentally friendly' and for any given purpose natural products are a priori assumed to be a more desirable alternative to synthetic chemicals. Obviously, this is per se not correct and many natural products are very toxic. For example botulinum toxin, a bacterially produced peptide with an LD_{50} of 1 ng kg^{-1} body weight is perhaps the most acutely toxic substance known (Fleming and Hunt 2000). In contrast, highly toxic inorganic arsenic has an LD_{50} (oral) of 763 mg kg^{-1} (http://ptcl.chem.ox.ac.uk/MSDS/AR/arsenic.html)

making it nearly 8×10^8 times less toxic than botulinum on a weight for weight basis. Nevertheless, living organisms, particularly plants, are brilliant synthetic chemists and produce a huge variety of physiologically active substances, thus providing an alternative to the combinatorial chemistry approach in the search for useful chemicals. To quote from a recent *Science* article: "Around half of the drugs currently in clinical use are of natural product origin." The authors go on to state that "Despite this statistic pharmaceutical companies have embraced the era of combinatorial chemistry, neglecting the development of natural products as potential drug candidates in favour of high-throughput synthesis of large compound libraries" (Peterson and Anderson 2005). This perhaps highlights a common perception among scientists that natural substances, per se are probably less effective than synthetic alternatives, or in a greater extreme that natural products are almost in the realm of esoterics and folk-lore. This can lead to a sceptical approach to each other's perspectives by both scientists and lay people and emphasizes the need for strict objectivity.

A potential advantage offered by natural products is that their effectiveness has been optimised by evolution for their particular task. In terms of plant protection this might be an antimicrobial, insecticidal or anti-feedant activity. Several microorganisms produce antibiotics, and many preformed and induced antimicrobial substances are known from plants (Mansfield 2000). These are obvious candidates to be considered for use in plant-protection strategies. Furthermore, many substances of natural origin which do not show direct antimicrobial activity might act as resistance inducers to prime systemic acquired resistance (SAR) relying on the plant's own defences (Goellner and Conrath 2007).

Although some structural components which are natural in origin, e.g. lignin or $CaCO_3$- or silica-containing shells, are very stable, natural products are generally easily biodegradable and after they have served their purpose do not tend to persist in the environment. This can also be a disadvantage, however, because the plant protectant has to be around for long enough to do its job before it is degraded and taking a natural product out of its cellular environment is usually de-stabilising.

Increasing interest in environmentally sustainable agriculture and horticulture, and organic farming, has

opened up a niche on the market for plant protection products compatible with regulations for labelling food as 'organic produce' and has forced the need to consider new alternatives. Thus, the use of copper-containing compounds, for example in combating potato blight, has traditionally been allowed in organic farming. Acknowledgement that the release of large amounts of this toxic heavy metal into the environment are not compatible with the ethos behind the organic farming movement has led to its phasing out as an allowed substance in the latest EU directives (Council Regulation 2092/91 on Organic Farming); however, an effective practical alternative has yet to be found.

Problems with natural substance development for plant protection

Substances honed by evolution for their function within the natural plant-pathogen context are not generally optimised for industrial production or external application. Thus, as mentioned above they may not have optimal stability for field applications and there may be a much cheaper synthetic alternative available which does the same job. Thus, early successes in the laboratory do not always transfer to the field situation. Furthermore, to be attractive to industry a product must be patentable and the status of many natural substances is unclear in this regard (see the example of neem products below). Nevertheless, as a starting point for derivatisation and formulation to enhance desirable and reduce undesirable properties, natural product structures can be an important starting point.

Examples

Some natural-product-inspired, natural-product-derived, or natural-product-similar plant protection chemicals have been important commercial successes:

Example (1)   The systemic benzimidazole fungicide Benomyl (Fig. 1a), which was released on to the market by the DuPont Company in 1968, has a heterocyclic ring structure reminiscent of benzoxazinones/benzoxazolinones which are weakly antifungal substances accumu-

lating in some grasses (e.g. wheat, rye, maize) (Fig. 1b). It seems that the biological activity of these compounds was not the inspiration that led to the development of Benomyl (Harvey Loux, personal communication); however, the structural similarity of Benomyl to these natural antifungals is clear. Benomyl interferes with microtubules and affects cell division and intracellular transport processes. Fungal microtubules seem particularly sensitive to benomyl which is probably the basis of its selective action. Although benomyl has such a low acute toxicity in mammals that it has not been possible to establish an LD_{50} for it (http://www.inchem.org/documents/ehc/ehc/ehc148.htm); concern about the effects of chronic exposure led to its phasing out and withdrawal from the market in 2001/2002.

Example (2)   Strobilurins are produced as natural antibiotics by the wood-rotting Basidiomycete fungus *Strobilurus tenacellus* whose fruiting bodies emerge from between pine-cone scales (Fig. 2). The fungus produces antibiotics in a 'chemical warfare strategy' to reduce competition for its habitat from other species. Strobilurins act at the outer ubiquinol binding site in the electron transport chain in aerobic respiration (cytochrome bc_1 complex, complex III) and are classed as Q_o inhibitors or 'Q_oI' (Grasso et al. 2006). Complex III is an integral component in mitochondrial electron transport in all eukaryotes and why strobilurins are selectively toxic to fungi is not understood. The basic strobilurin structure has been modified in the laboratory to improve characteristics for field application, such as UV-stability, and several analogues are marketed as successful fungicides. Interestingly, some novel strobilurin derivatives have been reported to have a resistance-inducing or 'priming' effect in

Fig. 1 The chemical structures of (**a**) the imidazole Benomyl, (**b**) 2-benzoxazolinone

a

b

the plant and also to stimulate plant growth and drought tolerance in addition to their direct antifungal activities and are thus beneficial for the plant even in the absence of any infection (Goellner and Conrath 2007).

Example (3) Neem oil/neem cake are products made from the seeds of the Neem tree (*Azadirachta indica*) a native of India and a member of the mahogany family (Meliaceae). Neem products have a long history of nutritional and medicinal uses by indigenous peoples and were the subject of an international dispute about patenting natural resources (Wolfgang 1995). Neem products are perhaps best known for their pesticidal and antifeedant activities but broad-range anti-mycotic properties have also been reported (Carpinella et al. 2003). The major insecticidal-active

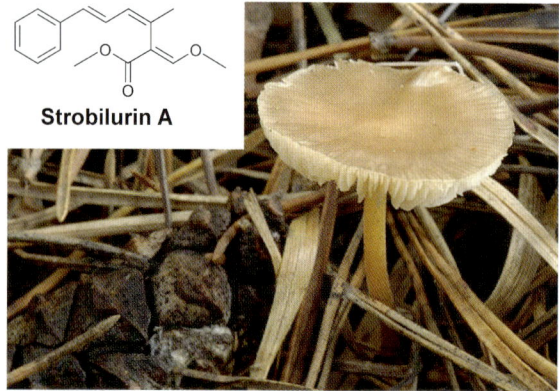

Strobilurin A

Fig. 2 A fruiting body of *Strobilurus tenacellus* (reproduced with the kind permission of Darek Karasinski, (http://grzyby. strefa.pl.). The inset shows the structure of strobilurin A

substance in Neem preparations is the triterpenoid azadirachtin (Butterworth and Morgan 1968). The structure is complex (Fig. 3) and despite early synthesis of the two sub-fragments of the molecule (decalin and a hydroxy furan), each of which shows independent insecticidal effects, total synthesis has remained elusive (Aldhous 1992; Ley 1994; Nicolaou et al. 2003). The exact mechanism of action is not well understood but azadirachtin acts as both a feeding deterrent and an insect-growth regulator. The molecule is acid and base-unstable and, because of the large number of double bonds, extremely UV-labile. More stable variants of the parent molecule have been developed but work is hampered because of the lack of an easy, cheap synthetic strategy (Aldhous 1992). Thus, the majority of uses employ preparations of or from neem seeds themselves. Azadirachtin is specifically listed as an acceptable plant protection substance for organic farming in EU directive 2092/91.

Perspectives

Certainly natural products are being considered in the search for plant-protection chemicals, as illustrated by the last two examples above. However, while many companies advertise their services for 'natural drug

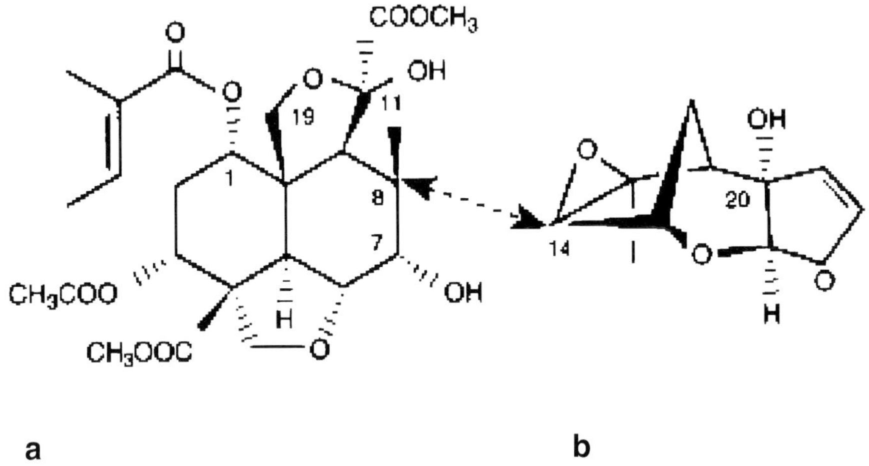

Fig. 3 The two component fragments of azadirachtin. (**a**) the decalin fragment and (**b**) the hydroxy furan fragment. From Aldhous (1992) reprinted by permission of AAAS

a

b

discovery,' the question remains as to whether plant protection will attract similar financial investment as for applications in human medicine. Since the use of raw extracts from plants is in many cases not economical for industrial scale applications due to the bulk of material needed, the future may well see the development of single molecules or mixtures that can serve as indicator structures or 'lead compounds' for derivatisation. Nevertheless, systematic scientific improvement of 'low-tech' solutions, where subsistence farmers might 'grow their own' plant protection, may be of real social value.

Although this review is focused on plant products it is pertinent at this point to mention microorganisms as sources of natural products for plant defence. The control of fireblight caused by the phytopathogenic bacterium *Erwinia amylovora* with the antibiotic streptomycin is a well known, if controversial, example. However, many groups of

bacteria and fungi have not been studied as a source of plant protection chemicals per se although there has been much progress in their direct use as antagonists.

Many phytoanticipins exist as precursors that need to be modified by enzymatic activity to achieve their anti-microbial potential. The future development of 'two-component' enzyme-substrate strategies for plant protection might therefore be productive. Work in this direction with the alliin-alliinase combination has already been published (Fry et al. 2005).

Case study: Allicin in garlic (*Allium sativum*)

When garlic is sliced or crushed it develops its characteristic odour because cellular damage leads to mixing of the vacuolar enzyme alliin lyase (E.C.4.4.1.4) and its cytosolic substrate alliin (S-allyl-L-cysteine

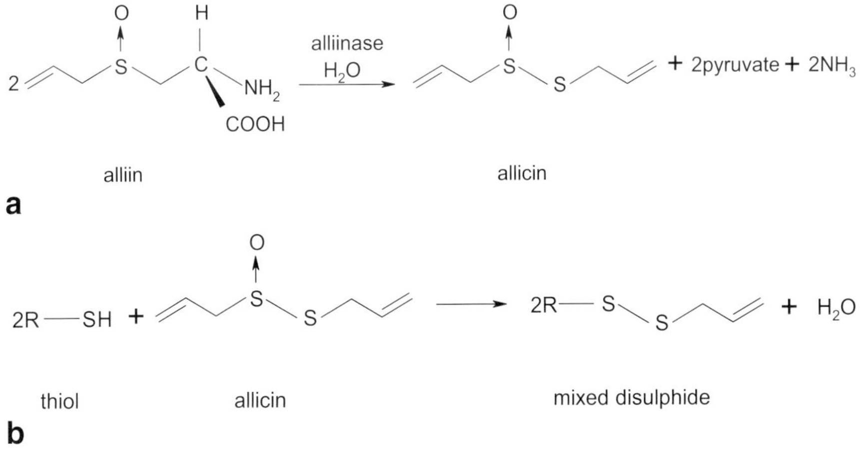

Fig. 4 (**a**) The production of allicin from alliin by alliin lyase, and (**b**) the thiol-disulphide exchange reaction with SH-compounds including amino acids in proteins thought to be the basis for allicin's biocidal activity

a

b

Eur J Plant Pathol (2008) 121:313–322

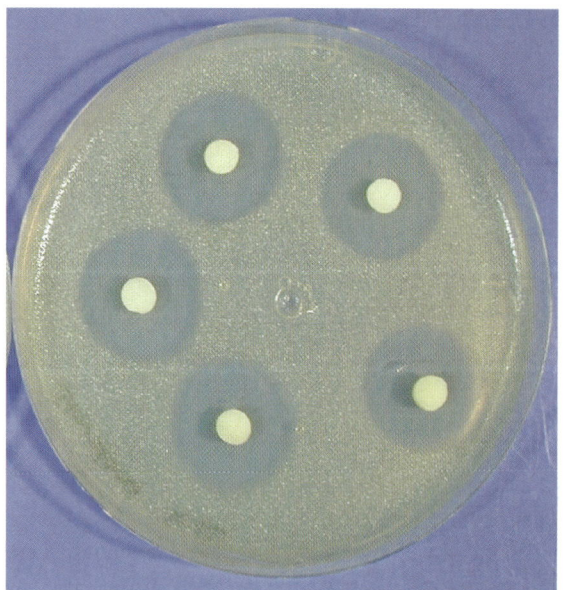

Fig. 5 Clear zones of inhibited growth of a wild-type *E. coli* K12 isolate around filter discs spotted with 20 μl of garlic juice containing a total of 90 μg allicin

sulphoxide). The immediate product is thiosulphenic acid which undergoes spontaneous dimerization to diallylthiosulphinate (allicin) (Fig. 4a). It is allicin that gives garlic its characteristic odour. Garlic has a long history of use in folk medicine. For instance the Codex Ebers, an Egyptian papyrus from 1800 BC, describes more than 800 medicinal preparations including 22 containing garlic (Block 1985). Allicin was shown to be the major antimicrobial substance in garlic by Cavallito and Bailey (1944) and the allicin metabolite ajoene shows potent anti-coagulent activity by inhibiting platelet aggregation (Jain and Apitz-Castro 1987). Allicin undergoes thiol-disulphide exchange reactions with free thiol groups in proteins (Fig. 4b) and it is thought that this, together with its ready membrane permeability (average LogP octanol:water$=1.52\pm0.80$, Tetko et al. 2005; http://www.vcclab.org), is the basis of its antimicrobial action (Miron et al. 2000; Rabinikow et al. 1998). Because of these attributes allicin has several potential targets within the cell and it is difficult for organisms to mutate to resistance. Allicin has been reported to be active against a broad-spectrum of taxonomically diverse organisms (Curtis et al. 2004; Portz et al. 2005 and references therein). Nevertheless, resistance to allicin is known and garlic is susceptible to *Puccinia porri*, which is presumeably insensitive to allicin or can colonise garlic without causing allicin

production. Similarly we have isolated a non-pathogenic allicin-resistant *Pseudomonas* isolate from fresh garlic cloves. However, the basis of the allicin resistance of this isolate is unclear.

The use of garlic preparations or allicin against plant pathogens has already been documented (Arya et al. 1995; Bianchi et al. 1997; Cao and vanBruggen 2001; Russell and Mussa 1977) and there are several preparations based on garlic compounds available commercially, although the latter are primarily aimed at controlling pests rather than pathogens.

Fig. 6 Control of leaf blight of tomato by spraying tomato plants with garlic juice 2 h before inoculation. Inoculation was done by spraying whole plants with a sporangial suspension of *P. infestans* (4–5×10^4 sporangia ml^{-1}). *Top panel*, inoculated plants; *middle panel*, inoculated and sprayed with diluted garlic juice containing 110 μg ml^{-1} allicin or, *bottom panel*, 85 μg ml^{-1} allicin

Fig. 7 An alginate formulation of garlic juice deposited on the soil surface in a pot test with *Phytophthora*-inoculated tomato seedlings

where bacterial growth has been inhibited (Fig. 5). When the concentration of bacteria suspended in the agar and the depth of agar in the plate are standardized, the diameter of the inhibition zone is highly reproducible between replicates. On this basis a Petri-plate bioassay to quantify the amount of allicin in crude garlic extracts was developed and originally calibrated by determining allicin using an approximate spectrophotometric assay (Curtis et al. 2004). The accuracy of this bioassay was subsequently improved by using an HPLC method to quantify a pure allicin standard (Krest and Keusgen 2002; Portz and Slusarenko unpublished).

Using a wild-type *E. coli* K12 isolate as an indicator strain, the diameter of the inhibition zone caused by 50 µg kanamycin was matched by 55 µg of allicin. On a molar basis this makes allicin approximately a quarter as potent as kanamycin. The quantity of the substrate alliin in garlic cloves varies but in our hands garlic purchased in the supermarket routinely yields approx. 2 mg allicin g^{-1} fresh weight. A single clove of garlic weighs around 5 g and a composite bulb around 50 g; this means 2 g allicin can be obtained from a kg (approximately 20 bulbs) of garlic. Thus, the antibiotic potential present in fresh garlic is considerable.

How potent an antibiotic is allicin?

Perhaps the best way to illustrate the potency of allicin is to make a comparison with a 'household' antibiotic like kanamycin which is used routinely in the laboratory in selective media. Spotting garlic juice or pure allicin to a Petri plate containing growth medium seeded with bacteria gives rise to clear halos

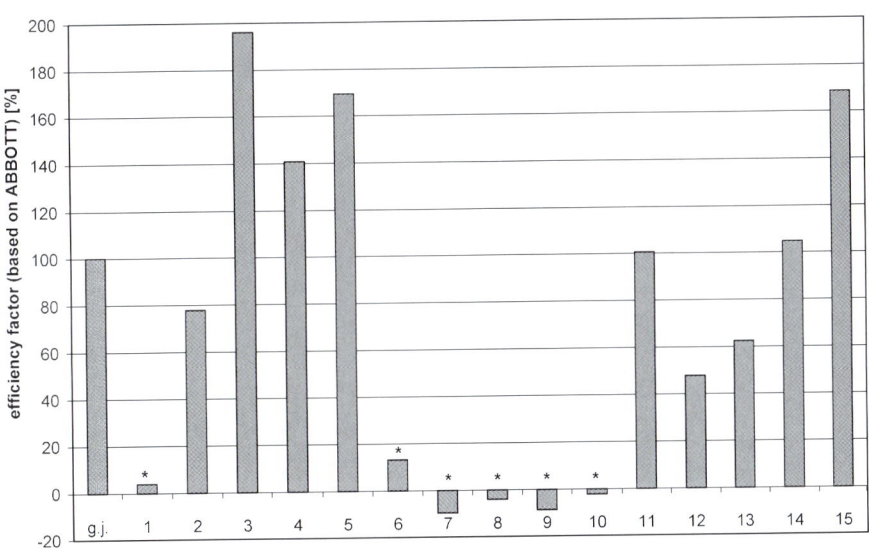

Fig. 8 The relative effectivity against *P. infestans* on tomato of applying 1.5 g of various garlic encapsulations onto the soil compared with a direct spray of 100 µg ml^{-1} allicin in garlic juice onto leaves. Both applications were done 2 h before whole plants were sprayed with sporangial suspensions of *P. infestans* (4–5×10^4 sporangia ml^{-1}). The data were treated according to the method of Abbott (1925). The asterisks indicate a significant difference to the treatment with diluted garlic juice (α=0.05, Dunn's Test). (g.j.=diluted garlic juice (100 µg ml^{-1} allicin), 1–15=different capsules)

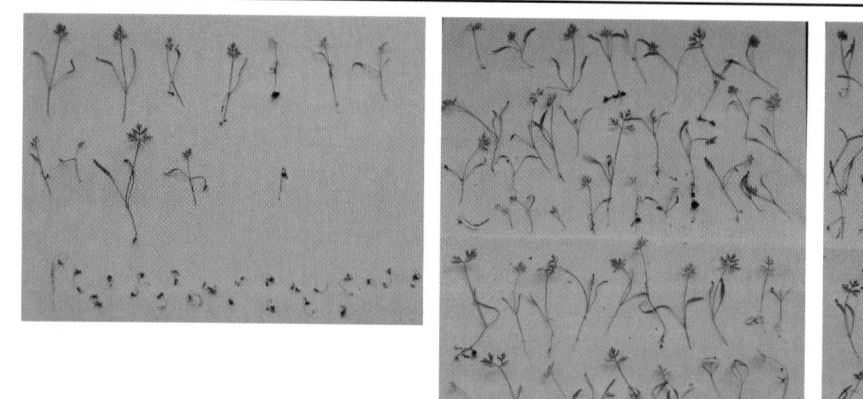
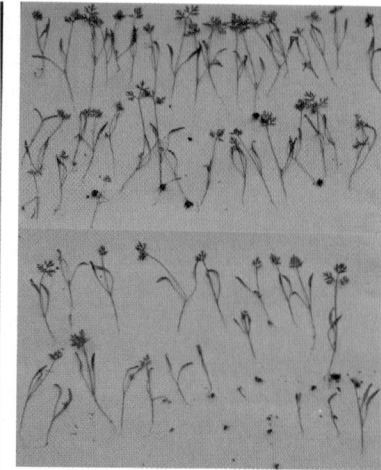

a   b   c

Fig. 9 Disinfection of *Alternaria*-infested carrot seed by imbibing with garlic juice. (**a**) Control seed without treatment, 12/100 seeds germinated, (**b**) Aatiram®-treated seed, 48/100 seeds germinated, (**c**) garlic juice treatment, 47/100 seeds germinated

Activity of allicin in garlic juice against plant pathogens in vitro and in planta

In vitro antibacterial, antifungal and anti-oomycete activity from garlic have been reported in many instances in the literature (see Curtis et al. 2004 and references therein). There are also reports of disease being reduced by treatment of infected plants and in the laboratory. We showed that garlic juice was able to reduce disease severity in several test pathosystems such as rice/*Magnaporthe oryzae*, *Arabidopsis*/*Hyaloperonospora parasitica*, and potato/*Phytophthora infestans* (*ibid*). In the latter case, tuber infection was investigated and control was achieved by applying allicin directly to the inoculation site as well as via the vapour phase in an enclosed space. This raises the possibility for developing fumigation protocols in special circumstances and relies on the volatile nature of the active principle allicin. Control of leaf blight of tomato after *P. infestans* inoculation was also achieved by spraying leaves with dilutions of garlic juice (Fig. 6).

The possibility that allicin might not only be acting directly against the pathogen but also by conditioning SAR was investigated in the *Arabidopsis*/*Hyaloperonospora* interaction by looking for accumulation of the SAR marker salicylic acid (SA) in local (treated) and systemic (untreated) leaves. No significant induction of either free or glycosylated SA was observed, leading us to conclude that garlic juice at the concentrations tested did not induce SAR directly (*ibid*).

Can allicin be transferred from the laboratory to the field?

The antibiotic potential and success in the laboratory of allicin/garlic treatments are clear. In a plant protection context, the big question is: can the small-scale effects in controlled environments be transferred at manageable cost to applications in the field? For this to happen several aspects must be considered.

Allicin has the reputation of being rather unstable. However, in a study on the degradation kinetics of allicin in different solvents Canizares et al. (2004) reported that allicin kept at 6°C retained its bacteriostatic activity against *Helicobacter pylori*, which causes stomach ulcers, for at least 10 months. In our hands the antimicrobial activity of garlic juice was destroyed after 10 min at 80°C but showed no loss of activity after 10 days at 4°C. Storage at room temperature led to a 50% reduction in the diameter of the inhibition zone against *E. coli in vitro* over the same period (Curtis et al. 2004). On a hobby gardener scale stability is not a problem because garlic juice can be freshly prepared and used immediately. The onus is on the plant protection industry to acknowledge the potency of this natural product and find ways to develop a suitable commercial product from it.

One approach to formulation of plant extracts is encapsulation which has been used successfully to stabilize and establish bio-pesticides in soil (Patel et al. 2004). For this reason we have carried out experiments using alginate and other formulations to encapsulate garlic juice and applying the capsules at various dosages to the soil around *Phytophthora*-inoculated tomato seedlings (Fig. 7). The results were promising in comparison to direct spraying of the plants. Thus, some formulations enhanced activity whereas others were clearly less effective (Fig. 8). Probably such factors as the rate of release and stabilisation of the garlic preparation were important. It seems there is scope for further optimisation in this area.

Whether allicin can be derivatized to improve its field qualities and still retain antimicrobial activity is unclear. A systematic investigation by synthetic chemists is needed to determine whether other related molecules or modification of allicin structure could lead to desirable plant protection chemicals.

Special applications of allicin: seed disinfection

In the EU-wide regulations governing organic farming (Council Regulation 2092/91 on Organic Farming) it is laid down that organic produce must be derived from the sowing of organically-produced seeds. Seed-borne diseases and seed hygiene are increasing problems in this sector and acceptable seed treatment procedures are urgently needed.

Commercial seed companies often employ a procedure called 'priming' to improve the germination rate and uniformity of germination of their seeds. Basically, seeds are allowed to imbibe for a period and are then dried down again to let them remain dormant (Bradford 1999; Gao et al. 1999). Priming procedures are generally empirically determined for particular seeds and each company has its own 'secret' protocol. By allowing seeds to imbibe allicin-containing preparations, followed by subsequent drying down, we have achieved improvements in the germination rate of *Alternaria*-infested carrot seed that are comparable to results obtained with the industrial seed dressing Aatiram® (active ingredient: 670 g kg^{-1} thiram) (Fig. 9). Whether these laboratory successes will transfer to commercial applications is yet unknown.

Concluding remarks

Whether the use the natural substance allicin, or garlic juice, for seed-disinfection or other plant protection strategies, conforms with organic farming practice must still be determined by the relevant regulatory body. Nevertheless, allicin seems to offer a promising alternative to the use of synthetic compounds and the analogy to the accepted use of azadirachtin and neem seed products is clear. Hopefully the future will see the increased development of successful plant protection strategies based on natural products.

Acknowledgements We are indebted to the kindness of the following people in providing accurate information and helpful comments on various parts of the manuscript: Charlie Delp, Jean-Luc Genet, Harvey Loux, David Morgan, Phil Russell, Nikolaus Schlaich.

References

Abbott, W. S. (1925). A method for computing the effectiveness of an insecticide. *Journal of Economon Entomology, 18*, 265–267.

Aldhous, P. (1992). Neem chemical: The pieces fall in place. *Science, 258*, 893.

Arya, A., Chauhan, R., & Arya, C. (1995). Effect of allicin and extracts of garlic and bignonia on two fungi. *Indian Journal of Mycology and Plant Pathology, 25*, 316–318.

Bianchi, A., Zambonelli, A., Zechini D'Aulerio, A., & Bellesia, F. (1997). Ultrastructural studies of the effects of *Allium sativum* on phytopathogenic fungi *in vitro*. *Plant Disease, 81*, 1241–1246.

Block, E. (1985). The chemistry of garlic and onions. *Scientific American, 252*, 94–99.

Bradford, K. J. (1999). Manipulation of seed water relations via osmotic priming to improve germination under stress conditions. *HortScience, 21*, 1105–1112.

Butterworth, J. H., & Morgan, E. D. (1968). Isolation of a substance that suppresses feeding in locusts. *Chemical Communications (London)*, 23–24.

Cao, K.-Q., & vanBruggen, A. H. C. (2001). Inhibitory efficacy of several plant extracts and plant products on *Phytophthora infestans*. *Journal of Agricultural University of Hebei, 24*, http://www.cipotato.org/gilb/Pubs/proceedings_easa/caokeqiang%2821%29.pdf.

Canizares, P., Gracia, I., Gomez, L. A., Garcia, A., de Argila, C. M., & Boixeda, D., et al. (2004). Thermal degradation of allicin in garlic extracts and its implication on the inhibition of the *in-vitro* growth of *Helicobacter pylori*. *Biotechnology Progress, 20*, 32–37.

Carpinella, M. C., Ferrayoli, C. G., & Palacios, M. S. (2003). Antimycotic Activity of the Members of Meliaceae. In M. Rai, & D. Mares (Eds.) *Plant-derived antimycotics* (pp. 81–115). New York: Food Products.

Cavallito, C. J., & Bailey, H. J. (1944). Allicin, the antibacterial principle of *Allium sativum* L. Isolation, physical properties

and antibacterial action. *Journal of the American Chemical Society, 66*, 1950–1951.

Curtis, H., Noll, U., Störmann, J., & Slusarenko, A. J. (2004). Broad-spectrum activity of the volatile phytoanticipin allicin in extracts of garlic (*Allium sativum* L.) against plant pathogenic bacteria, fungi and Oomycetes. *Physiological and Molecular Plant Pathology, 65*, 79–89.

Fleming, D. O., & Hunt, D. L. (Eds.) (2000). Biological safety: Principles and practices. (ASM Press) p. 267.

Fry, F. H., Okarter, N., Baynton-Smith, C., Kershaw, M. J., Talbot, N. J., & Jacob, C. (2005). Use of a substrate/alliinase combination to generate antifungal activity in situ. *Journal of Agricultural and Food Chemistry, 53*, 574–580.

Gao, Y.-P., Young, L., Bonham-Smith, P., & Gusta, L. V. (1999). Characterization and expression of plasma and tonoplast membrane aquaporins in primed seed of *Brassica napus* during germination under stress conditions. *Plant Molecular Biology, 40*, 635–644.

Goellner, K., & Conrath, U. (2007). Priming: It's all the world to induced disease resistance EJPP this issue.

Grasso, V., Palermo, S., Sierotzki, H., Garibaldi, A., & Gisi, U. (2006). Cytochrome *b* gene structure and consequences for resistance to Q_o inhibitor fungicides in plant pathogens. *Pest Management Science, 62*, 465–472.

Jain, M. K., & Apitz-Castro, R. (1987). Garlic: molecular basis of the putative 'vampire-repellant' action and other matters related to heart and blood. *Trends in Biochemical Science, 12*, 252–254.

Krest, I., & Keusgen, M. (2002). Biosensoric flow-through method for the determination of cysteine sulfoxides. *Analytica Chimica Acta, 469*, 155–164.

Ley, S. V. (1994). Synthesis and chemistry of the insect antifeedant azadirachtin. *Pure & Applied Chemistry, 66*, 2099–2102.

Mansfield, J. (2000). Antimicrobial compounds and resistance: The role of phytoalexins and phytoanticipins. In A. J. Slusarenko, R. S. S. Fraser, & L. C. van Loon (Eds.) *Mechanisms of resistance to plant diseases* (pp. 325–370). Dordrecht: Kluwer.

Miron, T., Rabinikov, A., Mirelman, D., Wilchek, M., & Weiner, L. (2000). The mode of action of allicin: its ready permeability through phospholipid membranes may contribute to its biological activity. *Biochimica et Biophysica Acta, 1463*, 20–30.

Nicolaou, K. C., Roecker, A. J., Monenschein, H., Guntupalli, P., & Follmann, M. (2003). Studies towards the synthesis of azadirachtin: Enantioselective entry into the azadirachtin framework through cascade reactions. *Angewandte Chemie International Edition, 42*, 3637–3642.

Patel, A. V., Slaats, B., Hallmann, J., Tilcher, R., Beitzen-Heineke, W., & Vorlop, K. D. (2004). Encapsulation and application of bacterial antagonists and a nematophagous fungus for biological pest control. In J. L. Pedraz, G. Orive, & D. Poncelet (Eds.), *Proceedings of the 12th International Workshop on Bioencapsulation, Vitoria, Spain, Servicio Editorial de la Universidad del Pais Vasco* (pp. 137–140). ISBN: 84-8373-649-7.

Peterson, I., & Anderson, E. A. (2005). The renaissance of natural products as drug candidates. *Science, 310*, 451–453.

Portz, D., Noll, U., & Slusarenko, A. J. (2005). Allicin from garlic (*Allium sativum* L.): A new look at an old story. In H.-W. Dehne, U. Gisi, K. H. Kuck, P. E. Russell, & H. Lyr (Eds.) *Proceedings of the 14th International Reinhardsbrunn Symposium, Modern Fungicides and Antifungal Compounds IV* (pp. 227–234). Alton, U.K.: British Crop Production Council.

Rabinikov, A., Miron, T., Konstantinovski, L., Wilchek, M., Mirelman, D., & Weiner, L. (1998). The mode of action of allicin: Trapping of radicals and interaction with thiol containing proteins. *Biochimica et Biophysica Acta, 1379*, 233–244.

Russell, P. E., & Mussa, A. E. A. (1977). The use of garlic (*Allium sativum*) extracts to control foot rot of *Phaseolus vulgaris* caused by *Fusarium solani* f.sp. *phaseoli*. *Annals of Applied Biology, 86*, 369–372.

Tetko, I. V., Gasteiger, J., Todeschini, R., Mauri, A., Livingstone, D., & Ertl, P., et al. (2005). Virtual computational chemistry laboratory—Design and description. *Journal of Computer-Aided Molecular Design, 19*, 453–463.

Wolfgang, L. (1995). Patents on native technology challenged. *Science, 269*, 1506.

Eur J Plant Pathol (2008) 121:323–330
DOI 10.1007/s10658-007-9238-1

REVIEW

Use of *Coniothyrium minitans* as a biocontrol agent and some molecular aspects of sclerotial mycoparasitism

J. M. Whipps · S. Sreenivasaprasad ·
S. Muthumeenakshi · C. W. Rogers · M. P. Challen

Received: 30 May 2007 / Accepted: 5 October 2007
© KNPV 2007

Abstract The use of the sclerotial mycoparasite *Coniothyrium minitans* as a biological control agent of diseases caused by sclerotium-forming pathogens especially *Sclerotinia sclerotiorum* is briefly reviewed. A number of studies have examined production and application methods, integrated control, ecology, and modes of action in order to understand the biology of the mycoparasite and enhance activity and reproducibility of use. Recently, development of a number of molecular-based techniques has begun to allow the examination of genes involved in mycoparasitism. Some of these procedures have been applied to identify pathogenicity genes involved in the infection of sclerotia of *S. sclerotiorum* by *C. minitans* and this work is discussed.

Keywords Biological control ·
Coniothyrium minitans · Mycoparasitism ·
Pathogenicity genes · Sclerotia · *Sclerotinia*

Introduction

Campbell (1947) first isolated the fungus *Coniothyrium minitans* from a sclerotium of *Sclerotinia sclerotium* in the USA and immediately recognised its potential as a biological control agent. It has subsequently been isolated from all continents except Antarctica, largely from sclerotia in soil (Sandys-Winsch et al. 1993; Monaco 1989). There are numerous phenotypes differing in colony morphology and a number of other biological characteristics (Sandys-Winsch et al. 1993; Jones and Stewart 2000; Grendene et al. 2002) but ability to parasitize sclerotia of *S. sclerotiorum* remains a key feature. Some isolates have begun to be characterised using molecular procedures such as random amplification of polymorphic DNA (RAPD) and simple sequence repeat (SSR) – PCR profiling, and rRNA gene sequencing (Goldstein et al. 2000; Ridgway and Stewart 2000; Muthumeenakshi et al. 2001; Grendene et al. 2002). Recently, based on anamorphic characteristics seen in vitro, and maximum parsimony analysis of ITS and SSU nrDNA sequences, *C. minitans* was reclassified as *Paraconiothyrium minitans* (Verkley et al. 2004). However, because of common usage, *Coniothyrium minitans* will continue to be applied throughout this paper.

Coniothyrium minitans as a biological control agent

Coniothyrium minitans has the ability to infect sclerotia of many ascomycetous fungi including *Sclerotinia minor*, *Sclerotinia sclerotiorum*, *Sclerotinia trifoliorum*, *Sclerotium cepivorum* but not apparently, those of

J. M. Whipps (✉) · S. Sreenivasaprasad ·
S. Muthumeenakshi · C. W. Rogers · M. P. Challen
Warwick HRI, University of Warwick,
Wellesbourne,
Warwick CV35 9EF, UK
e-mail: John.Whipps@warwick.ac.uk

 Springer

Ciborinia camellia (Van Toor et al. 2005), nor of any basidiomycetous sclerotia (Whipps and Gerlagh 1992). It has been used successfully when applied to soil in the glasshouse and field to control *S. sclerotiorum* in numerous crops including lettuce, celery, sunflower, bean and oilseed rape (Budge and Whipps 1991; Huang and Hoes 1980; Gerlagh et al. 1999; Luth 2001), *S. minor* on lettuce and peanut (Partridge et al. 2006b; Rabeendran et al. 2006) and *S. cepivorum* on onion (Ahmed and Tribe 1977; McLean and Stewart 2000). Conidia have also been sprayed onto foliage to prevent ascospore infection and disease development in alfalfa, beans and oilseed rape (Huang et al. 2000; Gerlagh et al. 2004; Li et al. 2003a, 2005a, 2006), to foliage to diminish sclerotial production and survival in rotations of several crops (Gerlagh et al. 1999) as well as to crop debris to decrease sclerotial carryover (Budge et al. 1995). It also survives well in soil for several years and can be responsible for suppressiveness to Sclerotinia disease in some areas (Huang and Kozub 1991; Huang and Erickson 2002; Jones and Whipps 2002). However, as with many biological control agents, there is evidence that reproducibility of control is sometimes variable depending on environmental and biological factors, especially disease pressure (Budge and Whipps 1991; McQuilken et al. 1995). Consequently, numerous studies have been carried out to characterise different properties of *C. minitans* to optimise its use in biological control.

Inoculum production and application

Once a screening exercise has been carried out to identify active isolates (Whipps and Budge 1990; Jones and Stewart 2000) it is important to optimise the quality and quantity of inoculum to be applied. Many experimental systems have applied solid substrate fermentation products directly to soil with concomitant arguments about cost effectiveness (Whipps and Gerlagh 1992). This process has also been used regularly to produce conidia which are then used in spray formulations (Gerlagh et al. 1999). Indeed, conidia produced by solid state fermentation are incorporated into the wettable granule formulation of the commercial *C. minitans* product, Contans® WG (de Vrije et al. 2001). Considerable efforts have been made to optimise the yield, viability, infectivity and surface characteristics of conidia in these systems with particular emphasis on physiological studies of substrate utilisation and influence of environmental conditions (Whipps and Gerlagh 1992; McQuilken et al. 1997b; Oostra et al. 2000; Ooijkaas et al. 1998, 1999; Smith et al. 1999; Jones et al. 2004b; Chen et al. 2005). Nevertheless, other workers have explored the potential to produce inoculum of *C. minitans* in liquid fermentation (McQuilken et al. 1997a; Cheng et al. 2003) and to improve infectivity by pre-germinating conidia prior to application procedures (Shi et al. 2004), and could be worth exploring further. There have also been numerous studies to compare application rates, timings and forms of inocula to obtain cost-effective disease control or infection of sclerotia (eg. Jones 2003a, 2004a; Gerlagh et al. 2003, 2004) and it is clear that these may be variable for each crop.

Integrated control

Another approach to improve efficacy has been to integrate *C. minitans* with other control treatments. For example, successful integration of *C. minitans* with the fungicide iprodione was obtained in glasshouse trials against Sclerotinia disease in lettuce (Budge and Whipps 2001). However, *C. minitans* is susceptible to numerous fungicides (Budge and Whipps 2001; Li et al. 2002; Partridge et al. 2006a) and so care must be taken with any strategy integrating fungicides with *C. minitans* unless a fungicide-tolerant isolate is available. In the study of Budge and Whipps (2001), use of a fungicide-tolerant isolate was not required as the mycoparasite in the soil was protected from the direct effects of foliar applied fungicide. *Coniothyrium minitans* has also been successfully integrated with combinations of *Trichoderma* spp. (Budge et al. 1995) for control of Sclerotinia disease in lettuce but under these glasshouse experiments, control reflected that of *C. minitans* rather than the *Trichoderma* spp. In this case, despite the fact that the *T. virens* used originated from a sclerotium of *S. minor*, the control reflected the temperature in the glasshouse and the temperature optima of the fungi involved. Thus, when temperatures in laboratory experiments were below 20°C only *C. minitans* was active in degrading sclerotia whereas when temperatures were increased above 25°C, *C. minitans* became inactive and *T. virens* became more active, dominating infection of sclerotia. Some small-scale studies have also explored the potential to integrate partial soil sterilisation (pasteurization) with

subsequent *C. minitans* application for longer-term control (Bennett et al. 2005).

Ecology

The ecology of *C. minitans* is gradually being revealed. It has long been known to have the ability to survive for several years in soil following application (Budge and Whipps 1991; Jones and Whipps 2002; Huang and Erickson 2007) but cannot grow through raw soil and utilise organic substrates in soil as a saprotroph (Williams 1996). Consequently, it must be viewed as an ecologically obligate mycoparasite. Tribe (1957) first proposed that the mycoparasite could survive for long periods of time in soil protected within sclerotia but only recently have a combination of experiments shown this hypothesis to be true. Firstly, studies using a strain genetically marked with GUS (β-glucuronidase (*uid A*)) and hygromycin resistance (hygromycin phosphotransferase (*hph*)) (Jones et al. 1999) showed that the mycoparasite infects a large proportion of sclerotia in soil rapidly from low population levels and that fungi colonising sclerotia already infected by *C. minitans* mask the detection of *C. minitans* rather than displacing it (Jones et al. 2003b). Indeed, only two conidia of *C. minitans* are needed to infect a sclerotium of *S. sclerotiorum* under ideal conditions (Gerlagh et al. 2003). Secondly, following colonisation of sclerotia of *S. sclerotiorum* by *C. minitans* in soil, the sclerotial medulla was largely converted to pycnidia of the mycoparasite with conidial droplets produced on the surface of a largely intact rind (Bennett et al. 2006). The pycnidia and dried conidial droplets were still present 6 months after infection and by 10 months approximately 13% of the conidia in dried droplets were still viable. Together these studies clearly show that *C. minitans* is a highly efficient sclerotial mycoparasite which uses sclerotia of *S. sclerotiorum* as reservoirs for survival.

Coniothyrium minitans exhibits no ability to infect healthy plants even on cut tissues (Gerlagh et al. 1996) although it can grow on foliage, petals and into stems precolonised by *S. sclerotiorum* (Gerlagh et al. 1994; Li et al. 2003a) and survive on petals of oilseed rape for several days (Yang et al. 2007). However, it is dispersed by water splash directly and as aerosols (Williams et al. 1998) and by animals including slugs, collembolans, mites and sciarid larvae (Whipps 2001).

It could be argued that despite its longevity in soil, *C. minitans* has few effects on the soil microbial population as it is an ecologically obligate mycoparasite. However, recent findings that *C. minitans* produces antimicrobial metabolites in culture (McQuilken et al. 2003; Li et al. 2005b) suggest this may not be the case and some studies to examine the impact of *C. minitans* on the soil microbial population have begun. Introduction of *C. minitans* at 10^3 or 10^6 colony forming units (cfu) g^{-1} soil resulted in no changes in culturable populations of bacteria over a 30-day period and only a small decrease (0.1 $\log_{10}$ cfu g^{-1}) in culturable fungal populations with the higher application rate (Bennett et al. 2003). Populations of *C. minitans* did not change over this period. Similarly, preliminary data using a genetically marked strain to enhance detection, indicate once again that *C. minitans* survives in soil for 6 months with little loss in cfus and that there is little impact on the microbial populations determined by microbial fingerprinting using PCR-denaturing gradient gel electrophoresis (Rogers and Whipps, unpublished). These data seem to confirm the concept that any effect of *C. minitans* on the natural microflora is minimal.

Modes of action

Although hyphal–hyphal interactions between *C. minitans* and numerous other fungi have been observed in vitro, the two major interactions in vivo reflect the competition or mycoparasitism that occurs between mycelium of *C. minitans* and *Sclerotinia* on petals, pollen or infected plant tissues (Li et al. 2002, 2003b), and the mycoparasitic attack of mycelium of *C. minitans* on host sclerotia (Whipps and Gerlagh 1992), the latter being most common. There have been no detailed microscopical or biochemical studies of the interactions on plant tissues and few studies of sclerotial mycoparasitism since the subject was last reviewed (Whipps and Gerlagh 1992). Cell wall degradation and host tissue collapse has long been associated with production of cell-wall degrading enzymes such as chitinases and glucanases by *C. minitans* as part of the mycoparasitic process and recently molecular studies have shown the increased expression of a β-1, 3 glucanase gene *cmg1* by *C. minitans* during infection of sclerotia of *S. sclerotiorum* (Giczey et al. 2001). *Coniothyrium minitans* has the ability to produce a wide spectrum of cell-wall degrading enzymes in culture when grown on complex

substrates such as purified fungal and Oomycete cell walls (Dahiya et al. 1998; Inglis and Kawchuk 2002; Kaur et al. 2005) and some mutants with enhanced β-glucanase activity have been obtained (Zantinge et al. 2003). Recently, the finding that *C. minitans* produces antimicrobial metabolites in culture (McQuilken et al. 2003; Li et al. 2005b) opens up the possibility that these compounds could be involved in the mycoparasitic or competitive processes occurring during these interactions in vivo. However, other than the study by Giczey et al. (2001), no good evidence for expression of specific enzymes, genes or metabolites during the sclerotial infection process by *C. minitans* has been obtained. The challenge to address this lack of knowledge using modern molecular techniques is considered in the next section.

Examining molecular aspects of sclerotial mycoparasitism

To date, there are less than 10 gene sequences of *C. minitans* available in public databases and the molecular tool-kits for gene discovery and gene function analysis in this challenging dual-ascomycete interaction system are only beginning to be developed. Recently, protocols for rapid and reliable transformation of *C. minitans* based on restriction enzyme mediated integration and *Agrobacterium tumefaciens*-mediated transformation were developed. These techniques were applied for insertional mutagenesis of *C. minitans* leading to the identification of 11 sclerotial mycoparasitism mutants (Rogers et al. 2004; Li et al. 2005b). Using inverse PCR, sequences flanking the site of insertion in some of the REMI mutants have been identified (Rogers et al. unpublished). For example, the sequence recovered in a non-mycoparasitic mutant R2427 showed high homology to the *PIF1* DNA helicase gene in yeasts and filamentous fungi. Functional complementation analysis by reintroduction of the wild-type *PIF1* gene into the mutant led to the restoration of sclerotial mycoparasitism. *PIF1* gene is likely to play a major role in maintaining mitochondrial integrity in response to oxidative stress, based on the recent work published in yeasts (Doudican et al. 2005).

Further, using degenerate PCR and macroarrays of a genomic cosmid library of *C. minitans*, *PKAC* and *PMK1* genes implicated in fungal pathogenicity and

signalling, have been isolated and fully sequenced with a view to characterising their role in signalling. In a separate study to identify the genes regulating key processes in sclerotial mycoparasitism, PCR-based suppression subtractive hybridisation (SSH) has been used (Muthumeenakshi et al. 2007). Following SSH between cDNA samples from *C. minitans* grown in culture and *C. minitans* colonising autoclaved sclerotia (simulated mycoparasitism), a cDNA library enriched for genes upregulated during this process was established. Sequencing of 672 cDNA clones and bioinformatic analysis led to the identification of 251 unisequences (putative genes) and their functional categorisation. The genes identified belonged to diverse functions such as signalling and cellular communication, cell wall degradation and hydrolysis of energy reserves, production of anti-microbial metabolites, detoxification, stress response and nutrient utilisation in *C. minitans* (Fig. 1). Nearly 35 putative genes encoding a whole range of hydrolytic enzymes were present among the *C. minitans* unisequences identified. Cell walls of the *S. sclerotiorum* sclerotia are a major barrier that the mycoparasite *C. minitans* has to overcome both to kill the host and to access the nutrients and a number of these hydrolytic enzymes are likely to play an active role in this process. It is also likely that some of these enzymes will play a role in remodelling the cells of *C. minitans*, as the mycoparasite grows through the sclerotial host. Further, Lu et al. (2004) expressed the *C. minitans* xylanase gene in *Arabidopsis* plants, providing a basis for investigat-

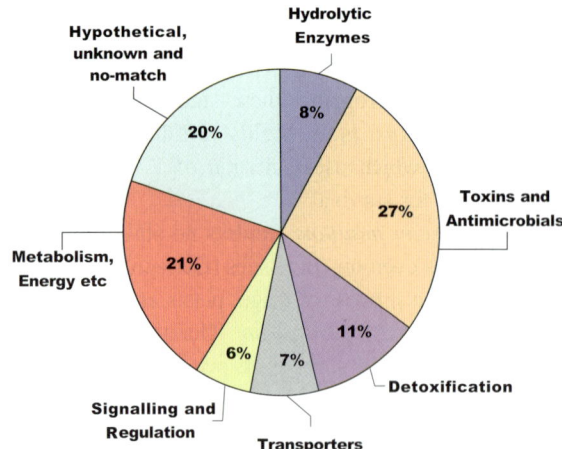

Fig. 1 Functional categorisation of gene transcripts expressed by *Coniothyrium minitans* during sclerotial mycoparasitism

ing transgenic resistance to *S. sclerotiorum* and other plant pathogens.

Various strategies were adopted to successfully perform expression analysis of *C. minitans* genes during sclerotial mycoparasitism (Muthumeenakshi et al. 2007). For example, specific probe sequences to avoid cross hybridisation were selected, by comparing the sequence of the *C. minitans* cDNAs with the genome sequence of *S. sclerotiorum*. Moreover, as the cDNA samples of *C. minitans* colonising live sclerotia yielded large quantities of RNA from the host fungus, slot blot analysis of known concentrations of the cDNA samples was carried out with different *C. minitans* probes and the cDNA ratio between the mycoparasite and the host was determined to be 1:6. Gene expression analysis carried out utilising dot blots and virtual Northerns of the unisequences, revealed different levels of upregulation of various *C. minitans* genes during sclerotial mycoparasitism. Experiments carried out with candidate genes suggest that RNAi and knock-out technologies are feasible for gene function analysis in *C. minitans*.

Overall, some of the genes identified in *C. minitans* such as those linked to signalling and detoxification belong to conserved gene families and pathways implicated in fungal-plant interactions. On the other-hand, the wide range of hydrolytic enzyme genes identified are likely to be important in fungal–fungal interactions and different to those expressed by fungal plant pathogens. Interestingly, less than 30% of the *C. minitans* genes identified showed homology to ESTs generated in *Trichoderma* spp. (Vizcaino et al. 2006). Moreover, nearly 20% of the genes are of unknown function suggesting that these could play novel roles in *C. minitans*, which shows a different eco-nutritional mode compared to other mycoparasites such as *Trichoderma* spp. These advances in gene discovery and gene function analysis in *C. minitans*, have laid a firm platform for comparative genomic investigations into the molecular mechanisms involved in sclerotial mycoparasitism. This knowledge is likely to provide strategies for a better exploitation of fungal biocontrol agents in future plant disease management.

Acknowledgements We would like to thank the BBSRC, Defra and the EU (Project: 2E-BCAs in crops) for financial support.

References

Ahmed, A. H. M., & Tribe, H. T. (1977). Biological control of white rot of onion (*Sclerotium cepivorum*) by *Coniothyrium minitans*. *Plant Pathology, 26*, 75–78.

Bennett, A. J., Leifert, C., & Whipps, J. M. (2003). Survival of the biocontrol agents *Coniothyrium minitans* and *Bacillus subtilis* MBI 600 introduced into pasteurised, sterilised and non-sterile soils. *Soil Biology & Biochemistry, 35*, 1565–1573.

Bennett, A. J., Leifert, C., & Whipps, J. M. (2005). Effect of combined treatment of pasteurisation and *Coniothyrium minitans* on sclerotia of *Sclerotinia sclerotiorum* in soil. *European Journal of Plant Pathology, 113*, 197–209.

Bennett, A. J., Leifert, C., & Whipps, J. M. (2006). Survival of *Coniothyrium minitans* associated with sclerotia of *Sclerotinia sclerotiorum* in soil. *Soil Biology & Biochemistry, 38*, 164–172.

Budge, S. P., McQuilken, M. P., Fenlon, J. S., & Whipps, J. M. (1995). Use of *Coniothyrium minitans* and *Gliocladium virens* for biological control of *Sclerotinia sclerotiorum* in glasshouse lettuce. *Biological Control, 5*, 513–522.

Budge, S. P., & Whipps, J. M. (1991). Glasshouse trials of *Coniothyrium minitans* and *Trichoderma* species for the biological control of *Sclerotinia sclerotiorum* in celery and lettuce. *Plant Pathology, 40*, 59–66.

Budge, S. P., & Whipps, J. M. (2001). Potential for integrated control of *Sclerotinia sclerotiorum* in glasshouse lettuce using *Coniothyrium minitans* and reduced fungicide application. *Phytopathology, 91*, 221–227.

Campbell, W. A. (1947). A new species of *Coniothyrium* parasitic on sclerotia. *Mycologia, 39*, 190–195.

Chen, X., Li, Y., Du, G. C., & Chen, J. (2005). Application of response surface methodology in medium optimization for spore production of *Coniothyrium minitans* in solid-state fermentation. *World Journal of Microbiology & Biotechnology, 21*, 593–599.

Cheng, J. S., Jiang, D. H., Yi, X. H., Fu, Y. P., Li, G. Q., & Whipps, J. M. (2003). Production, survival and efficacy of *Coniothyrium minitans* conidia produced in shaken liquid culture. *FEMS Microbiology Letters, 227*, 127–131.

Dahiya, J. S., Singh, D., & Nigam, P. (1998). Characterisation of laccase produced by *Coniothyrium minitans*. *Journal of Basic Microbiology, 38*, 349–359.

Doudican, N. A., Song, B., Shadel, G. S., & Doetsch, P. W. (2005). Oxidative DNA damage causes mitochondrial genomic instability in *Saccharomyces cerevisiae*. *Molecular and Cellular Biology, 25*, 5196–5204.

de Vrije, T., Antoine, N., Buitelaar, R. M., Bruckner, S., Dissevelt, M., & Durand, A., et al. (2001). The fungal biocontrol agent *Coniothyrium minitans*: Production by solid-state fermentation, application and marketing. *Applied Microbiology and Biotechnology, 56*, 58–68.

Gerlagh, M., Goossen-van de Geijn, H. M., Fokkema, N. J., & Vereijken, P. F. G. (1999). Long-term biosanitation by application of *Coniothyrium minitans* on *Sclerotinia sclerotiorum* infected crops. *Phytopathology, 89*, 141–147.

Gerlagh, M., Goossen-van de Geijn, H. M., Hoogland, A. E., & Vereijken, P. F. G. (2003). Quantitative aspects of

infection of *Sclerotinia sclerotiorum* sclerotia by *Coniothyrium minitans* – Timing of application, concentration and quality of conidial suspension of the mycoparasite. *European Journal of Plant Pathology, 109,* 489–502.

Gerlagh, M., Goossen-van de Geijn, H. M., Hoogland, A. E., Vereijken, P. F. G., Horsten, P. F. M., & de Haas, B. H. (2004). Effect of volume and concentration of conidial suspensions of *Coniothyrium minitans* on infection of *Sclerotinia sclerotiorum* sclerotia. *Biocontrol Science and Technology, 14,* 675–690.

Gerlagh, M., Kruse, M., Van de Geijn, H. M., & Whipps, J. M. (1994). Growth and survival of the mycoparasite *Coniothyrium minitans* on lettuce leaves in contact with soil in the presence or absence of *Sclerotinia sclerotiorum*. *European Journal of Plant Pathology, 100,* 55–59.

Gerlagh, M., Whipps, J. M., Budge, S. P., & Goossen van de Geijn, H. M. (1996). Efficiency of isolates of *Coniothyrium minitans* as mycoparasites of *Sclerotinia sclerotiorum, Sclerotium cepivorum* and *Botrytis cinerea* on tomato stem pieces. *European Journal of Plant Pathology, 102,* 787–793.

Giczey, G., Kerenyi, Z., Fulop, L., & Hornok, L. (2001). Expression of cmg1, an exo-beta-1,3-glucanase gene from *Coniothyrium minitans*, increases during sclerotial parasitism. *Applied and Environmental Microbiology, 67,* 865–871.

Goldstein, A. L., Carpenter, M. A., Crowhurst, R. N., & Stewart, A. (2000). Identification of *Coniothyrium minitans* isolates using PCR amplification of a dispersed repetitive element. *Mycologia, 92,* 46–53.

Grendene, A., Minardi, P., Giacomini, A., Squartini, A., & Marciano, P. (2002). Characterization of the mycoparasite *Coniothyrium minitans*: comparison between morphophysiological and molecular analyses. *Mycological Research, 106,* 796–807.

Huang, H. C., Bremer, E., Hynes, R. K., & Erickson, R. S. (2000). Foliar application of fungal biocontrol agents for the control of white mold of dry bean caused by *Sclerotinia sclerotiorum*. *Biological Control, 18,* 270–276.

Huang, H. C., & Erickson, R. S. (2002). Overwintering of *Coniothyrium minitans*, a mycoparasite of *Sclerotinia sclerotiorum*, on the Canadian prairies. *Australasian Plant Pathology, 31,* 291–293.

Huang, H. C., & Erickson, R. S. (2007). *Ulocladium atrum* as a biological control agent for white mold of bean caused by *Sclerotinia sclerotiorum*. *Phytoparasitica, 35,* 15–22.

Huang, H. C., & Hoes, J. A. (1980). Importance of plant spacing and sclerotial position to development of Sclerotinia wilt of sunflower. *Plant Disease, 64,* 81–84.

Huang, H. C., & Kozub, G. C. (1991). Monocropping to sunflower and decline of Sclerotinia wilt. *Botanical Bulletin of Academia Sinica, 32,* 163–170.

Inglis, G. D., & Kawchuk, L. M. (2002). Comparative degradation of oomycete, ascomycete, and basidiomycete cell walls by mycoparasitic and biocontrol fungi. *Canadian Journal of Microbiology, 48,* 60–70.

Jones, E., Carpenter, M., Fong, D., Goldstein, A., Thrush, A., & Crowhurst, R., et al. (1999). Co-transformation of the sclerotial mycoparasite *Coniothyrium minitans* with hygromycin B resistance and beta-glucuronidase markers. *Mycological Research, 103,* 929–937.

Jones, E. E., Mead, A., & Whipps, J. M. (2003a). Evaluation of different *Coniothyrium minitans* inoculum sources and application rates on apothecial production and infection of *Sclerotinia sclerotiorum* sclerotia. *Soil Biology & Biochemistry, 35,* 409–419.

Jones, E. E., Mead, A., & Whipps, J. M. (2004a). Effect of inoculum type and timing of application of *Coniothyrium minitans* on *Sclerotinia sclerotiorum*: Control of sclerotinia disease in glasshouse lettuce. *Plant Pathology, 53,* 611–620.

Jones, E. E., & Stewart, A. (2000). Selection of mycoparasites of sclerotia of *Sclerotinia sclerotiorum* isolated from New Zealand soils. *New Zealand Journal of Crop and Horticultural Science, 28,* 105–114.

Jones, E. E., Stewart, A., & Whipps, J. M. (2003b). Use of *Coniothyrium minitans* transformed with the hygromycin B resistance gene to study survival and infection of *Sclerotinia sclerotiorum* sclerotia in soil. *Mycological Research, 107,* 267–276.

Jones, E. E., Weber, F. J., Oostra, J., Rinzema, A., Mead, A., & Whipps, J. M. (2004b). Conidial quality of the biocontrol agent *Coniothyrium minitans* produced by solid-state cultivation in a packed-bed reactor. *Enzyme and Microbial Technology, 34,* 196–207.

Jones, E. E., & Whipps, J. M. (2002). Effect of inoculum rates and sources of *Coniothyrium minitans* on control of *Sclerotinia sclerotiorum* disease in glasshouse lettuce. *European Journal of Plant Pathology, 108,* 527–538.

Kaur, J., Munshi, G. D., Singh, R. S., & Koch, E. (2005). Effect of carbon source on production of lytic enzymes by the sclerotial parasites *Trichoderma atroviride* and *Coniothyrium minitans*. *Journal of Phytopathology, 153,* 274–279.

Li, M. X., Gong, X. Y., Zheng, J., Jiang, D. H., Fu, Y. P., & Hou, M. S. (2005b). Transformation of *Coniothyrium minitans*, a parasite of *Sclerotinia sclerotiorum*, with *Agrobacterium tumefaciens*. *FEMS Microbiology Letters, 243,* 323–329.

Li, G. Q., Huang, H. C., & Acharya, S. N. (2002). Sensitivity of *Ulocladium atrum, Coniothyrium minitans*, and *Sclerotinia sclerotiorum* to benomyl and vinclozolin. *Canadian Journal of Botany, 80,* 892–898.

Li, G. Q., Huang, H. C., & Acharya, S. N. (2003a). Importance of pollen and senescent petals in the suppression of alfalfa blossom blight (*Sclerotinia sclerotiorum*) by *Coniothyrium minitans*. *Biocontrol Science and Technology, 13,* 495–505.

Li, G. Q., Huang, H. C., & Acharya, S. N. (2003b). Antagonism and biocontrol potential of *Ulocladium atrum* on *Sclerotinia sclerotiorum*. *Biological Control, 28,* 11–18.

Li, G. Q., Huang, H. C., Acharya, S. N., & Erickson, R. S. (2005a). Effectiveness of *Coniothyrium minitans* and *Trichoderma atroviride* in suppression of sclerotinia blossom blight of alfalfa. *Plant Pathology, 54,* 204–211.

Li, G. Q., Huang, H. C., Miao, H. J., Erickson, R. S., Jiang, D. H., & Xiao, Y. N. (2006). Biological control of sclerotinia diseases of rapeseed by aerial applications of the mycoparasite *Coniothyrium minitans*. *European Journal of Plant Pathology, 114,* 345–355.

Lu, Z. X., Laroche, A., & Huang, H. C. (2004). Segregation patterns for integration and expression of *Coniothyrium minitans* xylanase gene in *Arabidopsis thaliana* transformants. *Botanical Bulletin of Academia Sinica, 45*, 23–31.

Luth, P. (2001). The control of *Sclerotinia* sp. and *Sclerotium cepivorum* with the biological fungicide Contans® WG – Experiences from field trials and commercial use. In C. S. Young, & K. J. D. Hughes (Eds.) *Proceedings of the XI International Sclerotinia workshop* (pp. 37–38). York: Central Science Laboratory.

McLean, K. L., & Stewart, A. (2000). Application strategies for control of onion white rot by fungal antagonists. *New Zealand Journal of Crop and Horticultural Science, 28*, 115–122.

McQuilken, M. P., Budge, S. P., & Whipps, J. M. (1997a). Production, survival and evaluation of liquid culture-produced inocula of *Coniothyrium minitans* against *Sclerotinia sclerotiorum*. *Biocontrol Science and Technology, 7*, 23–36.

McQuilken, M. P., Budge, S. P., & Whipps, J. M. (1997b). Effects of culture media and environmental factors on conidial germination, pycnidial production and hyphal extension of *Coniothyrium minitans*. *Mycological Research, 101*, 11–17.

McQuilken, M. P., Gemmell, J., Hill, R. A., & Whipps, J. M. (2003). Production of macrosphelide A by the mycoparasite *Coniothyrium minitans*. *FEMS Microbiology Letters, 219*, 27–31.

McQuilken, M. P., Mitchell, S. J., Budge, S. P., Whipps, J. M., Fenlon, J. S., & Archer, S. A. (1995). Effect of *Coniothyrium minitans* on sclerotial survival and apothecial production of *Sclerotinia sclerotiorum* in field-grown oilseed rape. *Plant Pathology, 44*, 883–896.

Monaco, C. (1989). Evaluacion de la eficiencia de micoparasitos sobre esclerocios de *Sclerotinia sclerotiorum* "in vitro." *Revista Facultad Agronomia, 65*, 67–73.

Muthumeenakshi, S., Goldstein, A. L., Stewart, A., & Whipps, J. M. (2001). Molecular studies on intraspecific diversity and phylogenetic position of *Coniothyrium minitans*. *Mycological Research, 105*, 1065–1074.

Muthumeenakshi, S., Sreenivasaprasad, S., Rogers, C. W., Challen, M. P., & Whipps, J. M. (2007). Analysis of cDNA transcripts from *Coniothyrium minitans* reveals a diverse array of genes involved in key processes during sclerotial mycoparasitism. *Fungal Genetics and Biology* (in press) DOI 10.1016/j.fgb.2007.07.011.

Ooijkaas, L. P., Ifeong, C. J., Tramper, J., & Buitelaar, R. M. (1998). Spore production of *Coniothyrium minitans* during solid-state fermentation on different nitrogen sources with glucose or starch as carbon source. *Biotechnology Letters, 20*, 785–788.

Ooijkaas, L. P., Wilkinson, E. C., Tramper, J., & Buitelaar, R. M. (1999). Medium optimization for spore production of *Coniothyrium minitans* using statistically-based experimental designs. *Biotechnology and Bioengineering, 64*, 92–100.

Oostra, J., Tramper, J., & Rinzema, A. (2000). Model-based bioreactor selection for large-scale solid-state cultivation of *Coniothyrium minitans* spores on oats. *Enzyme and Microbial Technology, 27*, 652–663.

Partridge, D. E., Sutton, T. B., & Jordan, D. L. (2006a). Effect of environmental factors and pesticides on mycoparasitism of *Sclerotinia minor* by *Coniothyrium minitans*. *Plant Disease, 90*, 1407–1412.

Partridge, D. E., Sutton, T. B., Jordan, D. L., & Curtis, V. L. (2006b). Management of Sclerotinia blight of peanut with the biological control agent *Coniothyrium minitans*. *Plant Disease, 90*, 957–963.

Rabeendran, N., Jones, E. E., Moot, D. J., & Stewart, A. (2006). Biocontrol of Sclerotinia lettuce drop by *Coniothyrium minitans* and *Trichoderma hamatum*. *Biological Control, 39*, 352–362.

Rogers, C. W., Challen, M. P., Green, J. R., & Whipps, J. M. (2004). Use of REMI and *Agrobacterium*-mediated transformation to identify pathogenicity mutants of the biocontrol fungus, *Coniothyrium minitans*. *FEMS Microbiology Letters, 241*, 207–214.

Ridgway, H. J., & Stewart, A. (2000). Molecular assisted detection of the mycoparasite *Coniothyrium minitans* A69 in soil. *New Zealand Plant Protection, 53*, 114–117.

Sandys-Winsch, C., Whipps, J. M., Gerlagh, M., & Kruse, M. (1993). World distribution of the sclerotial mycoparasite *Coniothyrium minitans*. *Mycological Research, 97*, 1175–1178.

Shi, J. L., Li, Y., Qian, H. L., Du, G. C., & Chen, J. (2004). Pregerminated conidia of *Coniothyrium minitans* enhances the foliar biological control of *Sclerotinia sclerotiorum*. *Biotechnology Letters, 26*, 1649–1652.

Smith, S. N., Armstrong, R. A., Barker, M., Bird, R. A., Chohan, R., & Hartell, N. A., et al. (1999). Determination of *Coniothyrium minitans* conidial and germling lectin avidity by flow cytometry and digital microscopy. *Mycological Research, 103*, 1533–1539.

Tribe, H. T. (1957). On the parasitism of *Sclerotinia trifoliorum* by *Coniothyrium minitans*. *Transactions of the British Mycological Society, 40*, 489–199.

Van Toor, R. F., Jaspers, M. V., & Stewart, A. (2005). Effect of soil microorganisms on viability of sclerotia of *Ciborinia camelliae*, the causal agent of camellia flower blight. *New Zealand Journal of Crop and Horticultural Science, 33*, 149–160.

Verkley, G. J. M., da Silva, M., Wicklow, D. T., & Crous, P. W. (2004). *Paraconiothyrium*, a new genus to accommodate the mycoparasite *Coniothyrium minitans*, anamorphs of *Paraphaeosphaeria*, and four new species. *Studies in Mycology, 50*, 323–335.

Vizcaino, J. A., Gonzalez, F. J., Suarez, M. B., Redondo, J., Heinrich, J., & Delgado-Jarana, J., et al. (2006). Generation, annotation and analysis of ESTs from Trichoderma harzianum CECT 2413. *BMC Genomics, 7*, 193.

Whipps, J. M. (2001). Ecological and biotechnological considerations in enhancing disease biocontrol. In M. Vurro, J. Gressel, T. Butt, G. E. Harman, A. Pilgeram, R. J. St. Leger, & D. L. Nuss (Eds.) *Enhancing biocontrol agents and handling risks* (pp. 43–51). Amsterdam: IOS Press.

Whipps, J. M., & Budge, S. P. (1990). Screening for sclerotial mycoparasites of *Sclerotinia sclerotiorum*. *Mycological Research, 94*, 607–612.

Whipps, J. M., & Gerlagh, M. (1992). Biology of *Coniothyrium minitans* and its potential for use in disease biocontrol. *Mycological Research, 96*, 897–907.

Williams, R. H. (1996). Dispersal of the mycoparasite *Coniothyrium minitans*. In Animal and Plant Sciences (pp 144). PhD thesis University of Sheffield, Sheffield.

Williams, R. H., Whipps, J. M., & Cooke, R. C. (1998). Splash dispersal of *Coniothyrium minitans* in the glasshouse. *Annals of Applied Biology, 132*, 77–90.

Yang, L., Miao, H. J., Li, G. Q., Yin, L. M., & Huang, H.-C. (2007). Survival of the mycoparasite *Coniothyrium minitans* on flower petals of oilseed rape under field conditions in central China. *Biological Control, 40*, 179–186.

Zantinge, J. L., Huang, H. C., & Cheng, K. J. (2003). Induction, screening and identification of *Coniothyrium minitans* mutants with enhanced beta-glucanase activity. *Enzyme and Microbial Technology, 32*, 224–230.

Eur J Plant Pathol (2008) 121:331–337
DOI 10.1007/s10658-007-9248-z

International standards for the diagnosis of regulated pests

Françoise Petter · Anne Sophie Roy · Ian Smith

Received: 14 May 2007 / Accepted: 29 October 2007
© KNPV 2007

Abstract For the last 10 years, the European and Mediterranean Plant Protection Organization (EPPO) has run a European Panel on diagnostics, which has developed regional standards on diagnostic protocols. Nearly 80 such standards have now been approved, and are in active use in EPPO countries. In 2003, the Commission for Phytosanitary Measures (CPM) of FAO, in reviewing global needs for International Standards for Phytosanitary Measures (ISPMs), recognized that there is a strong interest in developing diagnostic protocols for all contracting parties to the International Plant Protection Convention (IPPC). Such protocols would support the harmonization of detection and identification procedures worldwide, contribute to greater transparency and comparability in the diagnostics for regulated pests, and assist in the resolution of disputes between trading partners. In addition, such protocols would be very useful in technical assistance programmes. In 2004, the CPM adopted a mechanism for rapid development of ISPMs in specific areas, particularly suitable for diagnostic protocols. A Technical Panel was accordingly established to develop protocols for specific pests and meets on an annual basis. A format for international diagnostic protocols was adopted in 2006 and a list of priority pests was established. In 2003, EPPO initiated a new programme on quality management and accreditation for plant pest laboratories and Standards are now also being developed in this area. In 2006, a survey of existing diagnostic capacities in EPPO member countries was undertaken and a database on diagnostic expertise was created.

Keywords Diagnostic protocols · International Standards for phytosanitary measures · Regional Standards · Accreditation · Diagnostic expertise

F. Petter (✉) · A. S. Roy
European and Mediterranean Plant Protection Organization (EPPO),
1 rue Le Nôtre,
75016 Paris, France
e-mail: hq@eppo.fr

I. Smith
EPPO/OEPP,
Paris, France

Introduction

Global phytosanitary context

During the last century, movement of goods and persons across the world has increased considerably. Natural borders that were once effective barriers to the spread of pests are now under pressure from the increasing volume of international trade. As a consequence, the global community has developed cooperative mechanisms to protect plants and the environment from pests. The International Plant Protection Convention (IPPC) is an international treaty to protect plant health. The Convention, adopted in 1951, is deposited with the Director General of the Food and Agriculture Organization of the United Nations (FAO). The purpose of the IPPC is to secure common and effective action to prevent the spread and introduction of pests of plants and plant products and to promote appropriate measures for their control. Since the beginning of the 1990s, the international phytosanitary context has changed considerably. The IPPC was revised in 1997 (FAO 1997) to incorporate principles included in the 1994 SPS agreement (Agreement on the Application of Sanitary and Phytosanitary Measures) of the WTO (World Trade Organization). This agreement states in particular that "Members shall ensure that any sanitary or phytosanitary measure is applied only to the extent necessary to protect human, animal or plant life or health, is based on scientific principles and is not maintained without sufficient scientific evidence" and that "members shall base their SPS measures on international standards, guidelines and recommendations" (WTO 1994). The latter are defined as those 'developed under the auspices of the Secretariat of the IPPC in cooperation with regional organizations operating within the framework of the IPPC'. The revision of the IPPC in 1997 represented a major updating of the Convention. The changes were primarily intended to strengthen the IPPC, by provision of a mechanism for developing and adopting International Standards for Phytosanitary Measures (ISPMs) and by the creation of a Commission for Phytosanitary Measures (CPM) to promote the full implementation of the objectives of the Convention[1]. The New Revised Text of the IPPC came into force on 2005-10-02.

European context

The new revised text of the IPPC also provides for Regional Plant Protection Organizations (RPPOs) to develop internationally agreed standards on phytosanitary measures. In the IPPC context, they are known as 'regional standards'. The RPPO for Europe is the European and Mediterranean Plant Protection Organization (EPPO). This was created in 1951 to prevent the introduction of dangerous pests from

other parts of the world, and limit their spread within Europe if they were introduced. Today, 48 European and Mediterranean countries (including the 27 members of the European Union) are members of the Organization. Key partners of EPPO are National Plant Protection Organizations (NPPOs), i.e. the official services which are responsible for plant protection in each member country.

How are ISPMs developed by the CPM?

Global level

The process for developing an ISPM includes three stages: the preparation of a draft, a consultation stage and a formal approval stage (Fig. 1). Suggestions for topics for ISPMs can be made by the NPPOs, the IPPC Secretariat or the WTO-SPS Committee. Other organizations, such as the WTO, may also submit proposals for standards through the IPPC Secretariat. A Standards Committee oversees the standard-setting process and assists in the development of ISPMs by agreeing on the specifications for draft standards and checking the drafts before and after the consultation stage. This Committee comprises 25 members drawn from the seven FAO regions. The CPM is in charge, in particular, of establishing standard-setting priorities, and approves the final versions.

Regional level

For the European and Mediterranean region, EPPO has for many years had in place a Working Party on Phytosanitary Regulations which approves regionally harmonized texts on plant quarantine. These texts are developed by Panels of experts (e.g. the EPPO Panel on Diagnostics), are subject to a consultation procedure involving all Member Governments, and are finally approved by the Council of EPPO. When the IPPC introduced the concept of Regional Standards, all such EPPO documents were redefined as EPPO Standards, in a number of different series (for example, series PM 7 on diagnostic protocols). Regional standards are also developed by other RPPOs; in particular, the North American Plant Protection Organization and the 'Comité de Sanidad Vegetal del Cono Sur' in South America are also developing standards on diagnostic protocols. These regional initiatives have opened the

---

[1] Pending the entry into force of the text, an Interim Commission for Phytosanitary Measure was established and worked from 1998 to 2005. To facilitate the reading of the article it is called CPM throughout the text.

The standard-setting process

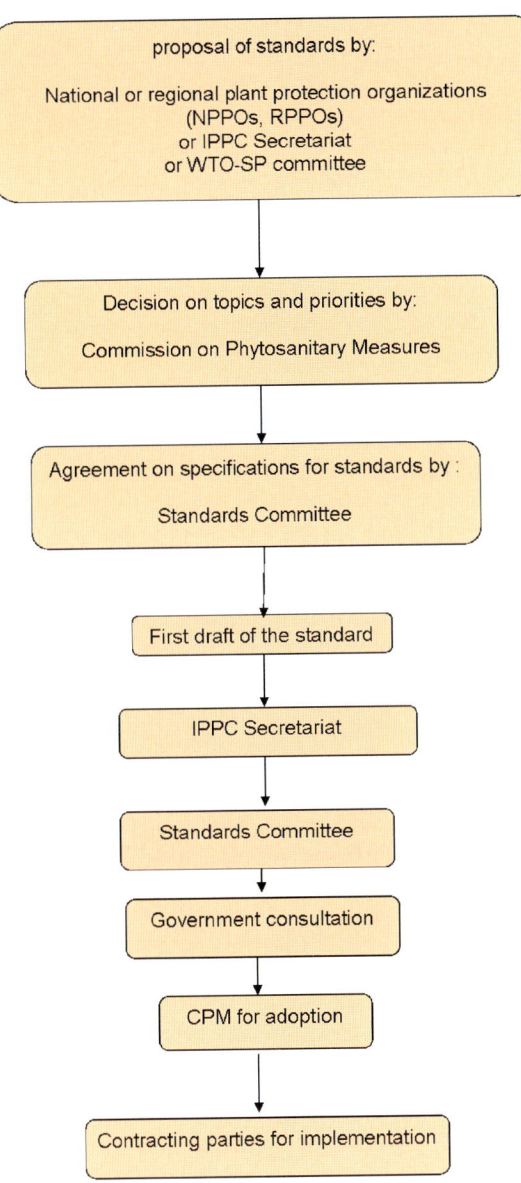

Fig. 1 The process for developing an ISPM. From the Guide to the International Plant Protection Convention (FAO 2002)

way for the development, at global level, of international diagnostic protocols.

Standards on diagnostic protocols

Regional level: EPPO Standards

The first initiatives in developing standards on diagnostic protocols were taken by EPPO. Standards

based on the work of the EPPO Panel on Diagnostics (Zlof et al. 2000) were published in 2001, and nearly 80 diagnostic protocols have now been approved (Table 1). Their preparation involved close collaboration with other EPPO Panels (Bacterial Diseases, Nematodes, Certification of Fruit Crops) and with the European Mycological Network. The EPPO system has also incorporated standards initially developed by the DIAGPRO Project of the Commission of the European Union (no. SMT 4-CT98–2252) (Harju et al. 2000). The objective of the EPPO programme was to develop internationally agreed diagnostic protocols for as many as possible of the regulated pests of the EPPO A1 and A2 lists and of the Annexes of EU Directive 2000/29 (of which there are about 340). The protocols are based on the many years of experience of EPPO experts. Each first draft is prepared by an individual expert, and written according to a 'common format and content of a diagnostic protocol' agreed by the Panel on Diagnostics, modified as necessary to fit individual pests. Each protocol is intended to contain all the information necessary to detect and positively identify a particular regulated pest.

Global level: IPPC diagnostic protocols

In reviewing global needs for plant protection in 2003, the CPM recognized that countries had a strong interest in international diagnostic protocols (FAO 2003). It considered in particular that international diagnostic protocols would provide harmonized detection and identification procedures worldwide, and contribute to greater transparency and comparability of diagnostics for regulated pests. Such protocols would be particularly useful in preventing disputes between trading partners. In addition, such protocols would constitute a very good basis for training and technical assistance.

In 2004, the CPM adopted a mechanism to speed up the development of ISPMs in specific areas, in particular diagnostic protocols (FAO 2004). A Technical Panel was accordingly established to develop diagnostic protocols for specific pests, with the following main tasks:

- identification of priorities for specific protocols to be developed and submitted to the Standards Committee. The Technical Panel should identify

Table 1 Published EPPO diagnostic protocols

Diagnostic protocol	
Published Bulletin OEPP/EPPO Bulletin 31(1), 2001	
PM 7/1 (1)	*Ceratocystis fagacearum*
PM 7/2 (1)	*Tobacco ringspot nepovirus*
PM 7/4 (1)	*Bursaphelenchus xylophilus*
PM 7/5 (1)	*Nacobbus aberrans*
Published Bulletin OEPP/EPPO Bulletin 32(2), 2002	
PM 7/6 (1)	*Chrysanthemum stunt pospiviroid*
PM 7/7 (1)	*Aleurocanthus spiniferus*
PM 7/8 (1)	*Aleurocanthus woglumi*
PM 7/9 (1)	*Cacoecimorpha pronubana*
PM 7/10 (1)	*Cacyreus marshalli*
PM 7/11 (1)	*Frankliniella occidentalis*
PM 7/12 (1)	*Parasaissetia nigra*
PM 7/13 (1)	*Trogoderma granarium*
Published Bulletin OEPP/EPPO Bulletin 33(2), 2003	
PM 7/14 (1)	*Ceratocystis fimbriata* f. sp. *platani*
PM 7/15 (1)	*Ciborinia camelliae*
PM 7/16 (1)	*Fusarium oxysporum* f. sp. *albedinis*
PM 7/17 (1)	*Guignardia citricarpa*
PM 7/18 (1)	*Monilinia fructicola*
PM 7/19 (1)	*Helicoverpa armigera*
Published Bulletin OEPP/EPPO Bulletin 34(2), 2004	
PM 7/20 (1)	*Erwinia amylovora*
PM 7/21 (1)	*Ralstonia solanacearum*
PM 7/22 (1)	*Xanthomonas arboricola* pv. *corylina*
PM 7/23 (1)	*Xanthomonas axonopodis* pv. *dieffenbachiae*
PM 7/24 (1)	*Xylella fastidiosa*
PM 7/25 (1)	*Glomerella acutata*
PM 7/26 (1)	*Phytophthora cinnamomi*
PM 7/27 (1)	*Puccinia horiana*
PM 7/28 (1)	*Synchytrium endobioticum*
PM 7/29 (1)	*Tilletia indica*
PM 7/31 (1)	*Citrus tristeza closterovirus*
PM 7/32 (1)	*Plum pox potyvirus*
PM 7/33 (1)	*Potato spindle tuber pospiviroid*
PM 7/34 (1)	*Tomato spotted wilt tospovirus*
PM 7/35 (1)	*Bemisia tabaci*
PM 7/36 (1)	*Diabrotica virgifera*
PM 7/37 (1)	*Thaumetopoea pityocamp*
PM 7/38 (1)	*Unaspis citri*
PM 7/39 (1)	*Aphelenchoides besseyi*
PM 7/40 (1)	*Globodera rostochiensis* and *Globodera pallida*
PM 7/41 (1)	*Meloidogyne chitwoodi* and *Meloidogyne fallax*
Published Bulletin OEPP/EPPO Bulletin 35(2), 2005	
PM 7/42 (1)	*Clavibacter michiganensis* subsp. *michiganensis*
PM 7/43 (1)	*Pseudomonas syringae* pv. *persicae*
PM 7/44 (1)	*Xanthomonas axonopodis* pv. *citri*
PM 7/45 (1)	*Cryphonectria parasitica*
PM 7/46 (1)	*Mycosphaerella dearnessii*

Table 1 (continued)

Diagnostic protocol	
PM 7/47 (1)	*Mycosphaerella pini*
PM 7/48 (1)	*Phoma tracheiphila*
PM 7/49 (1)	*Tomato ringspot nepovirus*
PM 7/50 (1)	*Tomato yellow leaf curl* and *Tomato mottle begomoviruses*
PM 7/51 (1)	*Aonidiella citrina*
PM 7/52 (1)	*Diaphorina citri*
PM 7/53 (1)	*Liriomyza* spp.
PM 7/54 (1)	*Lopholeucaspis japonica*
PM 7/55 (1)	*Rhizoecus hibisci*
PM 7/56 (1)	*Scirtothrips aurantii*, *Scirtothrips citri* and *Scirtothrips dorsalis*
PM 7/57 (1)	*Trioza erytreae*
Published Bulletin OEPP/EPPO Bulletin 36(1), 2006	
PM 7/58 (1)	*Burkholderia caryophylli*,
PM 7/58 (1)	*Clavibacter michiganensis* subsp. *sepedonicus*
PM 7/60 (1)	*Panthoea stewartii* subsp. *stewartii*,
PM 7/61 (1)	Candidatus *Phytoplasma aurantifoliae*,
PM 7/62 (1)	Candidatus *Phytoplasma mali*
PM 7/63 (1)	Candidatus *Phytoplasma pyri*
PM 7/64 (1)	*Xanthomonas arboricola* pv. *pruni*
PM 7/65 (1)	*Xanthomonas fragariae*
PM 7/66 (1)	*Phytophthora ramorum*
PM 7/67 (1)	*American plum line pattern ilarvirus*
PM 7/68 (1)	*Eotetranychus lewisi*
PM 7/69 (1)	*Lepidosaphes ussuriensis*
PM 7/70 (1)	*Maconellicoccus hirsutus*
PM 7/71 (1)	*Opogona sacchari*
PM 7/72 (1)	*Tecia solanivora*
PM 7/3 (2)	*Thrips palmi*
Published Bulletin OEPP/EPPO Bulletin 36(3), 2006	
PM 7/30 (2)	*Beet necrotic yellow vein benyvirus*
PM 7/73 (1)	*Gymnosporangium* spp. (non-European)
PM 7/74 (1)	*Popillia japonica*
PM 7/75 (1)	*Toxoptera citricidus*
PM 7/76 (1)	Use of EPPO diagnostic protocols
PM 7/77 (1)	Documentation and reporting on a diagnosis

the existing regional standards or protocols used by individual countries and consider suggestions for new protocols (i.e. those put forward by NPPOs, RPPOs, Expert Working Groups or other Technical Panels)

- identification of specialists to draft the Diagnostic Protocols
- production, or supervision of the production, of diagnostic protocols for specific pests

- submission to the Standards Committee of draft diagnostic protocols for specific pests and where necessary revisions of previously adopted protocols.

The Technical Panel is composed of eight diagnostic experts with at least one representing each taxonomic discipline: entomology/acarology, nematology, mycology, plant bacteriology, virology (including viroids and phytoplasmas), botany. An expert on quality assurance will also be included in the group. Between them, participants should have practical expertise in the use of morphological and molecular/biochemical diagnostic techniques, and in phytosanitary procedures. The membership of the Technical Panel is drawn from the seven FAO regions.

The Panel has already met three times since 2004. The Panel has prepared a standard explaining the scope, purpose and content of diagnostic protocols (IPPC 2006), and has recommended that specific protocols are added as annexes to this master standard (which was adopted by the CPM in April 2006 as ISPM no. 27 on *Diagnostic protocols for regulated pests*). The EPPO format was used as a basis for the development of this draft structure for diagnostics standards. The standard provides guidance on the scope and purpose of diagnostic protocols, but it does not address quality assurance issues. The Technical Panel recommended that experts drafting protocols should consider all appropriate methods for diagnosis of pests to ensure flexibility, and that the agreement of all Panel members should be needed for a method to be included in a protocol. Diagnostic protocols should take into account the fact that laboratories have differing capabilities and facilities. They should provide the essential requirements for reliable diagnosis of the pest, with alternative or supplementary methods and procedures to provide flexibility. The sensitivity, specificity and reliability of the methods should be indicated, so that NPPOs can determine the level of confidence given by each method or combination of methods. The Technical Panel also established a list of pests for which diagnostic protocols should individually be prepared (see Table 2). This list of priority pests was assembled on the basis of a list of existing regional diagnostic protocols and on pests for which requests for international diagnostic protocols had been made in 2003.

The IPPC Secretariat has collected names of experts from all parts of the world to take part in

Table 2 List of regulated pests for development of diagnostic protocols agreed in the IPPC work programme in 2007

Scientific name	Type of pest
Erwinia amylovora	Bacteria
Xylella fastidiosa	Bacteria
Liberibacter spp.	Bacteria
Xanthomonas fragariae	Bacteria
Xanthomonas axonopodis pv. *citri*	Bacteria
Phytophthora ramorum	Fungus-like
Tilletia indica/T. controversa	Fungi
Guignardia citricarpa	Fungi
Gymnosporangium spp.	Fungi
Gibberella circinata	Fungi
Puccinia psidii	Fungi
Anastrepha spp.	Insects
Thrips palmi	Insects
Anoplophora spp.	Insects
Trogoderma granarium	Insects
Bactrocera dorsalis complex	Insects
Dendroctonus ponderosae	Insects
Ips spp.	Insects
Liriomyza spp.	Insects
Bursaphelenchus xylophilus	Nematodes
Ditylenchus destructor/D.dipsaci	Nematodes
Xiphinema americanum	Nematodes
Aphelenchoides besseyi, A. ritzemabosi, and *A. fragariae*	Nematodes
Plum pox virus	Viruses
Tospoviruses (TSWV, INSV, WSMV)	Viruses
Citrus tristeza virus	Viruses
Potato spindle tuber viroid	Viruses
Viruses transmitted by *Bemisia tabaci*	Viruses
Phytoplasmas in general	Phytoplasmas

Expert Working Groups to prepare these diagnostic protocols. EPPO, acting as an intermediary in the process of selection of European experts, has communicated a list of names to the IPPC Secretariat, consisting mainly of experts who have been involved in the preparation of EPPO Diagnostic Protocols. The Expert Working Groups for drafting specific diagnostic protocols mostly work through e-mail discussions under the supervision of a member of the Technical Panel responsible for the appropriate discipline. The first diagnostic protocols were presented to the Standards Committee in May 2007.

Consequences for EPPO

EPPO with its long experience in coordinating the elaboration of regional diagnostic protocols will be

closely involved in the IPPC process. In particular, appropriate protocols already developed by EPPO may be available as working material for international standards. The work of the EPPO Panel will continue, but with new priorities determined in part by the progress in developing international diagnostic protocols. EPPO experts will concentrate on pests which are present in, and important for Europe, and not necessarily of great significance for other continents. EPPO expects that the international standards will in general fill the needs of EPPO countries for pests on which there is little practical experience.

Other EPPO initiatives in the field of diagnostics

Accreditation and quality management for plant health laboratories

Since 1999, there have been several discussions on the possibility that EPPO should assist national diagnostic laboratories in obtaining accreditation. It was concluded that EPPO can develop a quality assurance standard for diagnostic laboratories but that accreditation can be provided only by an official (national) outside body. In 2003, EPPO created an *ad hoc* Panel on Technical requirements for laboratories and began to develop guidelines on quality assurance. During the EPPO Conference on Diagnostics in Noordwijkerhout (NL) in 2004, accreditation of laboratories, quality assurance systems for laboratories, and EPPO's role in this area were discussed. It was proposed that the ad hoc Panel on Technical requirements for laboratories should identify the critical elements in ISO/IEC Standard 17025 on General requirements for the competence of testing and calibration laboratories (ISO, 2005), and develop an EPPO Standard which could serve as an interpretation of this Standard for plant pest identification. This would be particularly useful because national accreditation bodies currently interpret Standard ISO/IEC 17025 differently. An EPPO Standard on *Basic requirements for quality management in plant pest diagnosis laboratories* is being developed and will soon be adopted. Contacts have been made with the European Accreditation (EA) body to develop a standard which could be used as a guidance document for laboratories applying for accreditation, as well as for accreditation bodies.

EPPO database on diagnostic capacity

In 2004, EPPO Council stressed that the implementation of phytosanitary regulations for quarantine pests is being jeopardized by decreasing knowledge in plant protection and by a lack of well maintained collections. The Panel on Diagnostics consequently decided to identify practical actions which could be undertaken by EPPO to improve collaboration on diagnostics in Europe and to provide good scientific support for the diagnostic work of NPPOs. The EPPO Secretariat is accordingly making an inventory of the available expertise on diagnostics in Europe, of training capacities in diagnostics and of collections (including their maintenance), on the basis of a questionnaire to its member countries. Emphasis should be given to regulated pests and not to common pests which are widely distributed in Europe. The aim was to identify experts who could provide diagnosis of regulated species but also those who could help in the identification of new or unusual species for the EPPO region. In May and June 2006, this questionnaire was launched on the EPPO website. All NPPOs of EPPO member countries (and laboratories which were part of their diagnostic network) were invited to participate in this survey. As of May 2007, 27 countries have provided data about their diagnostic expertise (covering 118 diagnostic laboratories and 377 experts). These results are available in the form of a searchable database EPPO website. The database on collections is still under construction.

Conclusion

Activities on diagnostics have been and continue to be of prime importance for EPPO, and have been one of the main areas in which the Organization has made its mark in recent years, in its publications and on its website. A very valuable international resource has been created, and a basis has been laid for active cooperation between European countries in the provision of diagnostic services. This is reflected in the document on EPPO's mission, goals and strategy Period 2006–2009 stating that *"EPPO will expand its role in addressing diagnostic needs by supporting diagnostic laboratories, leading to the introduction of quality assurance systems and/or accreditation"*. More than 30 diagnostic protocols and two standards

on quality assurance are in a draft stage. More international cooperation on diagnostics is also expected with the IPPC initiative on Diagnostic protocols.

References

FAO (1997). *International Plant Protection Convention*. Rome (IT): FAO.

FAO (2002). *Guide to the International Plant Protection Convention*. Rome (IT): FAO.

FAO (2003) Report of the Fifth Interim Commission on Phytosanitary measures, Italy, 07–11 April 2004 (available at www.ippc.int).

FAO (2004) Report of the Sixth Interim Commission on Phytosanitary measures, Italy, 29 March-02 April 2004 (available at www.ippc.int).

Harju, V. A., Henry, C. M., Cambra, M., Janse, J., & Jeffries, C. (2000). Diagnostic protocols for organisms harmful to plants. DIAGPRO. *Bulletin OEPP/EPPO Bulletin*, *30*, 365–366.

IPPC (2006) ISPM no. 27 *Diagnostic protocols for regulated pests* in *International Standards for Phytosanitary Measures* pp. 335–345 IPPC Secretariat, FAO, Rome (IT). IPPC Secretariat, FAO, Rome (IT) available at www.ippc.int.

ISO (2005). General Requirements for the competence of testing and calibration laboratories. ISO, Geneva, Switzerland (available at www.iso.org).

WTO (1994). *Agreement on the application of sanitary and phytosanitary measures*. IN agreement establishing the World Trade Organization: Annex 1A: multilateral agreements on trade in goods. Geneva, Switzerland (available at www.wto.org).

Zlof, V., Smith, I. M., & McNamara, D. G. (2000). Protocols for the diagnosis of quarantine pests. *Bulletin OEPP/EPPO Bulletin*, *30*, 361–363.

Eur J Plant Pathol (2008) 121:339–346
DOI 10.1007/s10658-007-9247-0

Quality assurance in plant health diagnostics – the experience of the Danish Plant Directorate

Charlotte Thrane

Received: 26 April 2007 / Accepted: 29 October 2007
© KNPV 2007

Abstract Worldwide, there is increasing focus on implementation of Quality Assurance systems (QA-systems) in plant health diagnostic laboratories. Several laboratories are in the development or implementation phase and some laboratories have gained accreditation through approval by national accreditation boards. To initiate the process of developing and implementing QA-systems, management and staff need a strong motivation factor. First, because it is a time-consuming and demanding process to go through. Second, because plant health testing does not fit very well into the QA-systems that traditionally were developed for chemical or physical testing laboratories. External pressure is often the only way to generate this motivation factor amongst staff and management to initiate the development of QA-systems. The principal motivation factor in our laboratory was a national requirement that official testing laboratories should implement QA-systems. At the Danish Plant Directorate (PD) we have gained experiences with accreditation of plant health diagnostic methods during the past 5 years. The focus of this paper is a presentation of the consequences and the practical approach to comply with the require- ments of ISO 17025 in our plant health diagnostic laboratory. This includes the themes: staff competence and responsibilities, documentation and traceability, and continuous assessment and improvement of the QA-system.

Definitions

ISO 17025 is a specific standard for testing and calibrating laboratories. It is based on the certification standard *ISO 9001* (ISO/IEC 17025 standard 2005).

A laboratory can gain *accreditation* to do a certain process (method) by approval of an accreditation board after a *third part audit*.

Why are plant health diagnostic laboratories implementing ISO-accredited quality assurance-systems?

For any laboratory or other business, the main driver for implementing ISO-standard QA-systems, is a strong motivation factor. Customer complaints and customer requirements are usually strong motivation factors. Further, enforcing national administrative requirements, governments can place pressure on official testing laboratories to implement QA-systems. This was the case for our laboratory when the Danish government in the late 90s requested that official testing laboratories should implement QA-systems

C. Thrane (✉)
Laboratory of Diagnostic in Plants, Seed and Feed Stuff,
Danish Plant Directorate, Skovbrynet 20,
2800 Lyngby, Denmark
e-mail: cht@pdir.dk

and possibly gain accreditation to be competitive and to be preferred as a service to the government.

In general, development, implementation, and maintenance of QA-systems can be overwhelming to staff and management. Staff might fear increased bureaucracy, undesirable work routines, and management fear that there will be no added value compensating for the cost in money and time to work under ISO-standard conditions. However, when things go wrong, when the customer complains, or when questions arise during the trade of plant material, the advantages of a QA-system become more obvious.

When working under an ISO accredited quality system, documentation and traceability of the work processes is mandatory which is extremely useful if laboratory results are challenged by the customer. Within plant health in the increasingly enlarging EU and open markets, inspectorates are aware of the importance of international harmonisation of testing of plant products crossing borders. ISO 17025 accreditation together with validated testing methods are effective tools to increase the certainty of laboratory tests and subsequently, the quality of the products.

In our laboratory, the areas selected for accreditation have been tests with a large number of samples for routine screening and testing where findings have a high economical impact on the grower. This is the case for testing for quarantine bacteria of potatoes. Diagnostics of *Clavibacter michiganensis* ssp. *sepedonicus* causing ring rot of potatoes is the case for demonstrating a QA-system.

Impact of implementation of ISO 17025 in the diagnostic laboratory

Several years after gaining accreditation, the advantages of QA-systems have become apparent to laboratory staff and management (Thrane and Scheel 2005). In summary, the standardisation of testing, sample flow, and other processes have increased the quality of laboratory tasks. Because of the detailed descriptions of responsibilities and duties within the QA-system and the focus on training to develop or maintain their expertise, staff have increased confidence in their daily work and express higher satisfaction with their work. In addition, it is a valuable aspect when tests are offered to consumers.

The work in the diagnostic laboratory should be looked at as processes. The laboratory does certain processes from receiving samples to issuing the final test result to the customer (Fig. 1). The management of the quality is documented through process control at critical and measurable steps along the process. Such control steps can be calibration of equipment, use of controls (reference material) as part of the testing, blind samples, documenting the staff member that did the step etc. The in-built requirement to continuously improve the QA-system is one of the most important advantages of the ISO 17025 requirements (Fig. 2). The most important tools for the continual assessment and possibly improvement of the system are internal control (e.g. blind testing), recording of errors, internal and external audits, participation in proficiency testing, and finally, the yearly management review. The general quality of the diagnostic work has increased in the laboratory. But it is also obvious that the full implementation of the quality system depends on continual assessment of the system through internal and external audits.

Quality assurance in the diagnostic laboratory – examples on the practical approach to comply with ISO 17025

In all diagnostic laboratories, a certain level of quality assurance of the testing procedures and laboratory management is implemented. However, for accredited laboratories, the quality assurance has to be systematically approached and all points of the ISO 17025 Standard should be addressed and implemented accordingly. DANAK is the national accreditation board in Denmark.

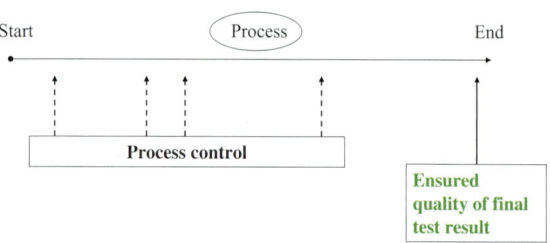

Fig. 1 All parts of a quality system should be subjected to continuous improvement. By actively responding to shortcomings of the system (errors, complaints etc), the quality system will improve

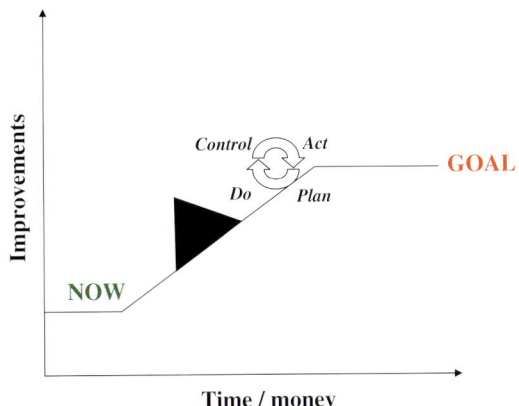

Fig. 2 Management of quality in laboratory testing through control checks at critical points of the process

Development and implementation

The resources required to develop and implement a QA-system depends on the scope for accreditation. We initiated our accreditation process for a limited part of the total laboratory activities. This served as a pilot exercise in assessing the process of implementing the various steps of an accredited QA system, thereby gaining valuable experience for a later scale-up. This initial exercise resulted in the following lesson being learned: rather than starting in a corner of the laboratory, it would have been valuable to involve all staff from the start and building up QA-systems for selected areas of testing that presented the different areas of diagnostics (virology, bacteriology, mycology, entomology, nematology). Now, 6 years after we started the process, all staff are involved with the QA-system at some level. Before, staff working under the ISO 17025 standard felt treated unfairly and isolated from the other staff whereas now all staff show a mutual perception of the QA-system. This development has increased the awareness and understanding of quality in their daily work and staff show an increased share of responsibilities for maintenance of the system.

The quality manual and document management

The quality manual is digital. A digital quality manual is less laborious to update than a paper-based manual. Thus, there is in general reduced bureaucracy in the document management of a digital quality manual than of a paper-based quality manual. Therefore there

is less risk of using outdated documents as long as staff make sure that they only use the current digital version. All staff have access to computers. Staff can make hard copies of the documents they need but it is the responsibility of the staff to make sure that the printed copy they use is the most recent version. Document control of the digital documents should be controlled by an employee with this designated responsibility to ensure that the documents cannot be changed without authorization. The current quality manual is an internet-based 'Qualiware'-system (Fig. 3). This system has many advanced applications and one of them is the extended use of flow charts to demonstrate the work processes and the decision trees. This is combined with a possibility to link to relevant documents. This facilitates the understanding of the processes and is especially useful when introducing new staff and for auditors. In all aspects of documentation it is important to minimize and restrict written text to only one or a few documents to prevent discrepancies and missing documentation when changes are made. In general, the staff doing a certain process are involved in developing the work flow and the necessary documents.

Staff competence

There are strong requirements in ISO 17025 that staff should be competent to do the tasks they are supposed to do. This includes specific training in different diagnostic tests and general training in sample flow, trouble shooting, and QA-requirements and principles. The competence of staff is ensured by training, evaluation of the acquired training, followed up by continual assessment of staff to perform their assigned duties. In general, internal training for a flexible period of time is required depending on the experience of the person and the complexity of the task they are going to do. Most frequently the most efficient way of training new staff is to let the responsible technician train the new staff in the specific task. If the effect of training can be evaluated objectively, this is done before approval of the person for the specific task (blind testing of samples or comparative testing between the new and the experienced staff). Sometimes specific courses (external) are needed later to increase the depth of knowledge enabling staff to trouble shoot in their work, to introduce new techniques etc.

The internal training for Ring rot testing includes training in all aspects of the testing including handling

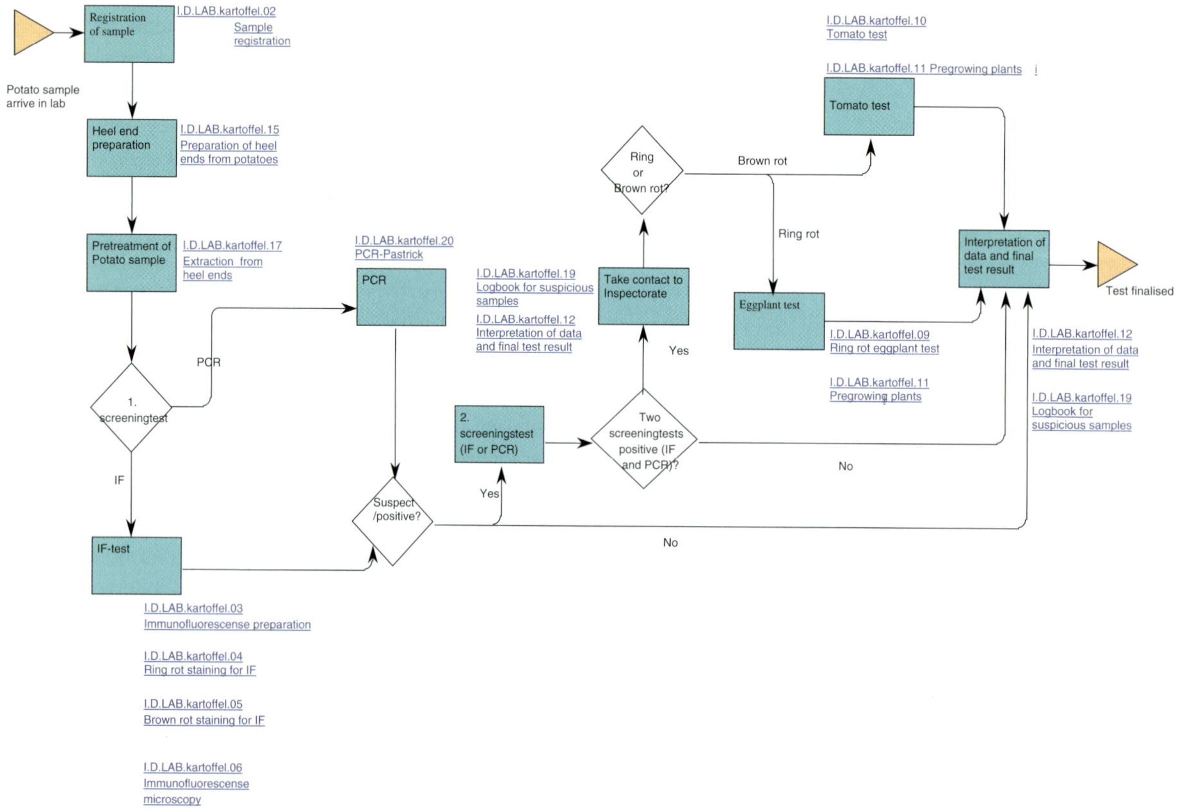

Fig. 3 Workflow of the potato testing scheme. **a** *Yellow triangles* start and end of process. **b** *Green boxes* activities. **c** *White diamonds* decision pathways. **d** *Blue underlined text* direct links to the written instructions of that part of the process (*I.D.LAB.kartoffel* Instruction Diagnostic Laboratory – potato.)

of samples as well as in special tasks as PCR, IF, egg plant testing. We have broken down the special tasks into separate training parts. Thus, one person can be approved for some parts of the testing only, or for the whole testing procedure. For the general handling of samples staff are trained under supervision for at least a month. This training is hard to assess objectively but in this case the responsible technician makes an evaluation where e.g. errors are recorded. For specific tasks as IF-microscopy and PCR, training might be longer and is always finalised with a test to ensure the competence of the trainee. No results can be issued by the trainee during the training period before the trainee has consulted the staff responsible.

Responsibilities and tasks

The QA-management at PD is an umbrella structure including three different laboratories. The overall re-

sponsibility of the quality management is the Controller-Unit (Office of Planning, Quality and International Projects) which is associated with the senior management of the organisation. The Controller-Unit prepares and issues the general procedures, approves laboratory specific procedures, and is responsible for the execution of internal and external audits as well as management reviews. In accredited laboratories, responsibilities and tasks should be clearly defined. This parameter is very important for the confidence of staff and subsequently for the quality of the work they do and the decisions they make. In the diagnostic laboratory, a database was developed for the staff to give an overview of the responsibilities and duties of all staff specified for the various tasks. One of the advantages of this database is to be able to foresee shortage of staff in specific areas of the laboratory tasks. For most tasks it is necessary that more than one person can do the work to make sure that the laboratory will be able to continuously meet customer requirements.

Documentation and traceability

Documentation is an essential part of QA-systems. Examples of documentation are the steps in the diagnostic work, evaluation of results from proficiency testing and ring tests, audit reports, records of staff training, records of equipment maintenance etc. Further, the documentation is necessary to keep track of the history of events in the laboratory. It is essential to do an initial assessment of which steps and parameters are necessary to document to ensure traceability in the work. And further, by doing this assessment it is possible to avoid non-essential documentation to reduce work load in order to be able to focus on the important parts of the QA-system. In general all documents are kept for six years. The long term storage of documents facilitates back-tracing in case errors or non-compliances are later identified. Samples are uniquely labelled by barcodes (Fig. 4). This ensures impartiality during testing, easy handling of samples, full traceability of all steps in the process, and thus fewer errors in the final diagnostic results.

Testing of equipment is a very important part of the process. Documentation of the characteristics and history of all equipment has a large focus in the ISO 17025 to reduce errors that otherwise could originate from malfunctional equipment (Fig. 5).

Non-compliances

The ISO 17025 standard determines that the laboratory should have a system for registration of errors and non-compliances. Specific forms are used for this purpose. When an employee identifies a problem, he or she describes it on the form. Then the impact of the non-compliances on the work is assessed. Together with the technical or quality-responsible staff, it is decided which corrective measures should be undertaken. The individual forms are evaluated on a regular basis to determine whether non-compliances appears to be systematic or only occurring accidentally, and whether the non-compliance is serious or of minor importance. It has been continuously necessary to emphasize the importance of identifying and registering non-compliances to staff and to assure staff that there is nobody personally to blame for errors occurring. The registration of non-compliances and implementation of corrective measures are very effective sources for improvements of testing quality.

Test methods

The European Plant Protection Organisation (EPPO) is one of the most important sources for improvements and standardisation of plant health diagnostic methods. For ring rot of potatoes the only official method accepted within EU, is stated in the ring rot control Directive (EU-Directive 2006/56/EEC). However, for most plant diagnostics, EU does not set the method for testing and laboratories need to find the best suitable method (preferably EPPO-methods or other validated or recognised methods). Unlike strict quantitative tests as in chemistry and physics, plant health diagnostics frequently implies interpretation of data from one or more tests before it is possible to provide a reliable test result to the customer. The more traditional methods in plant health diagnostics involve microscopy or morphological studies which require that staff are trained and experienced in the interpretation of results.

If the laboratory strictly follows official testing methods (validated methods), only verification of testing performance should be shown to ensure the quality of testing in the laboratory. It is useful to consult the guideline EA 4/10 for evaluation of testing quality of qualitative microbiological tests (European co-operation for Accreditation 2002). In case an official method is outdated or because the laboratory has another reason to deviate from the method, verification or even validation should be performed.

For morphological diagnostic methods, it is necessary to include some flexibility in the working instruction for the interpretation of results. Flexibility is necessary because of the biological variation between specimens of the same species and the similarity between different species (life stage of the pathogen, host plant, genetic variability, natural range of morphological variation between individuals etc). Because data can be hard to interpret it is:

- very important to standardise the process as much as possible,
- critical to develop decision trees of the diagnostic process,

Fig. 4 The use of bar coding for sample registration (**a**), sample processing (**b**), and entering the final test results (**c**) ensures a high degree of sample integrity and quality of the test result

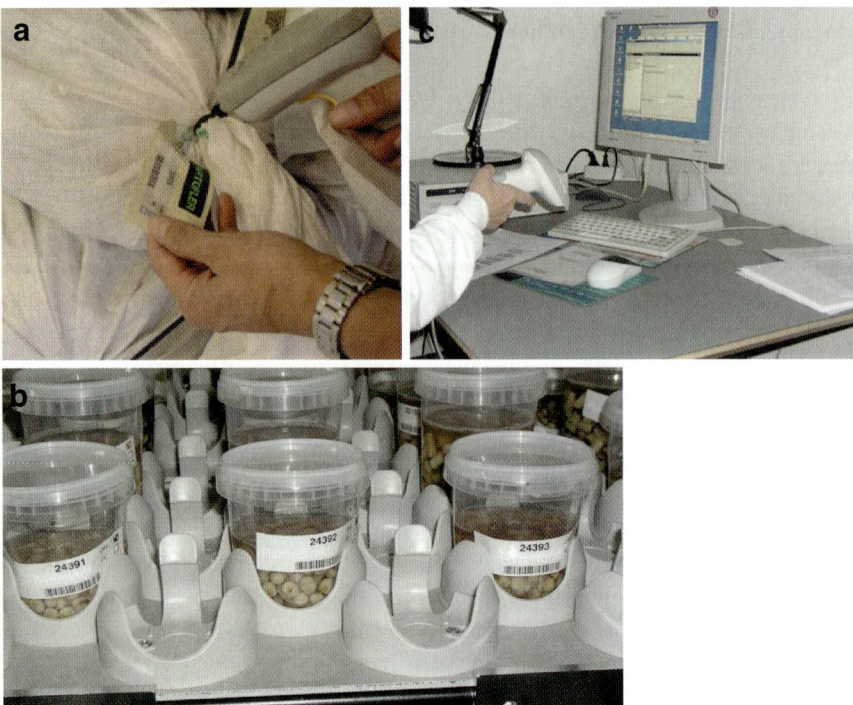

- necessary that staff are familiar with relevant literature and other sources of information,
- mandatory to document the findings on which the final judgment is made.

The use of molecular methods in plant health diagnostics has increased in recent years. First, technology has made this development possible. Second, molecular testing is often a fast way of testing samples and customers in general want results as quickly as possible. Third, younger staff prefer to work with molecular methods rather than traditional methods. Fourth, it is easier to train new staff in molecular methods than in classical diagnostic methods. And fifth, despite the requirements for technical skills, advanced equipment and facilities, the advan-

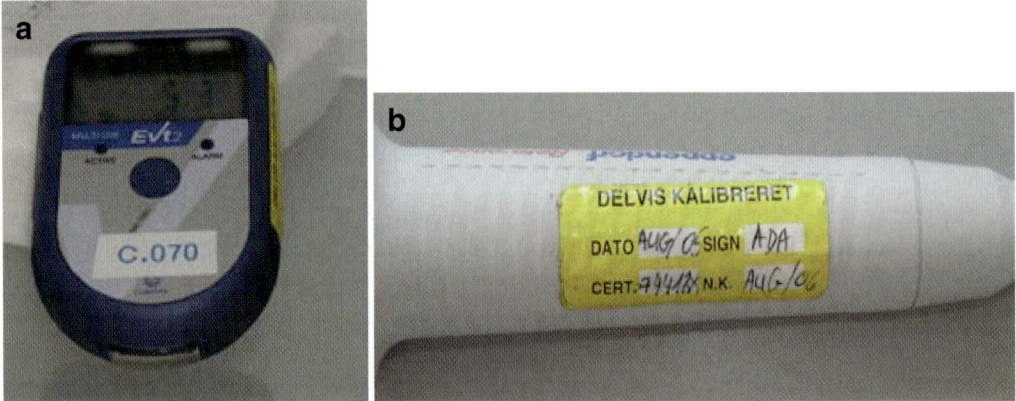

Fig. 5 Documentation and traceability of equipment history by unique labelling (**a**, thermo logger) and with calibration labels including certificate number to the reference material used for the calibration (**b**, pipette)

tage of molecular methods is that data interpretation does not depend as much on the skills of the analyst compared to morphological identification. Molecular methods can be standardised and the correct use of the methods can relatively easily be verified. Thus, it has in general been easier to become accredited for molecular tests than for classical morphological plant diagnostic tests. In particular, it can be difficult in morphological identification to:

- create standardised working instructions, and to
- demonstrate the competence of staff and laboratory.

Ring rot diagnostics

For ring rot testing, the method used is an official, validated, and robust method primarily developed for latent testing of the bacteria in tubers. EU member states have to follow the same standardised methods prescribed in the ring rot control EU-Directive. Frequently, new and improved methods are developed for diagnosis of plant pests. Such improvements of methods in general occur faster than updating of official methods. This is especially true when the official method is prescribed in an EU-Directive as it is the case for ring rot diagnostics. Thus, it is crucial in the preparation of official test methods that some flexibility is included, and that non-essential constraints on testing procedures are eliminated to make it possible to use new improved methods before they are published in the official document.

Assessment of testing quality

The ring rot test is a qualitative test and no strict statistical uncertainty measurement is possible and appropriate to evaluate the validity of test results. Thus, the components that can be easily evaluated should be under control to eliminate the uncertainties that could otherwise arise from the use of equipment etc. that is not well-maintained. Thus, uncertainties from the use of equipment should be regarded as minor in comparison with the uncertainties contributed by the subjectivity of analyst interpretation and the non-homogenous nature of the potato samples. The best way to assess the quality of ring rot testing is a top-down approach by some of the following initiatives; participation in proficiency testing, analysis of blind samples, comparative testing between analysts, and comparison of results obtained with different methods. External assessment of laboratory competence by proficiency schemes is currently being developed for plant health diagnostics by for example, the UK company, FAPAS.

Awareness of the risk of obtaining false positive and false negative results should be addressed in the choice of method for each test This is specifically important when testing non-symptomatic samples such as the ring rot survey. The use of controls during all testing is imperative and adds an essential component in verification of test results obtained. Threshold levels of detection should be addressed for all methods used. In ring rot testing, all suspicious samples are confirmed or refuted by alternative testing methods based on different biological principles according to the EU-directive. Based on positive screening tests, a preliminary result is issued to the customer. A positive finding in the screening tests for ring rot is further confirmed by Koch's postulates. This thorough testing before releasing a result of a positive finding, ensures a high quality of the issued test result.

Audits

DANAK performs external third part audits every 15 months on technical matters and on general quality management. Every 5 years DANAK performs a substantial audit and decides whether the laboratory can continue to be approved for accreditation. Internal audits are performed routinely and more frequently with assessment of equipment, test methods, and the general quality of management, respectively. There is no doubt that updating of the quality manual, organisational structure and general improvements of the QA-system largely depend on the accomplishment of internal and external audits.

Approval for accreditation

In many cases the national accreditation boards have only limited experience in auditing qualitative micro-

biology tests and especially in auditing plant health diagnostics. Thus, in general, the only feasible way to proceed is to use international experts in the specific fields. This can be expensive to the laboratory. However, it is crucial that the auditor knows the critical control points and possible problems in specific testing procedures in order to obtain a qualified assessment of the testing quality of the specific applicant.

References

EU-Directive (2006/56/EEC) for the control of potato ring rot.

European co-operation for Accreditation (2002). Accreditation for Microbiological Laboratories. EA - 4/10.

ISO/IEC 17025 standard (2005). General requirements for the competence of testing and calibration laboratories.

Thrane, C., & Scheel, C. (2005). Organization of quality assurance in diagnostics at the Danish NPPO. *EPPO Bulletin 35*(1).

Eur J Plant Pathol (2008) 121:347–353
DOI 10.1007/s10658-007-9228-3

REVIEW

Tracking fungi in soil with monoclonal antibodies

Christopher R. Thornton

Received: 30 March 2007 / Accepted: 27 September 2007 / Published online: 17 October 2007
© KNPV 2007

Abstract Species of the genus *Trichoderma* are ubiquitous soil-borne fungi that exhibit antagonism towards a number of economically important plant-pathogenic fungi and oomycetes. This review discusses recent developments in the use of monoclonal antibodies to detect these fungi in their natural soil environments and to quantify their population dynamics during antagonistic interactions with saprotrophic competitors in soil-based systems. Immunological approaches to detection and quantification are examined in relation to conventional plate enrichment techniques and to nucleic acid-based procedures. An example of recent research using a mAb-based assay to quantify the effects of saprotrophic competition on the growth of *Trichoderma* isolates in mixed species, soil-based, microcosms is presented. Future technological developments in immunoassays for tracking *Trichoderma* populations in soil are discussed and results presented showing the accurate detection and visualization of a plant growth-promoting isolate of *T. hamatum* in the rhizosphere of lettuce using mAb-based immunodiagnostic assays.

Keywords Monoclonal antibodies · Soil fungi · Quantification · Population dynamics · Competitive saprotrophic ability · Biological control · Lateral flow device

Historical perpectives

Soil fungi are a diverse group of eukaryotic organisms that encompass economically important root-infecting plant pathogens, beneficial mycorrhiza, root endophytes, nematode-trapping fungi and mycoparasites. As a consequence of their saprotrophic activities, soil fungi are the principal degraders of biomass in terrestrial ecosystems and are responsible for much of the organic re-cyling in the environment. As a group of organisms, they are arguably the most difficult to study since their natural habitat, soil, is chemically and biologically complex and remains largely uncharacterized. The close associations of fungi with other soil-borne micro-organisms such as bacteria, actinomycetes and oomycetes presents a significant challenge for detection and quantification of individual genera or species and their often copious production of asexual conidia causes significant restrictions in the accurate quantification of population dynamics.

Traditionally, detection of soil fungi has relied on plate enrichment techniques that incorporate selective or semi-selective media. Attempts have been made to replace these techniques with serological and nucleic-

C. R. Thornton (✉)
School of Biosciences, University of Exeter,
Geoffrey Pope Building, Stocker Road,
Exeter EX4 4QD, UK
e-mail: C.R.Thornton@ex.ac.uk

acid based procedures but it is worth noting that many such procedures still require a period of biological amplification through nutrient enrichment to allow the magnification of propagule numbers available for detection (Lees et al. 2002; Thornton et al. 1993, 2004). Biological amplification is not a suitable system for quantification of soil fungi. Colonies derived from asexual conidia lead to significant bias in quantification and mask ecologically important fluctuations in active hyphal biomass. Procedures have been developed for determining population dynamics based on measurements of colony-forming units (CFUs), but they are limited to experimental systems containing individual non-sporulating organisms such as the sterile root-infecting pathogen *Rhizoctonia solani* (Dewey et al. 1997; Thornton et al. 2004). For the quantification of population dynamics of spore-producing species, alternative methods needed to be developed that allowed discrimination between active hyphal growth and quiescent spore production. This has, so far, not been achieved using DNA-based techniques since they are unable to discriminate between spores and hyphae. Monoclonal antibody-based techniques do allow discrimination of biomass components and, provided that mAbs are raised against constitutively expressed antigens that are secreted during hyphal development, they can be used to quantify changes in active growth. The aim of this review is to examine techniques that have been developed to track fungi in soil, using as a case study species of the genus *Trichoderma*. Recent developments in immunological techniques that allow the accurate quantification of population dynamics of *Trichoderma* species during antagonistic interactions with other fungi in soil-based systems will be discussed.

Trichoderma—a case study

Species of the genus *Trichoderma* are cosmopolitan soil and compost-borne saprotrophic fungi. Certain strains of *T. harzianum* are noxious compost-borne pathogens of cultivated mushrooms (Seaby 1987), while a number of thermo-tolerant species have emerged as medically important opportunistic pathogens of healthy humans and of immuno-compromised patients (Walsh and Groll 1999). However, it is the ability of certain isolates to suppress plant disease (Whipps 1997, 2001) and to promote plant growth

(Harman et al. 2004) that has lead to intense and prolonged scientific interest in this important group of organisms.

The mycoparasitic activities displayed by certain *Trichoderma* strains has made them attractive candidates as biological control agents in the fight against plant diseases and considerable efforts have been made to promote their use for this purpose. Nevertheless, despite their unquestionable potential as ecologically sound alternatives to synthetic pesticides, their widespread application in the control of soil-borne pathogens has not been realized. Historically, biocontrol strains have been selected on the basis of activities exhibited under controlled laboratory conditions, conditions that are often far removed from those experienced in nature. Consequently, levels of disease control achieved in laboratory tests are rarely repeated following introduction of strains into alien soil environments. One reason for this is that artificially introduced strains are unable to compete for niches with soil saprotrophs already resident in soil ecosystems. It should be remembered that the mycoparasitic properties demonstrated by *Trichoderma* species (hyperparasitism, production of lytic enzymes, secondary metabolites and antibiotics) are also properties displayed many other species of soil fungi. Consequently, a biocontrol strain must possess both antagonistic activity and strong competitive saprotrophic ability (CSA), if it is to have a realistic chance of delivering levels of disease control comparable to conventional control methods. Determining whether a strain has a strong CSA in soil-based systems is problematic and requires techniques that allow saprotrophic activities to be accurately monitored during antagonistic interactions with other soil saprotrophs *in vivo*.

A number of procedures have been developed to track *Trichoderma* species in soil ranging from traditional techniques that employ selective media, through nucleic-acid based techniques that exploit polymerase chain reaction (PCR) to immunological techniques that incorporate specific monoclonal antibodies.

Conventional techniques

Conventional methods for detecting *Trichoderma* propagules, which employ soil dilution and plate-

enrichment techniques, have been used to recover isolates from naturally infested soils. The most specific medium for this purpose is TSM (*Trichoderma* Selective Medium)(Elad et al. 1981) which, when amended with compounds to inhibit fast growing oomycetes, is sufficient for general detection, although contamination by other fungi, particularly *Gliocladium* species, remains a problem. Assays based on selective isolation are also laborious and require taxonomic expertise to provide definitive identification of isolated fungi. The specificity of plate enrichment techniques can be improved by combining them with additional means of identification. For example, Thornton et al. (2002) used a *Trichoderma*-specific mAb-based enzyme-linked immunosorbent assay (ELISA) in combination with baiting and selective isolation to detect *Trichoderma* species in naturally infested composts. Isolates identified using the ELISA were recovered from the selective medium and their identity determined by analysis of the internally transcribed spacer (ITS) regions (ITS1-5.8S-ITS2) of the rRNA-encoding regions of the fungi.

A further limitation of plate-enrichment techniques is their inability to differentiate between colonies derived from hyphae and those derived from spores (conidia and chlamydopores). Consequently, selective media cannot be used to quantify the population dynamics of *Trichoderma* species based on active growth or measurements of CFUs.

Measurements of colony densities have also been shown to correlate poorly with other measures of biomass such as ATP, chitin, detection of respiration and ergosterol (Lumsden et al. 1990). Measures of ATP and chitin were regarded as suitable markers for estimates of *Trichoderma* biomass, but quantification base on amounts of chitin could only be used during interactions with oomycetes such as *Pythium* species that lack chitin in their cell walls (Lumsden et al. 1990), although these organisms have since been shown to have a chitinous cell wall component. In addition, ATP measurements can only be used to estimate total biomass because this technique is unable to discriminate between the metabolic activities of fungi and bacteria (Eiland 1985).

Nucleic acid-based techniques

Mutation of *Trichoderma* strains to benomyl tolerance and transformation of *Trichoderma* strains with β-glucuronidase and green-fluorescent protein-encoding genes have provided useful tools for ecological studies in the rhizosphere and bulk soil, but these studies have so far been constrained to individual fungicide-tolerant or recombinant isolates (Ahmad and Baker 1988; Bae and Knudsen 2000; Green and Jensen 1995; Hermosa et al. 2001; Pe'er et al. 1991; Thrane et al. 1995). An alternative approach adopted by Abbasi et al. (1999) employed a combination of sequence-characterized amplified region (SCAR) markers and dilution plating on a semi-selective medium to detect and enumerate propagules densities. However, the technique only permitted quantification of total propagule densities, because it was unable to differentiate between colonies derived from spores and from mycelium.

Immunological techniques

Hybridoma technology allows the production of monoclonal antibodies (mAbs) that are specific to individual genera, species or even isolates of fungi (Dewey and Thornton 1995) and are capable of discriminating between active growth and quiescent spore production (Thornton 2004; Thornton and Dewey 1996; Thornton et al. 1994). A monoclonal antibody (HD3) raised against a 20 kDa intracellular protein from *T. harzianum* was the first mAb produced using hybridoma technology that displayed a high degree of specificity to *Trichoderma* species (Thornton et al. 1994). Monoclonal antibodies have also been produced to the closely related fungus *Gliocladium roseum* using hybridoma technology (Breuil et al. 1992). The protein bound by mAb HD3 is only produced during active growth and, at the onset of sporulation, is no longer detected. The hyphal-specific nature of the antigen therefore represented a potentially useful target for the quantification of active growth of *Trichoderma* species in soil. However, subsequent studies showed that the low abundance of the antigen meant that it was not readily detected in soil extracts. Parallel studies using mAbs specific to *Rhizoctonia* species (Thornton and Gilligan 1999) showed that the most appropriate targets for the development of quantitative immunology assays were extracellular, constitutively expressed, antigens. Consequently, a mAb (MF2) was raised specific to *Trichoderma* species and the closely related fungi

Gliocladium viride, *Hypomyces chrysospermus*, *Sphaerostilbella* species and *Hypocrea* species (Thornton et al. 2002). It does not react with *Gliocladium catenulatum*, *G. roseum*, *Nectria ochroleuca* and *Clonostachys* species, or with a wide range of unrelated soil- and compost-borne fungi. Monoclonal antibody MF2 binds to a constitutive, extracellular, antigen secreted from the growing tip of hyphae and that, in certain species, is also released from conidia. In combination with PCR, Thornton et al. (2002) used mAb MF2 to develop a highly specific diagnostic procedure for the detection and identification of *Trichoderma* populations in naturally-infested peat-based composts (Thornton et al. 2002). Furthermore, studies showed that the mAb could be used to quantify the saprotrophic activity of biocontrol strains during antagonistic interactions with the plant patho-gen *R. solani* in peat and soil (Thornton 2004) and to visualize plant-growth-promoting *Trichoderma* spe-cies in the plant rhizosphere (Thornton and Talbot 2006).

The specificity of the quantitative assay also allows the accurate monitoring of the saprotrophic compe-tences of *Trichoderma* species and enables the competitive saprotrophic abilities of potential biocon-trol strains to be determined during competitive interactions with other saprotrophs in soil-based sys-tems. An example is shown in Fig. 1. Here, the population dynamics of *Trichoderma longibrachiatum* is quantified during antagonistic interactions with the abundant and widely distributed soil saprotroph *Aspergillus fumigatus*. Saprotrophic activity of *T. longibrachiatum* was quantified using the *Tricho-derma*-specific ELISA and standard calibration curves of biomass equivalents as described in Thornton (2004). Population dynamics of *T. longibrachiatum* were determined in single (*T. longibrachiatum* only) and mixed species (*T. longibrachiatum* and *A. fumiga-tus*) soil-based microcosms.

In microcosms containing *T. longibrachiatum* only (•), there was a rapid increase in active biomass of the fungus between days 1 and 2 post-inoculation. Thereafter, active biomass fluctuated, but there was an overall decline up to day 21 (see Fig. 1). Active growth dynamics of the fungus followed a similar trend in mixed species microcosms (○), but was significantly reduced by *A. fumigatus*. The aggressive saprotrophism exhibited by *A. fumigatus* was likely

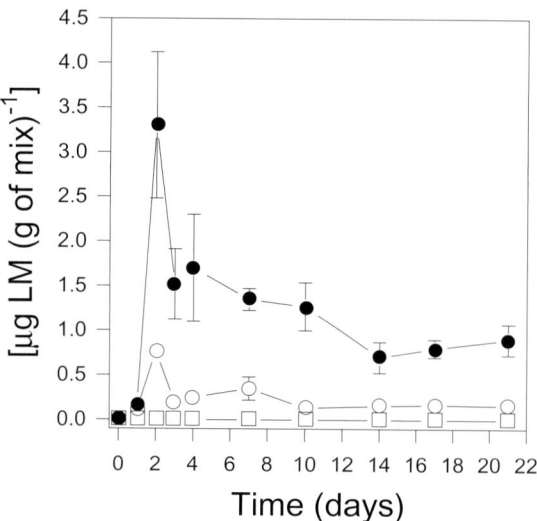

Fig. 1 Population dynamics of *T. longibrachiatum* in the presence (*empty circle*) and absence (*filled circle*) of *A. fumigatus*, quantified using the *Trichoderma*-specific mAb-based assay described in Thornton (2004). Absorbance values were converted to biomass equivalents {expressed as [µg LM (g of mix)$^{-1}$]} using a standard calibration curve. The mean ± SE was then calculated for each set of samples from the populations on each day of sampling. The specificity of the *Trichoderma*-specific mAb MF2 was shown using extracts from microcosms containing *A. fumigatus* only (*empty square*)

facilitated by the production of inhibitory secondary metabolites and antibiotics such as gliotoxin (Bok et al. 2006; Cramer et al. 2006) and the production of fungal cell wall-degrading enzymes such as chitinase (Escott et al. 1998), mechanisms that are similarly attributed to the antagonistic properties of *Tricho-derma* species (Harman and Kubicek 1998). Conse-quently, while *T. longibrachiatum* has been identified as a potential biocontrol agent for the control of oomycete plant pathogens (Sreenivasaprasad and Manibhushanrao 1990, 1993; Migheli *et al.* 1998), its efficacy might be significantly impaired in soil systems where aggressive soil saprotrophic fungi also reside. These types of considerations need to addressed if the biocontrol properties of *Trichoderma* species are to be fully exploited. In contrast to nucleic-based techniques that have been developed to detect individual *Trichoderma* strains, immunolog-ical techniques that incorporate genus-specific mAbs such as MF2 can be used to quantify the saprotrophic competence of multiple potential biocontrol strains.

Future technological developments

The use of immunological techniques for the detection of *Trichoderma* species is restricted to those laboratories that possess *Trichoderma*-specific mAbs. In order to expand the access of hybridoma technology to other workers interested in the biocontrol and plant-growth-promoting properties of these fungi, Thornton and co-workers have developed a 'user-friendly' mAb-based lateral flow device (LFD) for the specific detection of *Trichoderma* species in naturally infested soils and in the plant rhizosphere by adapting techniques previously developed for the detection of the soil-borne pathogen *R. solani* (Thornton et al. 2004). Briefly, the *Trichoderma*-specific mAb MF2 is immobilized to a defined capture zone on a porous nitrocellulose membrane, while the same mAb conjugated to colloidal gold particles serves as the detection reagent. Samples of solubilised antigens are added to a release pad containing the antibody-

Fig. 2 Detection of *T. hamatum* in the rhizosphere of lettuce. **a** Growth promotion of lettuce plants by the root-colonising *T. hamatum* strain GD12. The left hand microcosm shows plants grown in the presence of the fungus. Control plants grown in the absence of the fungus are shown on the right. Scale bar=1.7 cm. **b** Results of LFD tests using extracts prepared from root sections of lettuce plants grown in the presence (L$^+$) and absence (L$^-$) of GD12. **c** Scanning electron micrographs of a lettuce root showing colonisation of the root surface with mycelium of GD12. Scale bar=500 μM. **d** Immuno-enzymatic staining of the root surface of a lettuce plant grown in the presence of GD12, showing extensive colonisation by the fungus. Scale bar=20 μM

gold conjugate. The antibody-gold conjugate binds to the target antigen, passes along the porous membrane by capillary action, and binds to the mAb immobilized in the capture zone. Once an antigen extract is prepared and applied to the LFD, the test result is recorded within 10 min. Bound antigen-antibody-gold complex is seen as a red line with an intensity that is proportional to the antigen concentration. Anti-mouse immunoglobulin immobilized to the membrane in a separate zone acts as an internal control. In the absence of the *Trichoderma* antigen, no complex is formed in the zone containing solid-phase antibody, and a single internal control line is seen. In the presence of *Trichoderma* antigen, two lines are clearly visible. Antigens are extracted from soil samples using a simple 1-h extraction procedure, making the assay appropriate for use by personnel with limited access to microbiological facilities. The test can be used to monitor the presence of *Trichoderma* strains derived from commercial biocontrol and plant-growth-promoting formulations.

An example of the use of the LFD to track *Trichoderma* in the plant rhizosphere is shown in Fig. 2. Lettuce (*Lactuca sativa* cv. Webb's Wonderful) seeds were planted in peat containing inoculum of a plant growth-promoting strain of *T. hamatum* (strain GD12). After 2 weeks growth, seedlings were carefully removed at random from each tray and excess peat removed by shaking. Sections of root tissue 0–1.5 cm below the hypocotyl were excised, transferred to antigen extraction buffer and the roots crushed with a hand-held micro-pestle. Suspensions were centrifuged to pellet insoluble debris and supernatants tested for the presence of *Trichoderma* antigens using the LFD. Roots of plants were also examined under compound and scanning electron microscopes for colonisation by *T. hamatum*. Visualization of *T. hamatum* under the compound microscope was aided by immuno-localization of the fungus on the root surface using the *Trichoderma*-specific mAb MF2, according to the immuno-enzymatic staining procedure described in Thornton and Talbot (2006). Electron microscopy was carried out using a JEOL 6360LV scanning electron microscope.

Growth promotion of lettuce plants as a consequence of *T. hamatum* GD12 root colonization is shown in Fig. 2a. Positive results in LFD tests of extracts from lettuce roots were obtained with all plants grown in the presence of GD12 only. A typical

reaction of a positive sample is shown in Fig. 2b, where two lines in the LFD test shows the presence of *Trichoderma* antigens in root extracts. No antigens were detected in extracts from the roots of control plants grown in the absence of the fungus, indicated by a single line in the LFD test (Fig. 2b). Representative root sections were examined for the presence of the fungus using scanning electron microscopy and a *Trichoderma*-specific monoclonal antibody-based immuno-enzymatic staining procedure (Thornton and Talbot 2006). Representative sections from control plants all showed the absence of fungus in both SEM and IES tests. Sections of roots from plants grown in the presence of the fungus all showed the presence of *Trichoderma* mycelium both in SEM and IES tests (Fig. 2c and d respectively).

References

Abbasi, P. A., Miller, S. A., Meulia, T., Hoitink, H. A. J., & Kim, J.-M. (1999). Precise detection and tracing of *Trichoderma hamatum* 382 in compost-amended potting mixes by using molecular markers. *Applied and Environmental Microbiology*, 65, 5421–5426.

Ahmad, J. S., & Baker, R. (1988). Rhizosphere competence of benomyl tolerant mutants of *Trichoderma* species. *Canadian Journal of Microbiology*, 34, 694–696.

Bae, Y. S., & Knudsen, G. R. (2000). Co-transformation of *Trichoderma harzianum* with beta-glucuronidase and green fluorescent protein genes provides a useful tool for monitoring fungal growth and activity in natural soils. *Applied and Environmental Microbiology*, 66, 810–815.

Bok, J. W., Chung, D., Balajee, S. A., Marr, K. A., Andes, D., & Nielsen, K. F., et al. (2006). GliZ, a transcriptional regulator of Gliotxin biosynthesis, contributes to *Aspergillus fumigatus* virulence. *Infection and Immunity*, 74, 6761–6768.

Breuil, C., Luck, B. T., Rossignol, L., Little, J., Echeverri, C. J., & Banerjee, S., et al. (1992). Monoclonal antibodies to *Gliocladium roseum*, a potential biocontrol fungus of sap-staining fungi in wood. *Journal of General Microbiology*, 138, 2311–2319.

Cramer, R. A., Gamcsik, M. P., Brooking, R. M., Najvar, L. K., Kirkpatrick, W. R., & Patterson, T. F., et al. (2006). Disruption of a nonribosomal peptide synthetase in *Aspergillus fumigatus* eliminates gliotoxin production. *Eukaryotic Cell*, 5, 972–980.

Dewey, F. M., & Thornton, C. R. (1995). Detection of plant invading fungi by monoclonal antibodies. In J. H. Skerrit, & R. Appels (Eds.) *New diagnostics in crop sciences* (pp. 151–171). Oxford, UK: CABI.

Dewey, F. M., Thornton, C. R., & Gilligan, C. A. (1997). Use of monoclonal antibodies to detect, quantify and visualize fungi in soil. *Advances in Botanical Research Incorporating Advances in Plant Pathology*, 24, 275–308.

Eiland, F. (1985). Determination of adenosine triphosphate (ATP) and adenylate charge (AEC) in soil and use of adenosine nucleotides, as measures of soil microbial biomass and activity. Report no. S1777, Statens Planteavls Specialserie, Copenhagen, Denmark.

Elad, Y., Chet, I., & Henis, Y. (1981). A selective medium for improving quantitative isolation of *Trichoderma* spp. from soil. *Phytoparasitica*, *9*, 59–67.

Escott, G. M., Hearn, V. M., & Adams, D. J. (1998). Inducible chitinolytic system of *Aspergillus fumigatus*. *Microbiology*, *144*, 1575–1581.

Green, H., & Jensen, D. F. (1995). A tool for monitoring *Trichoderma harzianum*. 2. The use of a GUS transformant for ecological studies in the rhizosphere. *Phytopathology*, *85*, 1436–1440.

Harman, G. E., Howell, C. R., Viterbo, A., Chet, I., & Lorito, M. (2004). *Trichoderma* species—Opportunistic, avirulent plant symbionts. *Nature Reviews Microbiology*, *2*, 43–56.

Harman, G. E., & Kubicek, C. P. (Eds.) (1998). *Trichoderma* and *Gliocladium*: Enzymes, biological control and commercial applications. London: Taylor and Francis.

Hermosa, M. R., Grondona, I., Diaz-Minguez, J. M., Iturriaga, E. A., & Monte, E. (2001). Development of a strain-specific SCAR marker for the detection of *Trichoderma atroviride* 11, a biological control agent against soil-borne fungal plant pathogens. *Current Genetics*, *38*, 343–350.

Lees, A. K., Cullen, D. W., Sullivan, L., & Nicolson, M. J. (2002). Development of conventional and quantitative real-time PCR assays for the detection and quantification of *Rhizoctonia solani* AG-3 in potato and soil. *Plant Pathology*, *51*, 293–302.

Lumsden, R. D., Carter, J. P., Whipps, J. M., & Lynch, J. M. (1990). Comparison of biomass and viable propagule measurements in the antagonism of *Trichoderma harzianum* against *Pythium ultimum*. *Soil Biology and Biochemistry*, *22*, 187–194.

Migheli, Q., Gozalez-Candelas, L., Dealessi, L., Camponogara, A., & Ramon-Vidal, D. (1998). Transformants of *Trichoderma longibrachiatum* overexpressing the beta-1,4-endoglucanase gene *egl1* show enhanced biocontrol of *Pythium ultimum* on cucumber. *Phytopathology*, *88*, 673–677.

Pe'er, S., Barak, Z., Yarden, O., & Chet, I. (1991). Stability of *Trichoderma harzianum amdS* transformants in soil and rhizosphere. *Soil Biology and Biochemistry*, *23*, 1043–1046.

Seaby, D. A. (1987). Infection of mushroom compost by *Trichoderma* species. *Mushroom Journal*, *179*, 355–361.

Sreenivasaprasad, S., & Manibhushanrao, K. (1990). Biocontrol potential of fungal antagonists *Gliocladium virens* and *Trichoderma longibrachiatum*. *Journal of Plant Disease and Protection*, *97*, 570–579.

Sreenivasaprasad, S., & Manibhushanrao, K. (1993). Efficacy of *Gliocladium virens* and *Trichoderma longibrachiatum*

as biological control agents of groundnut root and stem rot diseases. *International Journal of Pest Management*, *39*, 167–171.

Thornton, C. R. (2004). An immunological approach to quantifying the saprotrophic growth dynamics of *Trichoderma* species during antagonistic interactions with *Rhizoctonia solani* in a soil-less mix. *Environmental Microbiology*, *6*, 323–334.

Thornton, C. R., & Dewey, F. M. (1996). Detection of phialoconidia of *Trichoderma harzianum* in peat-bran by monoclonal antibody-based enzyme-linked immunosorbent assay. *Mycological Research*, *100*, 217–222.

Thornton, C. R., Dewey, F. M., & Gilligan, C. A. (1993). Development of monoclonal antibody-based immunological assays for the detection of live propagules of *Rhizoctonia solani* in soil. *Plant Pathology*, *42*, 763–773.

Thornton, C. R., Dewey, F. M., & Gilligan, C. A. (1994). Development of a monoclonal antibody-based enzyme-linked immunosorbent assay for the detection of live propagules of *Trichoderma harzianum* in a peat-bran medium. *Soil Biology and Biochemistry*, *26*, 909–920.

Thornton, C. R., & Gilligan, C. A. (1999). Quantification of the effect of the hyperparasite *Trichoderma harzianum* on the saprotrophic growth dynamics of *Rhizoctonia solani* in compost using a monoclonal antibody-based ELISA. *Mycological Research*, *103*, 443–448.

Thornton, C. R., Groenhof, A. C., Forrest, R., & Lamotte, R. (2004). A one-step, Immunochromatographic lateral flow device specific to *Rhizoctonia solani* and certain related species, and its use to detect and quantify *R. solani* in soil. *Phytopathology*, *94*, 280–288.

Thornton, C. R., Pitt, D., Wakley, G. E., & Talbot, N. J. (2002). Production of a monoclonal antibody specific to the genus *Trichoderma* and closely related fungi, and its use to detect *Trichoderma* spp. in naturally infested composts. *Microbiology*, *148*, 1263–1279.

Thornton, C. R., & Talbot, N. J. (2006). Immunofluorescence microscopy and immunogold EM for investigating fungal infections of plants. *Nature Protocols*, *1*, 2506–2511.

Thrane, C., Lubeck, M., Green, H., Degefu, Y., Allerup, S., & Thrane, U., et al. (1995). A tool for monitoring *Trichoderma harzianum*. 1. Transformation with the GUS gene by protoplast technology. *Phytopathology*, *85*, 1428–1435.

Walsh, T. J., & Groll, A. H. (1999). Emerging fungal pathogens: evolving challenges to immunocompromised patients for the twenty-first century. *Transplant Infectious Disease*, *1*, 247–261.

Whipps, J. M. (1997). Developments in the biological control of soil-borne plant pathogens. *Advances in Botanical Research*, *26*, 1–34.

Whipps, J. M. (2001). Microbial interactions and biocontrol in the rhizosphere. *Journal of Experimental Botany*, *52*, 487–511.

Eur J Plant Pathol (2008) 121:355–363
DOI 10.1007/s10658-008-9284-3

Exploiting generic platform technologies for the detection and identification of plant pathogens

Neil Boonham · Rachel Glover ·
Jenny Tomlinson · Rick Mumford

Received: 24 July 2007 / Accepted: 31 January 2008

Abstract The detection and identification of plant pathogens currently relies upon a very diverse range of techniques and skills, from traditional culturing and taxonomic skills to modern molecular-based methods. The wide range of methods employed reflects the great diversity of plant pathogens and the hosts they infect. The well-documented decline in taxonomic expertise, along with the need to develop ever more rapid and sensitive diagnostic methods has provided an impetus to develop technologies that are both generic and able to complement traditional skills and techniques. Real-time polymerase chain reaction (PCR) is emerging as one such generic platform technology and one that is well suited to high-throughput detection of a limited number of known target pathogens. Real-time PCR is now exploited as a front line diagnostic screening tool in human health, animal health, homeland security, biosecurity as well as plant health. Progress with developing generic techniques for plant pathogen identification, particularly of unknown samples, has been less rapid. Diagnostic microarrays and direct nucleic acid sequencing (de novo sequencing) both have potential as generic methods for the identification of unknown plant pathogens but are unlikely to be suitable as high-throughput detection techniques. This paper will review the application of generic technologies in the routine laboratory as well as highlighting some new techniques and the trend towards multidisciplinary studies.

Keywords Plant health · Diagnostics · Detection · Real-time PCR · DNA barcoding · Microarrays · Isothermal amplification · LAMP · Direct tuber testing · Molecular diagnostics · TaqMan · De novo sequencing · Pyrosequencing

Introduction

A typical diagnostic laboratory usually maintains capacity to perform morphological identification using light or electron microscopy; methods for detecting proteins from the organisms such as enzyme-linked immunosorbent assay (ELISA) or electrophoresis; methods for detecting fatty acids; molecular methods identifying the nucleic acid of the organisms such as polymerase chain reaction (PCR), Reverse transcription PCR, real-time PCR, reverse polyacrylamide gel electrophoresis or nucleic acid hybridisation; and finally, traditional bioassays such as inoculation of test plants, indicators, or isolation on selective media followed by morphological identification. A decline in the availability of trained staff to perform traditional techniques is a significant issue for maintaining the critical mass required to deliver this type of service.

N. Boonham (✉) · R. Glover · J. Tomlinson · R. Mumford
Central Science Laboratory,
Sand Hutton,
York, UK
e-mail: n.boonham@csl.gov.uk

Generic methods that can be used for a number of different pests help with capacity maintenance and sustainability of small diagnostics laboratories. Using a smaller number of techniques in a laboratory means less expertise needs to be maintained and training new staff is more readily achievable; with robust techniques the staff do not need to be highly qualified, and this helps to keep running costs low, allows delivery of a cost-effective service, and makes sustainability more achievable.

Generic platform technologies

Generic methods are those that can be performed using the same basic technology in the same format, albeit with different reagents, to enable the detection of different pathogens. The techniques cut across disciplines whereby the same basic skills can be used for the detection of a range of pests of different types (insect, nematode, bacterial, fungal, or viral). Probably the first and most established of these is ELISA. Assays based on serological detection of virus proteins have been around since the 1950s, but the adoption of a generic platform on which to perform these assays (i.e. the 96-well plate) made the technology, including all the equipment needed, available to a wider market at an accessible cost. In addition, the ELISA test has proven to be very robust; it can be performed in almost any laboratory with a minimum of training. As a result, the technology remains virtually unchanged since it was first mooted for plant pathogens in the late 1970s (Clark and Adams 1977), and is used on a daily basis in almost all diagnostic laboratories world-wide for plant, clinical and veterinary targets.

Methods based on molecular biology have been developed since the early 1980s, usually in situations where ELISA was not suitable due to either low sensitivity, poor specificity, or lack of suitable antibodies. Significant uptake into the routine testing environment has been slow outside of niche applications, but 25 years on they are becoming more readily accepted and are even replacing traditional techniques in some laboratories. The reason for the recent uptake and proliferation of these techniques has been the generic nature and increased robustness brought on by advances in format. Initially, techniques were based on hybridisation or PCR (Puchta and Sanger 1989);

the latter proliferated primarily since it did not require probe synthesis or radioactive labels. Conventional gel-based PCR, however, has two major drawbacks; the first is susceptibility to contamination, and the second, perhaps more significantly, is a lack of robustness, especially in the hands of non-specialists in a routine setting.

Real-time PCR has started to become an established diagnostic technique since the format effectively 'turns PCR into ELISA', making it both robust and accessible. In diagnostics it has often been referred to as 'automated PCR' due to the lack of post-PCR manipulations (which also helps to prevent post-PCR contamination). However, the factor that has more significantly made an impact is the robustness of amplification. A similar level of skill is required to perform real-time PCR as is required for ELISA, and as such, unlike conventional PCR, it is not the preserve of experienced molecular biologists. The adoption of a common format (96-well and latterly 384-well plates), as with ELISA, has made the technology widely available to researchers as well as laboratories working in clinical, veterinary and plant diagnostics. This availability has driven down costs; the price of real-time instrumentation is now almost one tenth of that 10 years ago. These factors have made the technique well suited to high-throughput detection of known target pathogens. In some laboratories where the equipment is available, the generic nature of the technique has started to result in the replacement of ELISA due to the low start up costs when developing new assays, especially, though not exclusively, when antibodies/antiserum are not available. The laborious nature of nucleic acid extraction is now the main factor limiting the uptake of real-time PCR in routine plant pathogen detection.

Simplified detection

A drive to move diagnostics to the point of decision-making, rather than waiting for results returning from a laboratory has also had an impact on the methods being developed in a number of fields besides plant health. The development of in-field, point-of-care, and pen-side methods for plant, clinical, and veterinary diagnostics, respectively, is dominated by a single format: the serologically-based lateral flow device (LFD). Where appropriate reagents are avail-

able, and where serological assays give the level of sensitivity required, the LFD is the method of choice. The tests can be performed with limited training, have very few steps and give results in minutes. In fact, the technique is so dominant that it is often adopted and sold in very large numbers for targets (for example, *Mycobacterium tuberculosis*) where the limited sensitivity will often give false negative results.

To attain greater levels of sensitivity and specificity than can be achieved using LFDs, researchers have again turned to real-time PCR. The drive towards faster and more portable real-time PCR equipment designed with diagnostics in mind has been led by military and homeland security applications, although more recently platforms developed specifically for point-of-care (POC) clinical applications have been developed. The resulting equipment can be robust and rapid, such as the Smartcycler II (Cepheid) and Bioseeq (Smiths Detection); can have a simple user interface, such as the StepOne (Applied Biosystems); or can be completely self-contained, such as the GeneExpert (Cepheid), which is able to sequentially perform both extraction and reaction set up, thus requiring little or no user intervention. Although developed for in-field, remote POC applications, these techniques are also enabling technology for small laboratories, as the techniques are being developed specifically for diagnostics, require little training, and are very robust. This equipment has been deployed in a field situation for plant health applications (Fig. 1) for the detection of several pathogens in support of European Union policy.

Future technologies

There are several technologies on the horizon that might become valuable as generic diagnostic tools;

these are, at least in the short term, still in the research stage but worth discussing for the future.

Although the capital cost of real-time PCR has decreased in recent years, newer amplification chemistries may help to further reduce the cost of molecular testing. These include various isothermal amplification methods, such as nucleic acid sequence-based amplification (NASBA) (Leone et al. 1997); helicase-dependent isothermal amplification (Vincent et al. 2004); exponential nucleic acid amplification reaction (Tan et al. 2005); and methods based on rolling circle amplification (Yi et al. 2006; Zhang et al. 2006), all of which have the benefit of not requiring complex thermal cycling equipment, while retaining the specificity and sensitivity advantages of PCR-based methods. Such methods also have the potential to be combined with novel detection strategies such as the use of functionalised gold nanoparticles (Tan et al. 2005), bioluminescence (Gandelman et al. 2006), or biosensors (Jaffrezic-Renault et al. 2007), which could be less costly than the fluorescent detection required for real-time PCR. As an example, probably the most advanced isothermal amplification technology is loop-mediated amplification, which uses a highly processive strand displacing polymerase to separate the DNA strands, and primers that form loops to generate new priming sites (Notomi et al. 2000). In this case, the isothermal reaction generates so much DNA that a simple colour change can be used to discriminate positive and negative results, with a limit of detection approaching that of real-time PCR (Tomlinson et al. 2007).

Diagnostic microarrays and direct nucleic acid sequencing (DNA barcoding) both offer potential as generic methods for the detection and identification of unknown plant pests. For the diagnostic laboratory, microarrays offer potential as a generic method for detecting large numbers of known pathogens in a

Fig. 1 Real-time PCR being performed 'on-site' at the point of decision making (**a**) showing the SmartCycler II equipment in a car running on a generator and (**b**) for the testing of *Phytophthora ramorum* at a site remote from the laboratory

single test (Mumford et al. 2006; Boonham et al. 2007); in the UK Biochip project (www.bio-chip.co. uk) the generic nature extends beyond plant virus detection to the detection of animal and fish viruses also. Existing microarray methods are, however, complex and relatively insensitive, and a widely accepted diagnostic format has yet to be adopted.

More recently *de novo* sequencing methods have been used to good effect in identifying potential disease-causing agents. Recent reports (Cox-Foster et al. 2007; Ledford 2007) have shown that deep sequencing of cDNA and DNA generated using generic primer sets (e.g. 16S primers for bacteria) can be used to identify the presence of pathogen sequence. These massively parallel sequencing techniques employ sequencing-by-synthesis methods, for example, the Genome Sequencer FLX is based on pyrosequencing (Roche, formerly 454 Life Science) whilst the Genome Analyzer System (Illumina, formerly Solexa) uses a removable fluorescence-based chemistry. Both systems, however, can generate millions of sequences, yet require no *a priori* knowledge of the pathogens present in the host and offer for the first time a completely generic diagnostic tool regardless of host or pathogen. Currently, the sequencing is expensive and the bioinformatics required to deal with the large amount of data is in its infancy, but it is likely that it will be a method widely used in the future.

In the past, successful generic technologies for plant disease diagnosis have been developed based around commercially available platforms or formats; plant health diagnostics itself is not large enough to support novel platform development. This approach effectively 'piggybacks' onto widely available platforms used in other arenas, for example research, clinical diagnostics, or more recently, detection for homeland security. The wide availability of the platform then allows the generation of a commercial market for the specific reagents and consumables needed which helps to achieve both reduced costs and ultimately sustainability.

Applications

Although much has been done in developing modern nucleic acid-based 'molecular' tests over the last 20 years, much of the work has concentrated in niche

areas, often where other techniques (such as antibody-based ELISA) cannot be used. In the case of PCR detection, however, real-time PCR has changed this, and routine services based on this technology are not only being developed in niche areas, but are finally beginning to replace established main stream techniques. It is difficult to predict which other molecular techniques in the future will make an impact in the plant health arena, though with developments in sequencing technology moving at pace, the use of direct DNA sequencing almost certainly will. The following are examples of (1) an area in which a molecular biology technique is now established and is displacing the established technology, followed by (2) what will likely become an established technique in the next few years.

Direct tuber testing

The virus indexing of seed potatoes is one of the most widespread testing procedures performed by virologists. In general, a common approach is used based on taking eye cores from dormant tubers and growing these on in a greenhouse for several weeks, before testing the sprouts produced from these cores by ELISA. By testing at least 100 tubers, individually or in small batches, the test can be used to estimate the percentage of virus-infected tubers found in a particular seed stock, hence indicating its suitability for planting or the grade at which it should be classified. Over the last 30 years or more, this method has become almost universally adopted because of the advantages it offers: not only can it be used for most common potato viruses, it is also robust, simple to perform, and is well-suited to high-throughput testing. As a result, testing laboratories can routinely test hundreds or thousands of seed stocks in a season. For example, the Dutch General Inspection Service for agricultural seed and seed potatoes tests an average of 20,000 seed stocks every year, as part of official post-harvest control. This equates to one million ELISA tests for *Potato virus Y* alone (G.W. van den Bovenkamp, personal communication).

However, while the growing-on test has become the established method, it is not without problems. In addition to the requirement for a large amount of greenhouse space, the most obvious issue is the length of time the whole procedure takes. In general it takes at least 4 weeks for the eyes to break dormancy,

sprout and grow sufficiently to be tested. If the growing-on step could be removed, with tubers being tested directly, this would remove the need for glass-house space and reduce the amount of time taken. However, this cannot be achieved reliably using ELISA. Virus titres within dormant tubers are often very low, especially for late primary infections, and are often below the limit of detection for ELISA. This makes direct tuber testing using this method unreliable (Hill and Jackson 1984), hence the requirement to include growing-on as a bio-amplification step.

Given the limitations of the ELISA-based system, virologists have investigated more sensitive virus detection methods, to see if they might offer a reliable alternative for direct tuber testing. In the 1990s this work mainly focused on molecular methods; during this time, a range of different potato virus assays was designed including assays based on conventional PCR (Spiegel and Martin 1993; Mumford et al. 1994), the ligase chain reaction (O'Donnell et al. 1996), NASBA (Klerks et al. 2001; Leone et al. 1997), and real-time PCR (Schoen et al. 1996; Boonham et al. 2000). While these and other reports demonstrated that such methods could be used for the sensitive detection of potato viruses direct from tubers, it was also clear that much more work was required to turn this early progress into a useable, routine diagnostic service.

The first major issue was the need to validate any new system against the well-established, universally accepted growing-on test, in order to prove that a direct approach could be used as a reliable alternative. In particular, there were major issues related to the cost of performing this type of testing and its overall reliability. In the early stages of development, while it was relatively easy to prove the specificity of a particular assay and demonstrate it was comparable to ELISA, it was far harder to achieve such results when comparing the entire test (i.e. sample processing, RNA extraction, running assays and analysing results) to growing-on. In many cases, sample processing and RNA extraction were identified as the main problems and work focused on these aspects. At Central Science Laboratory (CSL), a combined approach was developed, taking small, uniform potato cores using specially-designed coring devices, homogenising with ball mill grinders and then using automated RNA extraction based on magnetic beads. Using this system, the consistency and reliability of extraction was greatly improved when compared with a more traditional approach using manual methods, which routinely had a failure rate of around 15–20% (Unpublished data). The use of extraction control assays (which detected endogenous plant genes such as *Cytochrome Oxidase 1*) also had a major effect on the overall quality of testing. By routinely using such controls, the performance of extraction, in terms of both quantity and quality of extracted RNA, could be closely monitored, and extractions repeated if necessary. In this way, the risk of false negative results due to extraction failure was virtually eliminated.

The second major issue with direct tuber testing, especially using a PCR-based approach, is the increased cost when compared to an ELISA-based growing-on system. Factors include an increase in the amount of hands-on staff time, the need for RNA extraction, relatively expensive reagents, and high capital equipment costs. Another contributing factor was the requirement to buy licences from the patent holders of the technologies used and the need to charge a royalty payment for every test performed. For example, a 15% royalty was charged for a PCR test. However, while increased costs were indeed a major factor in holding back the new approach, its significance has been reduced over the years. It has been possible to streamline the whole system, to ensure greater efficiency. For example, the use of real-time PCR has removed the need for post-PCR analysis and hence avoids the time and cost of gel running. The use of rapid coring and grinding, with automated extraction has also greatly reduced the amount of staff time required, while the use of liquid handling robotics has further reduced staff input and hence lowered per-sample costs. At the same time, reagent and equipment costs have also tumbled, especially in the real-time PCR arena where increased competition amongst suppliers has pushed down prices. Finally, the expiry of the PCR patents in 2006 has also reduced costs, ending the requirement for specific licences and royalties. While these will still be required for the various real-time technologies being used, the fact that there are different detection systems (e.g. TaqMan, Scorpions, Molecular Beacons, etc) means that competition is likely to keep the costs lower than before, in contrast to PCR where a monopoly existed.

Overall, these changes have meant that the costs related to a real-time PCR-based test are no longer prohibitive. It should also be remembered that, in

many laboratories, the true cost of performing growing-on tests is unclear, as the real costs involved with growing sprouted eye cores and the maintenance of the glasshouse facilities are often underestimated or totally overlooked. Indeed, if you were to establish from scratch a seed potato testing facility capable of dealing with moderate numbers of stocks, the capital costs of building a glasshouse facility of sufficient size and standard (e.g. insect-proof, with heating and lighting) would almost certainly outweigh the cost of equipping a laboratory capable of testing the same number of tubers by direct real-time PCR.

Ultimately, the success of any new diagnostic system must be measured by its adoption and use in routine testing. This has now been demonstrated for direct tuber testing. CSL has been offering a post-harvest tuber testing service based upon real-time PCR since 2001. Over that time the uptake by growers has grown significantly (Fig. 2), to the extent that in 2006–2007 direct tuber testing accounted for just under half of all the tuber stocks tested. These results clearly show that despite being more expensive, a direct testing approach is attractive to growers and they will be prepared to pay a premium price for a diagnostic test provided that it offers them real benefits (in this case speed). By getting rapid results, growers are able to make better, more timely decisions, for example, deciding which stocks to retain for planting as seed, before grading is carried out. As a result, they can often save large amounts of time, effort, and expense.

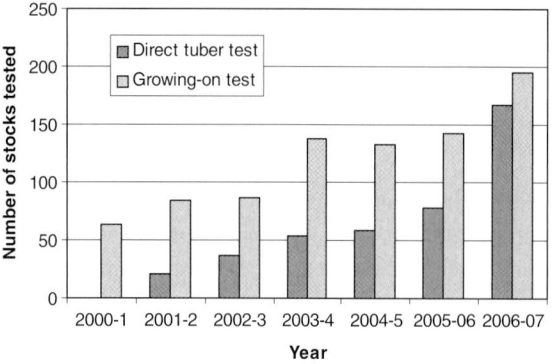

Fig. 2 A histogram showing the number of post-harvest tuber tests performed at CSL using either real-time PCR (direct tuber testing) or ELISA (the growing-on test)

DNA barcoding

DNA barcoding in an academic setting has two aims: firstly to assign unknown individuals to species, and secondly to enhance the discovery of new species (Hebert et al. 2003a; Stoeckle 2003; Blaxter 2003, 2004). The term 'DNA barcoding' originates from the idea of Universal Product Codes for manufactured goods being applied to DNA sequences for different species. Creation of the 'barcode' involves the PCR amplification and sequencing of a conserved gene sequence, typically around 600 bp of the mitochondrial *Cytochrome oxidase I* gene (*COI*). In simple terms, these sequences or barcodes are then aligned and a tree is produced; if a suitable gene has been used, the clusters should identify meaningful groups of individuals as distinct taxa.

Mitochondrial genes are particularly attractive for this application because of their lack of introns. Hebert et al. (2003b) have shown that in arthropods, most species have more than 50 substitutions in each 500 bp of their *COI* gene—more than enough for species identification. However, a common misconception is that each species has its own unique sequence for the entire 650 bp region of *COI*. Most species have a degree of intraspecific variation between individuals or populations, and therefore typically between five and ten individuals from each species are sequenced, in order to define the variation within the taxa studied.

The concept of sequencing a region of DNA and using it to identify species is not new; however, in recent years the scope and utilisation of DNA barcoding has expanded rapidly. The creation of the Consortium for the Barcode of Life (www.barcoding. si.edu) has led to the standardisation of protocols for DNA barcoding and many projects are ongoing. Recent projects have targeted birds (Hebert et al. 2004; Kerr et al. 2007), fish (Ward et al. 2005), bats (Clare et al. 2007), fungi (Seifert et al. 2007), and invertebrates (Ball et al. 2006).

DNA barcoding and plant health

Identifying invertebrate pests and fungal/bacterial pathogens to species-level using morphology alone can be time-consuming and requires specialist skills and knowledge. Many invertebrates can be morpho-

logically cryptic in their juvenile stages and may require culturing to gain a positive identification, a process that can take many weeks. Furthermore, an identification cannot be made if these samples are dead on arrival at the laboratory or die during culturing. In addition, there is a well-documented decline (e.g. Coomans 2002; Hopkins and Freckleton 2002) in the availability of experts in the field of morphological taxonomy, due at least in part to changing trends in teaching at Universities. Thus, maintaining a critical mass of expertise in these fields in order to provide a service is often difficult.

DNA barcoding could become a valuable tool in this arena. In addition to assigning unknown individuals to species and enhancing the discovery of new species, the technique can be used to identify unknown specimens. Given a validated dataset of sequences obtained from morphologically identified species, an unknown individual or juvenile may be identified by placing its sequence in the tree and seeing which species it clusters with. In contrast to the more traditional molecular diagnostic 'tests' (usually based on PCR) that only produce a yes/no answer (where the latter is often an unhelpful result) for the specific assay used, DNA barcoding can be thought of as a molecular identification tool. The technique also has a number of technological advantages since it is relatively simple, requiring only PCR (and access to the relevant primer sequences) and sequencing (a readily available service which is both rapid and inexpensive). The future of DNA barcoding as an application in the plant health arena will ultimately be determined by the availability of validated databases of sequences.

Many projects are currently underway to produce validated datasets of DNA barcodes. Their applications include forensics (Nelson et al. 2007) and elucidating cryptic species (Hulcr et al. 2007), in addition to identifying economically important species for biosecurity (Armstrong and Ball 2005; Ball and Armstrong 2006; Brunner et al. 2002). The Consortium for the Barcode of Life is also coordinating many bar coding projects. Of notable interest are the Mosquito Barcode Initiative, Tephritid Barcode Initiative and the International Network for Barcoding Invasive and Pest Species.

A further application of DNA barcoding data that is currently emerging is the use of short sections of the barcode as probes on microarrays (Pfunder et al. 2004;

Hajibabaei et al. 2007). This has the potential for large-scale microarrays containing probes for thousands of species on one slide; the future of this technique, however, lies in the development of inexpensive and validated arrays. The comparable cost of direct sequencing to even modest size arrays is weighted heavily in favour of the sequencing approach. Furthermore, whilst the sequencing approach generates the actual sequence data from which the identification is determined, the microarray approach effectively infers short stretches of the sequence on the basis of a hybridisation pattern. This in itself could be seen as a retrograde step.

One of the most important aspects of a DNA barcoding effort is the initial identification of material to be sequenced. The downstream identification of any unknown specimen is only as good as the data used for comparison. Web-based software produced by Ratnasingham and Hebert (2007) provides an identification engine that contains the DNA barcodes held to date, all of which were positively identified by morphology prior to sequencing. As the number of species, and the number of barcodes per species, in the public domain becomes larger, the power of the technique increases. However, beyond the academic aspects, it is important to consider the likely applications and potential practitioners of the method, and to engage with this community at an early stage. This community is the source of identified and validated material, without which the barcoding effort will be worthless. It is important that all the required information is captured for the material to be bar-coded and that this information is both relevant to the person using the approach to achieve an identification, and links to voucher specimens. The concept of DNA vouchers and also digital vouchers from which the DNA was extracted also need to be addressed.

Acknowledgements The authors would like to acknowledge funding from Plant Health Division and Chief Scientist Group of Defra (www.bio-chip.co.uk), and also the European Union for funding under the fifth framework programme (www.diagchip.co.uk) and also the sixth framework programme (www.portcheck.eu.com) project (SSPE-CT-2004-502348). In addition the authors would like to acknowledge the help and support provided through the COST project 'Agricultural Biomarkers for Array Technology' (http://www.cost853.ch/).

References

Armstrong, K. F., & Ball, S. L. (2005). DNA barcodes for biosecurity: invasive species identification. *Philosophical Transactions of the Royal Society B, 360,* 1813–1823.

Ball, S. L., & Armstrong, K. F. (2006). DNA barcodes for insect pest identification: a test case with tussock moths (Lepidoptera: Lymantriidae). *Canadian Journal of Forest Research, 36,* 337–350.

Blaxter, M. (2003). Molecular systematics: counting angels with DNA. *Nature, 421,* 122–124.

Blaxter, M. (2004). The promise of a DNA taxonomy. *Philosophical Transactions of the Royal Society B, 359,* 669–679.

Boonham, N., Tomlinson, J., & Mumford, R. (2007). Micro-arrays for rapid identification of plant viruses. *Annual review of Phytopathology, 45,* 307–328.

Boonham, N., Walsh, K., Mumford, R. A., & Barker, I. (2000). The use of multiplex real-time PCR (TaqMan®) for the detection of potato viruses. *EPPO Bulletin, 30,* 427–430.

Brunner, P. C., Fleming, C., & Frey, J. E. (2002). A molecular identification key for economically important thrips species (Thysanoptera: Thripidae) using direct sequencing and a PCR-RFLP-based approach. *Agricultural and Forest Entomology, 4,* 127–136.

Clare, E. L., Lim, B. K., Engstron, M. D., Eger, J. L., & Hebert, P. D. N. (2007). DNA bar coding of neotropical bats: species identification and discovery within Guyana. *Molecular Ecology Notes, 7,* 184–190.

Clark, M. F., & Adams, A. N. (1977). Characteristics of the microplate method of Enzyme-Linked Immunosorbent Assay for the detection of plant viruses. *Journal of General Virology, 34,* 475–483.

Coomans, A. (2002). Present status and future of nematode systematics. *Nematology, 4,* 573–582.

Cox-Foster, D. L., Conlan, S., Holmes, E. C., Palacios, G., Evans, J. D., Moran, N. A., et al. (2007). A metagenomic survey of microbes in honey bee colony collapse disorder. *Science, 318*(5848), 283–287, Oct 12, Epub 2007 Sep 6.

Gandelman, O., Church, V. L., Moore, C., Carne, C., Jalal, H., Murray, J. A. H., et al. (2006). Bioluminescent alternative to real-time PCR (BART). *Luminescence, 21,* 276–277.

Hajibabaei, M., Singer, G. A. C., Clare, E. L., & Hebert, P. D. N. (2007). Design and applicability of DNA arrays and DNA barcodes in biodiversity monitoring. *BMC Biology, 5,* 24.

Hebert, P. D. N., Cywinska, A., Ball, S. L., & deWaard, J. R. (2003a). Biological identification through DNA barcodes. *Proceedings of the Royal Society London B, 270,* 313–321.

Hebert, P. D. N., Ratnasingham, S., & deWaard, J. R. (2003b). Bar coding animal life: Cytochrome C oxidase subunit 1 divergences among closely related species. *Proceedings of the Royal Society London B, 270,* S96–S99.

Hebert, P. D. N., Stoeckle, M. Y., Zemlak, T. S., & Francis, C. M. (2004). Identification of birds through DNA barcodes. *PloS Biology, 2,* e312.

Hill, S. A., & Jackson, E. A. (1984). An investigation of the reliability of ELISA as a practical test for detecting *Potato leaf roll virus* and *Potato virus Y* in tubers. *Plant Pathology, 33,* 21–26.

Hopkins, G. W., & Freckleton, R. P. (2002). Declines in the numbers of amateur and professional taxonomists: implications for conservation. *Animal Conservation, 5,* 245–249.

Hulcr, J., Miller, S. E., Setliff, G. P., Darrow, K., Mueller, N. D., Hebert, P. D. N., et al. (2007). DNA bar coding confirms polyphagy in a generalist moth, *Homona mermerodes* (Lepidoptera: Tortricidae). *Molecular Ecology Notes, 7,* 549–557.

Jaffrezic-Renault, N., Martelet, C., Chevolot, Y., & Cloarec, J. P. (2007). Biosensors and bio-bar code assays based on biofunctionalized magnetic microbeads. *Sensors, 7,* 589–614.

Kerr, K. C. R., Stoeckle, M. Y., Dove, C. J., Weight, L. A., Francis, C. M., & Hebert, P. D. N. (2007). Comprehensive DNA barcode coverage of North American Birds. *Molecular Ecology Notes, 7,* 535–543.

Klerks, M. M., Leone, G. O. M., Verbeek, M., van den Heuvel, J. F. J. M., & Schoen, C. D. (2001). Development of a multiplex AmpliDet RNA for the simultaneous detection of *Potato leafroll virus* and *Potato virus Y* in potato tubers. *Journal of Virological Methods, 93,* 115–125.

Leone, G., van Schijndel, H. B., van Gemen, B., & Schoen, C. D. (1997). Direct detection of *Potato leafroll virus* in potato tubers by immunocapture and the isothermal nucleic acid amplification method NASBA. *Journal of Virological Methods, 66,* 19–27.

Ledford, H. (2007). Rapid sequencer puts virus in the frame for deaths. *Nature, 447,* 12–13.

Mumford, R. A., Barker, I., & Wood, K. R. (1994). The detection of Tomato Spotted Wilt Virus using the polymerase chain reaction. *Journal of Virological Methods, 46*(3), 303–311.

Mumford, R., Boonham, N., Tomlinson, J., & Barker, I. (2006). Advances in molecular phytodiagnostics - new solutions for old problems. *European Journal of Plant Pathology, 116,* 1–19.

Nelson, L. A., Wallman, J. F., & Dowton, M. (2007). Using COI barcodes to identify forensically and medically important blowflies. *Medical and Veterinary Entomology, 21,* 44–52.

Notomi, T., Okayama, H., Masubuchi, H., Yonekawa, T., Watanabe, K., Amino, N., & Hase, T. (2000). Loop-mediated isothermal amplification of DNA. *Nucleic Acids Research, 28,* e63.

O'Donnell, K. J., Canning, E., & Young, L. G. A. (1996). Detection of *Potato virus Y* using ligase chain reaction (LCR), in combination with a microtitre plate based method for product detection. In G. Marshall (Ed.) *Diagnostics in Crop Production: BCPC Symposium Proceedings No. 65* (pp. 187–192). Farnham, UK: British Crop Protection Council.

Pfunder, M., Holzgana, O., & Frey, J. E. (2004). Development of microarray-based diagnostics of voles and shrews for use in biodiversity monitoring studies, and evaluation of mitochondrial cytochrome oxidase I vs. cytochrome b as genetic markers. *Molecular Ecology, 13,* 1277–1286.

Puchta, H., & Sanger, H. L. (1989). Sequence analysis of minute amounts of viroid RNA using the polymerase chain reaction (PCR). *Archives of Virology, 106,* 335–340.

Ratnasingham, S., & Hebert, P. D. N. (2007). BOLD: The barcode of life data system. *Molecular Ecology Notes, 7,* 355–364. (www.barcodinglife.org).

Schoen, C. D., Knorr, D., & Leone, G. (1996). Detection of *Potato leafroll virus* in dormant potato tubers by immunocapture and a fluorogenic 5ô nuclease RT-PCR assay. *Phytopathology, 86*, 993–999.

Seifert, K. A., Samson, R. A., deWaard, J. R., Houbraken, J., Levesque, C. A., Moncalvo, J. M., et al. (2007). Prospects for fungus identification using CO1 DNA barcodes, with *Penicillium* as a test case. *Proceedings of the National Academy of Sciences USA, 104*, 3901–3906.

Spiegel, S., & Martin, R. R. (1993). Improved detection of *Potato leafroll virus* in dormant potato tubers and microtubers by the polymerase chain reaction and ELISA. *Annals of Applied Biology, 122*, 493–500.

Stoeckle, M. (2003). Taxonomy, DNA and the barcode of life. *BioScience, 53*, 2, 3.

Tan, E., Wong, J., Nguyen, D., Zhang, Y., Erwin, B., Van Ness, L. K., et al. (2005). Isothermal DNA amplification coupled with DNA nanosphere-based colorimetric detection. *Analytical Chemistry, 77*, 7984–7992.

Tomlinson, J. A., Barker, I., & Boonham, N. (2007). Faster, simpler, more specific methods for improved molecular detection of *Phytophthora ramorum* in the field. *Applied and Environmental Microbiology, 73*, 4040–4047.

Vincent, M., Xu, Y., & Kong, H. M. (2004). Helicase-dependent isothermal DNA amplification. *EMBO Reports, 5*, 795–800.

Ward, R. D., Zemlak, T. S., Innes, B. H., Last, P. R., & Hebert, P. D. N. (2005). DNA bar coding Australia's fish species. *Philosophical Transactions of The Royal Society of London B, 360*, 1847–1857.

Yi, J. Z., Zhang, W. D., & Zhang, D. Y. (2006). Molecular Zipper: A fluorescent probe for real-time isothermal DNA amplification. *Nucleic Acids Research, 34*, e81.

Zhang, D., Wu, J., Ye, F., Feng, T., Lee, I., & Yin, B. J. (2006). Amplification of circularizable probes for the detection of target nucleic acids and proteins. *Clinica Chimica Acta, 363*, 61–70.

Eur J Plant Pathol (2008) 121:365–375
DOI 10.1007/s10658-008-9311-4

The challenge of providing plant pest diagnostic services for Africa

Julian J. Smith · Jeff Waage ·
James W. Woodhall · Sam J. Bishop ·
Nicola J. Spence

Received: 6 July 2007 / Accepted: 20 March 2008
© KNPV 2008

Abstract The consequences of a globalisation of trade and climate change present an increased threat from first-entry pests and a challenge to plant health authorities. In this paper, pest reporting for the continents of Africa and Europe are discussed, and argued as a barometer of effective Plant Pest Diagnostic Services (PPDS) in terms of human capacity, infrastructure and policy-culture for phytosanitary issues. To illustrate particular areas of concern, case studies are presented on recent pest events which include outbreaks of *Ralstonia solanacearum* on *Pelargonium*, *Xanthomonas campestris* pv. *musacearum* on banana (banana Xanthomonas wilt) and *Puccinia graminis* f.sp. *tritici* race Ug99 on wheat (black stem rust). Examples are given of some recent initiatives to invigorate diagnostic capacity in East Africa, spanning state-of-the-art centres of excellence, traditional capacity building and networking projects, and grass-root level 'going-public' pest surveillance initiatives. Discussion is presented on the provision of PPDS and the impact of technology, institutional factors, the private sector, accreditation of services and policy. Emphasis is placed on the role of PPDS in support of regulatory policy. In recognising the precarious nature of many African cropping systems, the argument is made for a more consolidated approach to PPDS in and for Africa. The paper is presented from the perspective of European practitioners in pest diagnostic and risk analysis.

Keywords Climate change · Globalisation · Plant Pest Diagnostic Services · Centres of excellence · Reference laboratory · Capacity building

J. J. Smith (✉) · J. W. Woodhall · S. J. Bishop ·
N. J. Spence
Plant Health Group, Central Science Laboratory,
York YO41 1LZ, UK
e-mail: julian.j.smith@csl.gov.uk
URL: http://www.csl.gov.uk/

J. Waage
Centre for Environmental Policy, Imperial College London,
London SW7 2AZ, UK

Present address:
J. Waage
London International Development Centre,
36 Gordon Square,
London WC1H 0PD, UK

Introduction

Additive consequences of an increase in global trade (Josling et al. 2003), climate change (Garrett et al. 2006; Harwell 2002) and evolving plant pest pathogenic capacities (Brazier 2001; Zhou et al. 1997) that may be exacerbated by new host encounters due to global trade and adaptation to climate change (Garrett et al. 2006) identify an increased risk from plant pests for agriculture and the environment across the globe. A recent review on the causality of emerging infectious diseases of plants has recently been undertaken

(Anderson et al. 2004). Against this background, it is a widely accepted view that developing nations, and the more vulnerable sectors of these nations, are at the greatest risk from the predicted events of climate change and globalisation (Chancellor and Kubiriba 2006, Waage et al. 2006; Anon 2006).

Recent and high impact examples of animal and plant pest introductions in the UK, such as foot and mouth disease, Newcastle disease, Avian influenza, horse chestnut miner, potato brown and ring rot and *Phytophthora ramorum* and *P. kernoviae* exemplify the concern (see www.defra.gov.uk). Likewise, in Africa the spread of plant pests such as banana Xanthomonas wilt (BXW; Tushemereirwe et al. 2004), coffee wilt (Rutherford 2006) and cassava mosaic disease (Zhou et al. 1997) across East Africa have presented an additional and serious risk to the livelihoods of already vulnerable households. The impact of these events has prompted an increased and global vision of the threat presented by pest introductions, focusing research on future technologies for pest diagnostics (Barker et al. 2006) and more 'biosecure' quarantine systems (Waage and Mumford 2008). The function of Plant Pest Diagnostic Services (PPDS) is at the nexus of policy-based phytosanitary systems, trade, breeding programmes, integrated pest management and farmer advice.

Further evidence from a recent UK government 'foresight study' focusing on future infectious disease risks (Brownlie et al. 2006) has highlighted the global nature of these threats, drawing comparisons for plant diseases between Africa and the UK in terms of the impact of climate change (Chancellor and Kubiriba 2006), prospects for future disease control (Barker et al. 2006), and governance issues which may affect these (Quinlan et al. 2006). More specifically the foresight study presented a preliminary analysis of plant disease introductions (viruses, fungi and bacteria) into Africa and Europe based on first reports (Waage et al. 2006). These data suggest that Europe has experienced a growing rate of new plant pest introductions over the past century, while introductions in Africa (whilst greater overall, possibly due to greater crop diversity), appear to have peaked and declined during that same period. The study concludes that this may reflect differences in recent international trade by Europe and Africa, but that it may, alternatively, reflect differences in national plant protection capacity. That is, a recent lack of plant protection capacity, infrastruc-

ture and effort may cause a decline in reporting of new diseases in Africa, in contrast to continuing, strong European systems.

A further observation from the study by Waage et al. (2006) was the differences over time in the rates of first plant disease records for bacteria, fungi and viruses. In this comparison, it is observed that a marked increase has occurred for viruses for both continents, but particularly Europe. Most probably this trend is explained by recent advances in technology for diagnosis, such as the application of PCR, and a rapidly increasing database resource of nucleic acid sequences of diagnostic value that have increased our ability to detect and identify non-culturable organisms amongst field crops and in the environment. First disease reports of phytoplasmas would be expected to mirror this trend. The facility for molecular identification can be expected to increase further because of the myriad of research projects, such as the Bar Coding of Life Project (http://www.barcoding.si.edu), that continue to add to the sequence database available (e.g. GenBank). The consequences of advanced technologies for pest diagnostics on PPDS may be profound and present a particular challenge for African nations that often have neither the trained personnel nor the fiscal resources to take advantage of such change. Consequently the technology gap between north and south, which is already significant, may grow further over the coming years unless major effort is sought to address these capacity and infrastructural needs.

Finally, although it is beyond the scope of this paper to look in depth at the interface of National Plant Protection Organisations (NPPO) and how these are aligned to regional (e.g. European Plant Protection Organisation (EPPO) for the European Union and Inter African Phytosanitary Council (IAPSC) for Africa) and international treaties (e.g. World Trade Organisation and the Sanitary and Phytosanitary (SPS) Agreement, International Plant Protection Convention (IPPC), *Codex Alimentarius* (CODEX), Office International des Epizooties (OIE)), it is important to note that collectively these provide the overarching framework for safeguarding the health of humans, animals and plants against adverse effects through the international movement of people and traded goods. For example, under the IPPC it is recognised for contracting parties 'to the best of their ability' to have updated lists of regulated plant pests known to occur within their territories and to conduct pest surveillance as supports non-records of presence. The incomplete and poorly verified nature of national pest lists amongst many African nations is recognised as a key constraint in negotiations on trade.

In the sections below, case studies for three pathogens of recent significance within East Africa and one methodology for extension and surveillance are discussed that identify with particular technology, trade, and institution mandates. Emphasis is given on the role of PPDS in support of related regulatory policy. This paper does not include provision of services for food, such as mycotoxin, pesticide residue and GMO testing, as extend to private sector compliances of GlobalGap and EurepGAP (Labuschagne 2006). However, the opportunity of linkages with these services is brought into some of the discussion in the context of the value of generic technologies, broader laboratory function and the engagement of the private sector for the commercialisation of diagnostic services.

Case studies

Case study 1: *Ralstonia solanacearum*
on *Pelargonium* within the private sector of Kenya

In 2003 the *Pelargonium* industry of Kenya and USA suffered significant economic losses due to an outbreak of *Pelargonium* wilt caused by *Ralstonia solanacearum* biovar 2a, a pathogen normally associated with potato and of quarantine status in Europe, USA and many other countries (e.g. EU Council Directive 98/57/EC). In the USA, the outbreak resulted in the contamination of 27 *Pelargonium*-growing greenhouses in 13 states and the quarantine of over a further 800 glasshouses (Anon 2003). Immediate economic cost to the flower industry was estimated at several million US dollars, with longer-term consequences to trade. Interceptions of the pathogen were also made in Europe (Janse et al. 2004).

The presence of *R. solanacearum* biovar 2a in *Pelargonium* was first confirmed in Europe in 1999 and 2000, and the source of infection was traced to Kenya (Janse et al. 2004). Although this was not a first report for this pathogen on *Pelargonium*, these were the first confirmed reports of biovar 2a affecting *Pelargonium*. Earlier reports of *R. solanacearum* on *Pelargonium* from the USA did not substantiate the race or biovar status (Strider et al. 1981); however, it

was assumed that these were most probably race 1 strains as this race is known to have the greatest propensity for extending its host range (race 1 has the largest host range of the five races of *R. solancearum*) and biovar 2a (synonymous with race 3) is not known to occur in the USA (Anon 2004).

At the time of the confirmed reports of *R. solanacearum* biovar 2a in Europe, preventative actions were introduced by Europe on imported material that have, in conjunction with better practices within the industry in Kenya, to date prevented further introduction. These better private sector practices, though not accredited to any recognised international criteria, have been developed in partnership with CSL specialists, and include the use of rapid diagnostic kits (Pocket Diagnostics® lateral flow devices) for *R. solanacearum* and *Xanthomonas hortorum* pv. *pelargonii*, a pathogen of *Pelargonium* with similar symptoms. Likewise, the trade of *Pelargonium* cuttings with the USA is now governed by a 'Minimum Sanitation Protocols for Offshore Geranium Cutting Production' that operates as a certification protocol to ensure freeness of the pathogen (Anon 2007d). The formal nature of the *Pelargonium* industry in Kenya, which is a consequence of its private sector basis, has markedly assisted in the positioning of pest preventative measures and the strategic monitoring of biovar 2a through the use of diagnostic kits.

Case study 2: banana Xanthomonas wilt (BXW) within the informal cropping systems of East Africa

Having been a relatively minor disease of enset and banana in Ethiopia since the mid 1960s (Yirgou and Bradbury 1968), *Xanthomonas campestris* pv. *musacearum*, the causal organism of BXW, has recently been reported in Uganda in 2001 (Tushemereirwe et al. 2006) and, subsequently, other nations of the Great Lakes region of East Africa (Reeder et al. 2007). Damage attributed to the disease on banana within the Great Lakes region has been significant (Karamura 2006). Prior to the occurrence of the disease in Uganda the risk presented by this disease had not been formally considered. However, the Global Plant Clinic (GPC) on identifying the bacterium as a new pest record for Uganda advised on a precautionary approach, stating the risk posed by the disease was unknown for Uganda and the Great Lakes region and that the more intense and contiguous banana cultivations of the

region could identify with a higher risk of spread and impact than seen in Ethiopia. Events have shown this to be true (Smith 2007) and, although a formal Pest Risk Analysis is now available (Smith et al. 2008), the absence of a contingency plan to manage an outbreak of BXW may have contributed to delays in action that allowed the disease to become more established.

From a PPDS perspective the outbreaks of BXW have identified various issues in national capacity to survey for and identify the causal organism (Smith 2007; Aritua et al. 2008). Notably, for some nations the evidence from farmers and national authorities on the date of introduction of BXW has been inconsistent, with farmers reporting awareness of the disease prior to the national authorities (Reeder et al. 2007). Whilst such observations are anecdotal, they suggest that for these nations national surveillance mechanisms for emerging plant diseases are inadequate. Moreover, confirmation of the disease in all the affected nations has required the support of the Global Plant Clinic (GPC: www.globalplantclinic. org) as an external northern partner. Indeed, in the case of Burundi, confusion has persisted as to the disease status (present or absent) when reports of disease had been based on symptoms only and not supported by successful isolation and identification. In this instance the uncertainty as to disease presence (which has been confirmed by the GPC) diverted attention and action away from implementing control measures that must have been to the detriment of the control effort. Only in Kenya was a successful isolation of the bacterium onto culture medium performed as a step towards identification, with these cultures then transported to the GPC in the UK for formal identification. From these events it is evident that across East Africa, and most likely for other developing African nations, there is a general low level of expertise and infrastructure available in bacteriology; spanning the capacity to undertake isolation of the bacterium from infected plants onto culture medium through to the identification that frequently requires the use of sophisticated analytical methods.

In efforts to control BXW, much debate has centred on the use of a diagnostic tool (Smith 2007; Aritua et al. 2008; Aritua et al. submitted), and if such a tool had been available at the time of the first outbreak would a different scenario of disease spread have occurred. In considering this question, it is necessary to appreciate the type of detection technol-

ogy, the receiving environment of the test (the banana cropping system of Uganda and the Great Lakes region) and the entry points for the technology along the farm-fork continuum, from producing banana plantlets for farmers, to production and marketing of the banana itself, to the disposing of crop waste. An analysis quickly concludes that within the informal banana cropping systems typical of the Great Lakes region no obvious entry point for detection technology, be it laboratory or field-based, is readily identifiable as would allow for a critical intervention to reduce the rate of spread of the pathogen (Smith 2007). In practical terms, any diagnostic tool, without a critical entry-point from which to operate, will only have limited application in confirming and 'chasing' the disease. This is not to conclude that we should not invest in diagnostic technologies, for example, the prosaic ambition of achieving a rapid and accurate identification has value in mobilising contingency efforts, but it would be a mistake to see such tools as a panacea on their own. The point is made that in development of such tools, effective entry-points for the technology must be recognised, and for Africa and food this most notably requires a greater degree of formalisation of agricultural production (private sector players) and the strengthening of the policy environment that oversees such practices. The potential and use of such technologies for Africa have been considered by the Foresight Project (Barker et al. 2006, Quinlan et al. 2006).

Case study 3: black stem rust and a new virulence

A new variant of black stem rust, caused by *Puccinia graminis* f.sp. *tritici* race Ug99, was identified in Uganda in 1999 that had evolved the capacity to overcome resistance genes present in the majority of globally cultivated varieties (Pretorious et al. 2000; Singh et al. 2006). The risk posed by this variant within Africa and globally has been described by the Consultative Group for International Agricultural Research (CGIAR) centre, CIMMYT (International Centre for the Improvement of Maize and Wheat), as significant and requiring rapid action, and that the response to the threat to date has been inadequate. In an article to the New Scientist Environment (Anon 2007b) it is argued that in previous times the routine surveillance for variants of the pathogen that was, but is now no longer due to financial constraints,

undertaken by CIMMYT would have allowed for the earlier detection of the variant and a more effective response. The disease has now spread across East Africa and spores have been detected in Yemen and Sudan, making the link to Egypt, Turkey the Middle East and beyond highly likely (Brown and Hovmoller 2002). A parallel can be drawn with the UK Cereal Pathogen Virulence Survey (UKCPVS) that screens for changes in virulence of named UK cereal pathogens (yellow and brown rust of wheat, brown rust of barley, powdery mildew of wheat and barley and *Rhynchosporium* of barley), and determines the consequences of any changes observed (Anon 2007f). This surveillance has been undertaken by National Institute of Agricultural Botany (NIAB) since 1967 and continues today, with the costs accepted by UK Government (Department for Environment, Food and Rural Affairs) and the private sector cereal levy board, Home-Grown Cereals Authority (HGCA). Amongst developing nations, where governmental and private sector support for pest surveillance is historically weak, such a role of surveillance for pathogens known to present a high risk (through introduction, emergence and/or through having a known predisposition to evolving new virulence) is appropriate for the CGIAR (or any other body with the capacity to provide such a service). Some complementarity with the primary mandate of the CGIAR as a custodian of germplasms seems evident. The example of cassava mosaic disease and the recent new virulent form that has swept across East Africa would also fall within this category (Zhou et al. 1997).

Case study 4: pest surveillance, field level diagnosis and provision of advice to farmers (extension services)

The delivery of extension services is a highly challenging ideal and historically, for Africa, one that has met with limited success. Even within developed nations, many such services have been radically reduced where these have relied on government support, with a transfer of service delivery provided by the private sector. A positive innovation in this area has been the example of 'Going Public' and mobile 'Plant Health Clinics', as initiated under the Global Plant Clinic (GPC) (Bentley et al. 2004; www.global-plantclinic.org). Going Public (on a particular disease) and Plant Health Clinics (for general plant health) are

not extension services in the typical form, but rely on trained diagnosticians attending public gatherings with relatively limited and simple props and making themselves available. These approaches have shown the effectiveness of being 'with the people' in the two-way communication of plant health news between researchers and farmers. Advice can be provided and new disease concerns can be quickly reported and followed-up on, with the option of taking samples for laboratory analysis if necessary. For example, Going Public has been effective in raising awareness on identifying and controlling napier grass stunt in Kenya (Anon 2005a) and the concept of mobile Plant Health Clinics has been enthusiastically received within Uganda (Anon 2005b). Under the GPC these modes of extension have been piloted with significant success in numerous countries, including Uganda, DR Congo, Tanzania and Kenya within East Africa. As a mechanism for communicating to, and gathering information from, grass-root level stakeholders on 'today's' pest concerns, along with other issues such as food quality and marketing constraints, these approaches command significant merit (Boa 2007).

Functionalising diagnostic capacity (PPDS) in Africa

Over the past few decades, significant external investment has been made in African agricultural research and phytosanitary institutions. For example in Kenya, in the 1980s and 1990s, the Kenya Agricultural Research Institute was substantially supported under the National Agricultural Research Programme (NARP I and NARP II), which aimed to improve infrastructure and human capacity through building and equipping new premises, *ad hoc* training and PhD programmes. Similarly, Uganda has been a focus for major external investment. Consequently, amongst East African nations, Kenya and Uganda support the highest number of national agriculture-related PhDs and have some of the best-equipped laboratories. Yet, despite the marked progress achieved through such investment, by example of the impact of diseases such as **Cassava mosaic virus** (CMV), coffee wilt and BXW, even these better-equipped nations remain vulnerable to new and emerging pests. Perhaps reflecting this concern, in recent years support by

development partners has moved towards more grass-root level activity, notably in working with Non-Government Organisation (NGOs) and in promoting private sector entrepreneurialship and the market values of agriculture. The crux of a sustainable PPDS is in connecting institutions, farmers, the private sector and consumers within a policy framework, where all the stakeholders of a food chain equitably share risks and benefits of investment.

Within Europe, the discussion on how to support PPDS is relatively well advanced. The starting point is not strongly comparable with the challenges that beset Africa, particularly with respect to private sector activity and the opportunity this affords through levy boards and other industry platforms as, in part, characterise developed nations. However, some aspects can be paralleled.

Capacity and advances in technology for PPDS

The maintenance of plant pest taxonomic expertise for identification purposes in Europe increasingly resides with a few institutes. For the UK this is primarily with the Central Science Laboratory (CSL) that serves as a national laboratory for the identification of plant pests. This diagnostic capacity is in the main supported by the UK government, through the Department for Environment, Food and Rural Affairs (Defra), with a further value realised (approx. 10% for 2006) with commercial customers. Defra further supports CSL to undertake underpinning research on pests of strategic importance that builds from and adds synergy to the taxonomy and diagnostic resource. Notably, taxonomy and diagnostic capacity are not seen as a stand-alone resource and are integrated with other areas of research and development. It is the fullness of this understanding between CSL and Defra, and Defra's responsibility (policy position) to industry and the public that is important in maintaining a credible PPDS at CSL.

It is recognised that PCR-based and generic nucleic acid platform diagnostic technologies (micro-arrays) have the potential to 'reduce' diagnostic procedures for the divergent disciplines of bacteriology, mycology, nematology, virology and even entomology to a much simplified non-technical testing format (Mumford et al. 2006; Boonham et al. 2007). Thus, a move towards the use of advanced approaches questions the need for expertise in classical pest identification and presents an opportunity to deliver a service with fewer and less

qualified staff. Requirements for ISO-accreditation in pest diagnostics are also more attainable by advanced methods that avoid some of the subjective assessments required by classical taxonomy and diagnostics. Such accreditation of PPDS is becoming increasingly important with clients. Thus, on a fiscal basis, cost-saving and greater business opportunity may be envisaged by embracing modern methods. It is widely recognised that expertise in classical taxonomy for plant pests is on the decline and the high costs of maintaining this is often seen as a core factor. In this context, advanced molecular methods for pest diagnostics are seen positively. However, it would be interesting to establish the truths of this statement, especially in the context of Europe and Africa, noting the respective positions of high salary, low consumable costs and low salary, high consumables costs for these continents, respectively. Moreover, there is also a view that classical taxonomy and diagnostics should not be overlooked as, whilst it seems expedient to embrace technology, there is a risk of the diagnosis process becoming detached from the deeper knowledge of the organism inherent to classical approaches. The consequences of diluting expertise in taxonomy, if this leads to a less effective research and development programme for controlling a particular pest, might prove to be to a greater detriment than the investment needed to maintain such capacity.

Organisation of PPDS

Whilst for most nations the provision of PPDS is considered a national responsibility, at the European level some discussions are in train on the value of identifying Community Reference Laboratories (CRLs) that may serve a regional function. This may seem attractive especially for those nations without a critical mass in PPDS. For example and on a bilateral basis, CSL provides an identification service to the Swedish government for plant pests. An extension of this idea may see institutes recognised at the EU level to provide regional PPDS for particular organisms for which they are recognised experts, with others pests aligned to different institutions within the same or another EU country. This approach to 'partitioning' expertise is already evident within veterinary diagnostics of the EU where CRLs are recognised for the key pests (e.g. the CRL for Avian influenza is the

Central Veterinary Laboratory, Weybridge, UK; Dir 2005/94/EC) that serve to backstop National Reference Laboratories (NRL; Anon 2007h). In this context the CRL will provide resources to the region that are unique for that pest, such as reference strains or highly specialised diagnostic tests that are not available to the NRL. A further example is also evident with mycotoxin and GMO testing in food and feed where EC regulations recognise NRLs and a single CRL, the Joint Research Centre Institute for Reference Materials and Measurements (Anon 2007i, j).

For Europe a dispersed regional facility in the area of plant health, akin to that with veterinary diagnostics, would be achievable. However, such partitioning of expertise may risk fragmenting the synergies that operate between taxonomic disciplines, synergies that are increasingly apparent with the generic nature of many of today's emerging nucleic acid-based technologies, and would not allow for the necessary institutional planning (succession of staff and infrastructure updates) to maintain a critical mass in diagnostic capacity. Thus, whilst a credible service can be identified today this may not be the case in 5 to 10 years if a disparately located PPDS was realised. The opportunities and concerns in setting up plant health CRLs are being discussed by the European plant health community (Giltrap, N., personal communication).

Aspirations of CRL in Europe have not extended to the establishment of a single or limited number of regional 'centres of excellence' that aim to provide a complete PPDS for the region. However, with the continuing advances being made in molecular diagnostics, e.g. microarrays (Boonham et al. 2007) there may well be greater motivation towards such an holistic approach in the future. Currently for the EU, the relative strengths and long histories of many national laboratories makes this scenario unlikely. However, in an African context, where there is not such a strong existing plant pest diagnostic capacity and the investment needed to raise the standard of many nations appears prohibitive, such an argument for a 'one-stop' centre of excellence with a regional mandate may be more viable. Another view may suggest it runs against political will to 'outsource' services to another nation when this may lead to an erosion of national expertise and a reduced-competitiveness in bidding for research and service contracts, an argument that is also applicable with the development of CRLs. Ideas of consolidating institutions further, as might be suggested when

generic diagnostic technologies allow for plant, animal and human health pest diagnostic services to be joined, would present a further political dimension.

These arguments are very relevant to Africa where the scarcity of human capacity and infrastructure strongly identifies with the logic of identifying a regional capability, located centrally or virtually through a network.

Examples of pest diagnostic initiatives in Africa

Examples of initiatives to invigorate pest diagnostic capacity in Africa are evident as training projects, networks and regional 'centres of excellence'. To note a few within the East Africa region:

Bioscience East and Central Africa (BeCA; Anon 2007a) is based in Kenya on the grounds of the CGIAR centre ILRI (International Livestock Research Institute) and aims to serve as a research centre of excellence for the East Africa region that could extend to PPDS. As a recently formed centre, BeCA is still at an early stage of development. A key to its success will be the complementarities realised with the regional national research institutions with which BeCA must make effective partnerships.

The International Plant Diagnostic Network (IPDN) is a USAID-supported initiative that aims to establish regional networks for plant disease diagnostics through a 'hub (main laboratory) and spoke (neighbouring country laboratory)' structure (Anon 2007c). Currently, three regional networks are in development, two in Africa (West and East Africa) and one in Central America. With this initiative, strong emphasis is placed on communication and data-networked systems that extend to partners in the USA and Europe. The placement of such networks with regional bodies of Africa such as COMESA in the context of sanitary and phytosanitary measures and the realising of robust data sets for national/regional pests as complies with the WTO and facilitates region/export trade would be of substantial value.

The Nematode Initiative for East and Southern Africa (NIESA) is ostensibly a capacity-building programme for expertise in nematology that is funded by the Gatsby Charitable Foundation. The primary aim of the project is to develop a 'Nematology in Africa Platform' for accessing and sharing information on all aspects of plant parasitic nematodes (Anon 2007e).

The Global Plant Clinic, an ostensibly DfID-funded plant health diagnostic clinic, has for many years provided a service for pest diagnosis for developing nations that is free on a request basis and, significantly, in recent years has extended to field support. The work of the GPC represents an example of on the ground surveillance for pests, through linking field and market observations of farmers and consumers to extension-styled diagnostics and on-the-spot provision of control recommendations, with referrals to technical laboratories as required (see case study 4).

Implementing policy standards

How a plant health standard is implemented (policy position) represents a key challenge that goes to the core of functionalising institutions that either implement or oversee standards. In Europe provision to ensure its new member states are proficient in plant health has been addressed through twinning projects tailored to address specific policy positions. For example, in the twinning project on plant health in Estonia (Anon 2007g), provision has been made to address EU regulations on ring rot of potato that has involved the purchase of a TaqMan PCR machine, capacity building and the implementation of Standard Operating Procedures (SOPs) that are accredited to an EU-recognised body. Such projects are akin to 'learning alliances' that simply partner like-mandated institutions between old and new member states for purposes of infrastructure and capacity building. However, within these projects limited consideration is given to the sustainability of the capacity building, and the assumption is made that either the linkages of the PPDS to the private sector or the national government are sufficient to provide viability. Thus whilst 'learning alliances' may present effective models for institutional and policy strengthening, for African nations, where there is a weak private sector pull for services and a limited expectation for support from governments, more is needed to establish the linkages to production pathways to allow for the associated costs of PPDS to be positioned.

Cost sharing of PDDS

Amongst the farm-to-fork continuum for agricultural produce and the stakeholders therein (farm, institution, private sector and consumer, etc) is the requirement to pay for PPDS. Arguably this presents the greatest challenge as it identifies with the transformation of human capacity, infrastructure and appropriate policy positions into functional systems that deliver services in a cost-effective and timely way such as supports the interests of, and is valued by, all stakeholders. The challenge is truly a difficult one to resolve, as the provision of quality PPDS, especially if accredited, requires quality science, training, infrastructure and processes of monitoring and evaluation that are high cost. When these services are to be placed within informal food chains that are often of low value, as is typical for Africa and its staple crops, how such costs can be supported is not obvious. Evidently, the exception is with high-value products that attract or support private sector activity. However, the majority of crops grown in Africa are destined for local markets of low value where the necessary market structures upon which PPDS can be positioned are not in place.

Examples of paying by the private sector for PPDS in Africa are evident. The example of *Pelargonium* in addressing the disease threat by *R. solanacearum* has been presented in case study 1. Such examples tend to identify with export and higher value crops, where the engagement with PPDS is driven primarily by the needs of consumers of the importing nation and the standards for food safety and agricultural practices therein. Moreover, many of these standards are inclusive of non-regulatory good agricultural practices that have been set by the private sector for, by example, organic, ethical, fairtrade and EurepGAP standards, and as such builds value in the receiving markets with consumers.

A key realisation for Africa will be the positioning of plant health standards that stimulates the engagement of the private sector and a greater volume of formal local and regional trade. In these instances, arguably, standards that are equivalent for export are not appropriate for Africa and it might be advantageous to set less exacting standards, that whilst secure and safe, are more tolerant of pests and their products, and therefore more achievable in the near-term. Such standards may in time act as a stepping-stone towards future higher private sector-led standards that are the

equivalent of the exporting standards set by the EU and USA. An example of the constraint imposed by setting overly exacting standards is with seed potato in Kenya and Uganda and the setting of a phytosanitary standard at zero for *R. solanacearum* (cause of bacterial wilt). For these nations where the prevalence of bacterial wilt within farmer's fields is high, the setting of zero tolerance for bacterial wilt in seed-tubers has largely precluded all smallholder-derived potatoes being officially sold as seed. Consequently, the majority of certified seed-tubers available are those produced by the respective national agricultural research stations, with a private sector venture unlikely. It is a circular argument, but it is only when a greater demand exists for certified seed-tubers that a significant private sector interest for seed potato production can be justified, and ware farmers remain to be convinced of the value of good quality seed because of its limited availability and high price. Thus, part of the solution lies with the education of stakeholders as to the need for quality, be this with seed or agrochemical inputs or food, as ultimately the consumers will drive standards.

A further key point to realise is that PPDS and other diagnostic tests that encompass food and feed quality and authenticity are not equal in their commercial value and therefore support from the private sector. Many of the diagnostic services reside with food and feed and are driven as much by the private sector and consumer interface as by regulatory need, and demand a higher volume of testing. Tests for pesticide minimum residue levels (MRLs), mycotoxins and GMO contamination are integral to food and feed-testing regimes for Europe and many developed nations, and potentially these tests can provide the bulk of commercial diagnostic service income for a laboratory. This is the experience of CSL. The corollary of this position is in the provision of diagnostic services that hold little direct value to the private sector, yet are nonetheless recognised as essential to the overall wellbeing of an industry. In these cases, alternative ways of support are required. An indirect mechanism for private sector engagement may be achievable through private sector trade boards that use part of their industry levy to support the industry through provision of services that are most effectively implemented by a higher body. It is worth noting that nations with developed PPDS tend to have established industry bodies, such as the British Potato

Council and Potatoes South Africa. In case study 3 the role of the HGCA in cost-sharing with the UK government for virulence surveillance of cereal pathogens is described.

However, it is not the case that the private sector can support all PPDS and, whilst the example of Europe and other developed nations may testify to an increased importance for engaging the private sector in PPDS, a role for government is nonetheless also evident. As in the example of CSL and its relationship with Defra, there remains a key role for government to support the position on regulatory pest diagnosis, risk analysis and contingency planning. Not all pest diagnostic needs can be packaged for the private sector to pick up.

Concluding remarks

Based on these arguments its seems unavoidable that the development of a pest diagnostic capacity for Africa, that aims to serve the majority of farmers' crops and notably the low value staples, must be part of a broad PPDS that is substantially linked to government or regional support mechanisms of national/regional bio-security. However, were viable, the role and buy-in of the private sector must be identified and strengthened. This would also seem to be the objective of developed nations of Europe and USA that continue to evolve through mult-national arrangements (for the EU), address of homeland security and increasing private sector activity.

The porosity of the borders between African nations to pests identifies with endeavours to promote regional initiatives that address PPDS, especially in the context of phytosanitary policy. Such a remit sits well with the aims and objectives of the New Partnerships for Africa's Development (NEPAD) through its Comprehensive Africa Agricultural Development Programme (CAADP) and its advocacy body, The Forum for Agricultural Research for Africa (FARA), in the facilitation of regionally coordinated actions. The extent to which such an African PPDS capacity is national or regional, networked, with resources duplicated, unique, disparate or centralised and linked with partners of advanced research institutes/centres of excellence in Europe, South Africa or USA requires debate, along with the necessary links to the private sector.

There is however, a clear need for better pest diagnostic capacity, as without such tools any ambitions for pest detection, identification and monitoring will remain unrealised. Costing PPDS serves is a key issue, and noting that policy and private sector-based sanitary and phytosanitary standards shape production and trade and protect the national interest, efforts to develop and implement PPDS that bridge private sector and mandated institutional interests, may provide the vehicle to catalyse broad-based development. The inclusion 'under one roof' of food and feed diagnostic services alongside PPDS will present greater opportunity to maximise processing efficiency, reduce outlay for infrastructure and human capacity and realise revenue through private sector interest.

References

Anderson, P. K., Cunningham, A. A., Patel, N. G., Morales, F. J., Epstein, P. R., & Daszak, P. (2004). Emerging infectious diseases of plants: Pathogen pollution, climate change and agrotechnological drivers. *Trends in Ecology and Evolution, 19*, 535–544.

Anon (2003). Status of *Ralstonia solanacearum* race 3 biovar 2 in U.S. greenhouses. North American Plant Protection Organisation, Phytosanitary Alert System. www.pestalert.org.

Anon (2004). Questions and answers on *Ralstonia solanacearum* race 3 biovar 2. *APHIS Plant Protection and Quarantine, Fact Sheet*. April 2004.

Anon (2005a). Grass stunt draws a crowd. http://www.new-ag.info/05

Anon (2005b). Uganda's mobile 'plant clinics' answer emergency call. http://www.scidev.net/News/index.cfm?fuseaction=readNews&itemid=2259&language=1.

Anon (2006). Adapting to climate change in developing countries. *The Parliamentary Office of Science and Technology, Postnote 269*, www.parliament.uk/post.

Anon (2007a). BECA. http://www.biosciencesafrica.org/background.htm.

Anon (2007b). Billions at risk from wheat super-blight, New Scientist; <http://environment.newscientist.com/channel/earth/mg19425983.700>;).

Anon (2007c). International Pest Diagnostic Network. http://www.oardc.ohio-state.edu/ipdn/about.asp.

Anon (2007d). Minimum Sanitation Protocols for Offshore Geranium Cutting production. http://www.aphis.usda.gov/plant_health/plant_pest_info/ralstonia/downloads/ralstoniaworkplan.pdf.

Anon (2007e). Nematode Initiative for East and Southern Africa (NIESA), http://www.africannematology.info/index.asp.

Anon (2007f). The UK Cereal Pathogen Virulence Survey. http://www.niab.com.

Anon (2007g). Upgrading of Functional Capability on testing of Harmful Organisms in Estonia, Twinning Contract No EE04-IB-AG-01, http://pmk.agri.ee/twinningproject.

Anon (2007h). http://ec.europa.eu/food/animal/diseases/laboratories/index_en.htm.

Anon (2007i). http://www.irmm.jrb.be/html/CRLs/crl_mycotoxins/imdex.htm.

Anon (2007j). http://gmo-crl.jrc.it/legalbasis.htm.

Aritua, V., Parkinson, N., Thwaites, R., Heeney, J. V., Jones, D. R., Tushemereirwe, W., et al. (2008). Characterisation of the *Xanthomonas* sp. causing wilt of enset and banana reveals it is a strain of *X. vasicola*. *Plant Pathology, 57*, 170–177.

Barker, I., Bokanga, M., Lenne, J., Otim-Nape, W., & Spence, N. (2006). Future control of infectious diseases in plants with emphasis on sub-Sahara Africa. In: Foresight 2006, UK Government Office of Science and Innovation, Infectious Diseases: Preparing for the future, D3.1.

Bentley, J. W., Boa, E. R., Van-Mele, P., Almanza, J., Vasques, D., & Eguino, S. (2004). Going Public: A new extension method. *International Journal of Agricultural Sustainability, 1*(2), 108–123.

Boa, E. (2007). Plant healthcare for poor farmers: An introduction to the work of the Global Plant Clinic. *APSnet*, (in press). http://www.apsnet.org/online/feature/clinic.

Boonham, N., Tomlinson, J., & Mumford, R. (2007). Microarrays for rapid identification of plant viruses. *Annual Review of Phytopathology, 45*, 307–328.

Brasier, C. M. (2001). Rapid evolution of plant pathogens via interspecific hybridization. *Bioscience, 51*, 123–133.

Brown, J. K. M., & Hovmoller, M. S. (2002). Aerial dispersal of pathogens on the global and continental scales and its impact on plant disease. *Science, 297*, 537–541.

Brownlie, J., Peckham, C., Waage, J., Woolhouse, M., Lyall, C., Meagher, L., et al. (2006). In *Foresight 2006. Infectious Diseases: Preparing for the Future Threats* (pp. 1–83). London: Office of Science and Innovation.

Chancellor, T., & Kubiriba, J. (2006). The effect of climate change on infectious diseases of plants. In: Foresight 2006, UK Office of Science and Innovation, Infectious diseases: Preparing for the future. Study T7.2

Garrett, K. A., Dendy, S. P., Frank, E. E., Rouse, M. N., & Travers, S. E. (2006). Climate change effects on plant disease: Genomes and ecosystems. *Annual Review of Phytopathology, 44*, 489–509.

Harwell, C. D. (2002). Climate warming and disease risks for terrestrial and marine biota. *Science, 296*, 158–162.

Janse, J. D., van den Beld, H. E., Elphinstone, J., Simpkin, S., Tjou-Tam-Sin, N. N. A., & van Vaerenbergh, J. (2004). Introduction to Europe of *Ralstonia solanacearum* Biovar 2, Race 3 in *Pelargonium zonale* cuttings. *Journal of Plant Pathology, 86*(2), 147–155.

Josling, T., Roberts, D., & Orden, D. (2003). *Food regulation and trade. Towards a safe and open global system*. Washington, D.C: Institute for International Economics.

Karamura, E. (2006). Assessing the impact of the Banana Bacterial Wilt, *Xanthomonas campestris* pv. *musacearum*, on household livelihoods in East Africa. Final Technical report: DfID RNRRS CPP R8437, UK.

Labuschagne, L. (2006). Audits, pesticides and IPM—explained. Worldgrower, 23rd Nov.

Mumford, R. A., Tomlinson, J., Barker, I., & Boonham, N. (2006). Advances in molecular phytodiagnostics—new solutions for old problems. *European Journal of Plant Pathology, 116*, 1–19.

Pretorious, Z. A., Singh, R. P., Wagoire, W. W., & Payne, T. S. (2000). Detection of virulence to wheat stem rust of resistance gene *Sr31* in *Puccinia graminis* f.sp. *tritici* in Uganda. *Plant Disease, 84*, 203.

Quinlan, M. M., Phiri, N., Zhang, F., & Wang, X. (2006). The influence of culture and governance on the detection, identification and monitoring of infectious diseases. In: Foresight 2006, UK Office of Science and Innovation, Infectious diseases: Preparing for the future, Study D4.1.

Reeder, R. H., Muhinyuza, J. B., Opolot, O., Aritua, V., Crozier, J., & Smith, J. (2007). Presence of banana bacterial wilt (*Xanthomonas campestris* pv. *musacearum*) in Rwanda. *Plant Pathology, 56*(6), 1038.

Rutherford, M. (2006). Current knowledge of Coffee Wilt Disease, a major constraint to coffee production in Africa. *Phytopathology, 96*, 663–666.

Singh, R., Hodson, D. P., Jin, Y., Huerta-Espino, J., Kinyua, M. G., Wanyyera, R., et al. (2006). Current status, likely migration and strategies to mitigate the threat to wheat production from race Ug99 (TTKS) of stem rust pathogen. *CAB Reviews: Perspectives in Agriculture, Veterinary Science, Nutrition and Natural Resources* 1, No.054.

Smith, J. (2007). Summary of a workshop entitled 'Expert Consultation on Progressing the Road Map for the Control of Banana Xanthomonas Wilt in Uganda and across East Africa'; hosted at Central Science Laboratory, York UK, 24–27th July 2006.

Smith, J. J., Jones, D. R., Karamura, E., Blomme, G., & Turyagyenda, FL. (2008). An analysis of the risk from **Xanthomonas campestris** pv. **musacearum** to banana cultivation in Eastern, Central and Southern Africa. InfoMus@, www.promusa.org, http://www.promusa.org/index.php?option=com_content&task=view&id=66.

Strider, D. L., Jones, R. K., & Haygood, R. A. (1981). Southern bacterial wilt of geranium caused by *Pseudomonas solanacearum*. *Plant Disease, 65*, 52–53.

Tushemereirwe, W., Kangire, A., Ssekiwoko, F., Offord, L. C., Crozier, J., Boa, E., et al. (2004). First report of *Xanthomonas campestris* pv. *musacearum* on banana in Uganda. *Plant Pathology, 53*, 802.

Tushemereirwe, W. K., Nankinga, G. K., Okaasai, O., Kubiriba, J., Masanza, M., & Odoi, N. (2006). Status of banana bacterial wilt disease (Kiwotoka) in Uganda and a synthesis of successes towards its control. In G. Saddler, J. Elphinstone, & J. Smith (Eds.), *Programme and Abstract Book of the 4th International Bacterial Wilt Symposium, 17th–20th July 2006, The Lakeland Conference Centre, Central Science laboratory, York, UK,* (p. 56).

Waage, J. K., & Mumford, J. D. (2008). Agricultural biosecurity. *Philosophical Transactions of the Royal Society B, 363*, 863–876.

Waage, J. K., Woodhall, J. W., Bishop, S. J., Jones D. R, Smith, J. J., & Spence, N. J. (2006). Patterns of new disease spread: A plant pathogen database analysis. In: Foresight 2006, UK Government Office of Science and Innovation, Infectious Diseases: Preparing for the future, Study T15.pdf.

Yirgou, D., & Bradbury, J. F. (1968). Bacterial wilt of enset (*Ensete ventricosum*) incited by *Xanthomonas musacearum* sp. n. *Phytopathology, 58*, 111–112.

Zhou, X., Liu, Y., Calvert, L., Munoz, C., Otim-Nape, G. W., Robinson, D. J., et al. (1997). Evidence that DNA-A of a geminivirus associated with severe cassava mosaic disease in Uganda has arisen by interspecific recombination. *Journal of General Virology, 78*, 2101–2111.

Eur J Plant Pathol (2008) 121:377–385
DOI 10.1007/s10658-008-9303-4

Application of pathogen surveys, disease nurseries and varietal resistance characteristics in an IPM approach for the control of wheat yellow rust

Mogens S. Hovmøller · Karen E. Henriksen

Received: 8 October 2007 / Accepted: 3 March 2008
© KNPV 2008

Abstract The present paper presents the rationale for the use of pathogen surveys, inoculated and non-inoculated disease nurseries and varietal resistance characteristics in an integrated approach to control wheat yellow rust in Denmark. The non-inoculated disease observation plots, which gave valuable information about yellow rust at the year, site and variety level, served as the primary sample source for the pathogen survey revealing pathogen virulence dynamics. This survey was also the main source for isolates of new pathotypes, a prerequisite for the assessment of the resistance characteristics of varieties and breeding lines in inoculated nurseries, and the postulation of race-specific resistance genes. A simple grouping of varieties into four categories with respect to resistance to the current yellow rust population proved robust, and this grouping was used as a determinant in a web-based decision support system for pesticide applications in cereals, Crop Protection On-line (CPO). The interplay between the different research and survey activities in the integrated pest management (IPM) approach demonstrated the need for a coherent and long-term involvement at all stages from plant breeding to the official variety approval

M. S. Hovmøller (✉) · K. E. Henriksen
Faculty of Agricultural Sciences,
Department of Integrated Pest Management,
University of Aarhus,
Flakkebjerg,
4200 Slagelse, Denmark
e-mail: mogens.hovmoller@agrsci.dk

system, extension service and research in disease epidemiology and resistance genetics.

Keywords *Puccinia striiformis* f. sp. *tritici* ·
Integrated pest management · Disease resistance ·
Pathotype

Introduction

Yellow rust or stripe rust, caused by *Puccinia striiformis* f. sp. *tritici*, is a common disease of wheat in temperate regions of Europe and elsewhere (Zadoks 1961; Chen 2005). The pathogen is a biotrophic basidiomycete, which does not complete sexual reproduction, apparently due to the lack of an appropriate alternate host (Stubbs 1985). It has been recognized as harmful to wheat since Theophrastus (371–286 BC), one of the classical Greek authors, who made remarks on the susceptibility of wheat to cereal rusts and that disease development was affected by weather (Hermansen 1968). The first comprehensive review on the biology and host plant symptoms at different growth stages was published as early as the late nineteenth century (Eriksson and Henning 1896).

In recent years, the control of yellow rust by host resistance has been generally successful in NW Europe (Johnson 1992), although the emergence of new virulence phenotypes has sometimes caused disease epidemics on varieties with race-specific resistance genes (Hovmøller 2001; Enjalbert et al. 2005) resulting in subsequent fungicide sprays (Schmidt 2003). Resistance to *P. striiformis* may be based on genes which are expressed at all wheat growth stages or genes which are effective mainly at the post-seedling and adult plant stages; the latter is denoted as 'adult plant resistance' (APR) (Johnson 1984). Genes providing effective disease control at the seedling stage generally follow race-specificity (Manners 1988), although this may also be the case for some APR genes (Johnson 1992). Thus, the ratio of virulence/avirulence in the current pathogen population has a great influence on the expected disease severity on wheat varieties carrying different sources of *P. striiformis* resistance (Priestley et al. 1984; Hovmøller 2001). Virulence is defined as the qualitative ability to cause disease on plant genotypes possessing a particular resistance gene (Flor 1971; Brown 2003), i.e. virulence represents the lack of recognition between a specific pathogen avirulence gene and the corresponding resistance gene in the host (Flor 1971). Depending on the context, terms such as 'pathotype', 'race' and 'virulence phenotype' are often used synonymously, just as 'emergence of virulence' and 'loss of avirulence'.

Forecasts on expected yellow rust severity at the crop level is a challenging task due to variable disease loads in different years and regions, influenced by seasonal weather conditions as well as prior winter survival (Hovmøller 2001) and the patchiness by which the disease often appears (Buiel et al. 1989). Assessments of average or maximum disease severity for specific varieties, when challenged by specific pathogen isolates with well-defined virulence/avirulence characteristics under disease conducive field conditions, may therefore be valuable in estimating average and worst case scenarios for such varieties (Priestley et al. 1984; Pinnschmidt et al. 2006).

In the context of a wish to reduce the pesticide use in Danish crop production (Jørgensen and Kudsk 2006), the challenging features of the yellow rust—wheat system were tackled through an integrated pest management (IPM) strategy by a sequence of research and survey activities: disease observation plots and virulence surveys (Hovmøller 2001), assessment of host resistance genes (Hovmøller 2007), national and European wheat nurseries using yellow rust pathotypes typical for each area (Bayles et al. 2000), and the application of decision support systems for pesticide use (Hagelskjær and Jørgensen 2003) and choice of varieties (http://www.sortinfo.dk/). Al-though most pages on internet sites referred to in this paper were in Danish, they provided summaries in English. This paper presents the rationale for an IPM strategy for yellow rust on wheat in Denmark, illustrated by examples of data sources and how data were generated.

Materials and methods

Survey data were generated through ongoing activities in the official cereal variety approval system (Anonymous 2007a; http://www.pdir.dk/), the Danish Agricultural Advisory Service (http://www.lr.dk/) and pesticide efficacy evaluations (e.g., Jørgensen 2006).

Disease observation plots subject to natural disease occurrence

All varieties and advanced breeding lines considered for approval in Denmark, were grown in non-fungicide treated, non-inoculated disease observation plots distributed at 15–20 locations across Denmark (Pinnschmidt et al. 2006). The naturally occurring diseases on each variety were generally assessed visually once or twice during the growing season, when the disease severity on the most susceptible varieties had reached a substantial level. Disease scores were recorded as percent leaf area covered by symptoms of the individual disease, assessed visually using a scale comprising the following steps (in percent): 0–0.1–0.5–1–5–10–25–50–75–100. Mean values of two adjacent steps were used when a score was not assigned to a specific category with sufficient confidence.

Pathogen virulence surveys

Representative, single-lesion samples of yellow rust were collected from susceptible varieties as well as previously yellow rust-resistant varieties in disease observation plots (Hovmøller 2001). Samples were multiplied on seedlings of the standard susceptible variety 'Cartago' or the variety from which the sample was collected, and covered by spore-proof cellophane bags to minimize the risk of contamination. Seedlings with sporulating lesions were gently rubbed onto 12–16 day-old seedlings of a set of 20–30 differential wheat varieties and lines. Typically,

these comprised 15 varieties from the 'world' and 'European' differential sets (Johnson et al. 1972), a range of near-isogenic wheat lines in the Avocet background (Wellings 2007), which duplicated some of the considered resistance genes, and supplementary commercial varieties of particular interest. Disease reactions were scored using a 0–9 scale (McNeal et al. 1971) in which infection types up to six indicated varying levels of incompatibility between host and pathogen, whereas types 7–9 were interpreted as compatible interactions (Hovmøller and Justesen 2007).

Assessment of sources of resistance with major effect at the seedling stage

Wheat varieties were assessed for race-specific resistance by analysing the patterns of disease reaction (qualitatively defined infection types) on seedlings when challenged by 10–16 yellow rust isolates of diverse origin and pathotype. A comparison with patterns of infection types on differential varieties with defined resistance genes (Johnson et al. 1972), pedigree information and occasional information on molecular markers in specific varieties (Christiansen et al. 2006), allowed an interpretation of data in terms of presence/absence of specific *Yr*-resistance genes in more than 150 European varieties (Hovmøller 2007).

Inoculated yellow rust nurseries at the adult plant stage

A range of yellow rust isolates obtained via the pathogen survey and selected according to pathotype, were used to investigate yellow rust resistance characteristics under field conditions. Most varieties were challenged by three to six individual isolates in multiple years, sown in six-rowed, 1 m² plots

consisting of two varieties per plot and two rows of an isolate-specific, susceptible spreader (Hovmøller 2007). The spreaders were inoculated at GS 30–32 by gently rubbing seedlings with fresh spores, often twice at 2-day intervals, preferably prior to natural dew formation or light rain. Main plots were surrounded by winter barley for minimizing unintentional mix of isolates in the field. Disease scores were recorded as percent leaf area covered by lesions (scale shown above). First assessment in a trial was done at first symptom appearance, typically three weeks after inoculation, and then once for each subsequent yellow rust generation until senescence of leaves made reliable scoring impossible.

Comparative analysis of seedling data and adult plant data allowed conclusions about presence of components of resistance to *P. striiformis*, which were mainly expressed at the adult growth stages, and denoted as APR (Johnson 1984). Thus, the level of APR was not accessible for varieties being resistant to all isolates at the seedling stage.

Results and discussion

The general yellow rust disease level in Denmark, measured by yellow rust severity on a susceptible check, was highly variable across years (Fig. 1). Previous research has shown that these dynamics were much influenced by pathogen winter survival as well as the distribution of host varieties with specific resistance characteristics within Denmark (Hovmøller 2001). Random samples, as well as targeted sampling from varieties previously resistant to yellow rust, were obtained from these observations plots and from farmer's fields and assayed for pathotypes as described.

A total of 14 new pathotypes have emerged since 1997, i.e., an average rate of approximately one per

Fig. 1 Yellow rust attack on var. 'Anja' (susceptible check) in non-fungicide treated disease observation plots in Denmark, average of eight to ten locations at GS 65–71

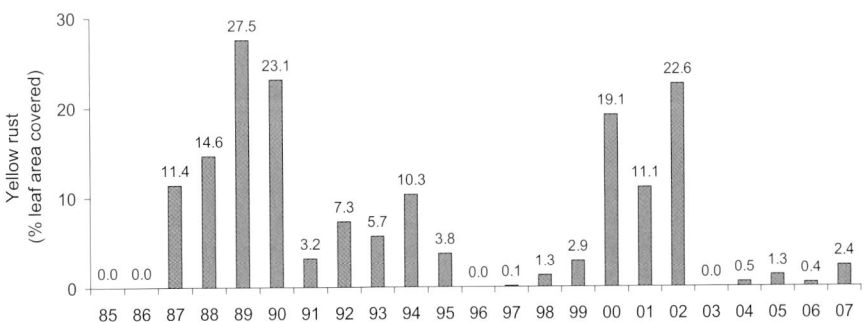

year (Table 1). However, the time of first appearance, and virulence characteristics of new pathotypes were highly unpredictable. For instance, the most frequent pathotype in Denmark in 2007, which was virulent on host varieties possessing $Yr17$ and $Yr32$, had not been observed in previous years. The same occurrence pattern of this particular pathotype applied for France (de Vallavieille-Pope, personal communication), although it had been present in moderate to high frequencies in the UK since 2000 (Bayles et al. 2004). This new pathotype may be of immediate agronomic relevance for Danish farmers because $Yr17$ and $Yr32$ were commonly present in NW-European breeding programmes (Hovmøller 2007). A similar chain of events has been observed several times in the past in northern Europe (e.g., Bayles et al. 2000).

Isolates of pathotypes XI and XII (Table 1), which represent the first yellow rust detected in a wheat crop in Denmark in 2007, were observed as randomly dispersed single lesions at the field scale, approximately ten lesions per square meter without focal development in the area, i.e., a clear indication of random spore dispersal by wind from distant source(s). This hypothesis was confirmed by highly atypical pathotypes as compared to the resistance characteristics of wheat varieties currently grown in Denmark. In fact, a range of 'atypical' pathotypes has been observed in recent years in Denmark (Hovmøller and Justesen 2007), suggesting relatively frequent, long-distance spread of airborne yellow rust spores from exotic sources into northern Europe.

The implementation of variety and pathogen characteristics in an IPM approach is illustrated in Table 2, where the resistance groupings in Crop Protection On-line (CPO) are presented for five varieties. CPO is a web-based decision support system (DSS) for pesticide use in cereals (Hagelskjær and Jørgensen 2003), which is widely used by farmers and extension service in Denmark (Jørgensen et al. 2007). The resistance grouping has immediate influence on the CPO-recommended disease management strategy. Most of the information in Table 2 is accessible online through a DSS for choice of variety in cereals, 'SortInfo' (http://www.sortinfo.dk). This system contains all public available information about growth and resistance characteristics of cereals, which have been marketed in Denmark since 1995.

The varieties Penta and Ambition were both in CPO group 0, i.e., sources of resistance expected to be effective for yellow rust control in the year considered. No yellow rust was detected in these varieties in the observation plots, or in inoculated field trials, although var. Penta was susceptible to several, frequent pathotypes at the seedling stage (Hovmøller 2007). Therefore, var. Penta was classified as a variety with a highly effective APR resistance. A similar conclusion could not be made for var. Ambition, in which the source(s) of seedling resistance has not yet been resolved due to the absence of virulent isolates.

Variety Robigus was initially grouped as 0 based on results in the disease observation plots, 2003. The grouping changed to 1 based on 2004 results in inoculated nurseries (>10% diseased leaf area on average during the epidemic caused by 71/93, a $Yr32$-virulent isolate). If $Yr32$-pathotypes had been present in the natural Danish *P. striiformis* population at that time, the variety would have been grouped as 2. Group 1 indicates 'partial susceptibility', i.e., there is a risk from yellow rust if a corresponding virulence emerges. $Yr32$-pathotypes reappeared in Denmark in 2007 (Table 1). The level of APR was classified as low based on disease severity levels when challenged by virulent isolates (>10% leaf area covered, seasonal average).

In recent years, vars Bill and Blixen were grouped as 2 and 3, respectively, as reflected by moderate to high levels of disease in inoculated nurseries. The APR in these varieties was classified as medium and very low, respectively. A moderate to low frequency of pathotypes virulent on var. Blixen may, in part, explain the annual fluctuations in disease severity on this variety in disease observation plots (0–7% leaf area covered).

The CPO-groupings for a larger set of varieties, which have covered a substantial part of the Danish wheat area in recent years, had changed for most varieties since 2002 (Table 3). The groupings were defined according to the procedures above, and generally based on results in a previous year. However, group changes may occur within the growing season, e.g., if a new pathotype suddenly emerges at a high frequency. In general, the groupings were robust in the sense that yellow rust was never observed to any significant extent on a non-inoculated variety with a 0-grouping. The relative distribution of the varieties, e.g. the low proportion of the highly resistant var. Penta, suggests that the choice of variety is a multifactor

Table 1 Frequency of *Puccinia striiformis* f. sp. *tritici* pathotypes in Denmark, 2000–2007

Number	Pathotype															Pathotype frequency (2000–2007)								First observation (DK)
	1	2	3	4	6	7	8	9	10	15	17	32	Sd	Su	An	2000 n=61	2001 n=38	2002 n=72	2003 n=22	2004 n=12	2005 n=12	2006 n=6	2007 n=51	
I	–	2	3	4	6	–	–	9	–	–	17	–	Sd	Su	An	13	13	4	9	0	0	0	0	1994
II	–	2	3	–	–	–	–	–	–	–	17	–	Sd	–	An	44	34	47	9	0	0	0	0	1997
III	–	2	3	4	–	–	–	9	–	–	17	–	Sd	Su	An	41	42	6	5	0	0	0	4	1997
IV	–	2	3	4	–	7	–	9	–	–	17	–	Sd	Su	An	2	0	0	0	0	0	0	0	2000
V	–	(2)	–	–	–	–	8	9	–	–	–	–	(Sd)	–	An	0	11	8	14	8	0	0	0	2001
VI	–	2	3	–	6	7	–	9	–	–	17	–	Sd	–	An	0	0	29	18	50	42	0	0	2002
VII	–	2	3	–	6	–	–	9	–	15	17	–	Sd	–	An	0	0	6	0	0	0	0	0	2002
VIII	–	*	–	–	–	–	–	–	–	–	*	–	*	–	An	0	0	0	36	17	0	0	0	2003
IX	–	(2)	–	–	6	7	8	–	–	–	–	–	*	–	An	0	0	0	0	17	42	67	2	2004
X	–	–	–	–	6	7	8	–	–	–	–	–	–	–	–	0	0	0	0	0	0	33	0	2006
XI	–	–	–	–	6	–	–	–	–	–	–	–	–	Su	An	0	0	0	0	0	0	0	4	2007
XII	–	–	3	–	6	–	–	–	–	–	–	(32)	(Sd)	Su	An	0	0	0	0	0	0	0	4	2007
XIII	1	2	3	4	6	–	–	9	–	–	17	32	Sd	Su	An	0	0	0	0	0	0	0	78	2007
XIV	–	(2)	–	–	6	7	8	–	10	–	–	–	(Sd)	–	An	0	0	0	0	0	0	0	2	2007
XV	–	(2)	–	–	6	7	8	–	10	–	–	–	*	–	–	0	0	0	0	0	0	0	4	2007
																100	100	100	100	100	100	100	100	

Figures and symbols defining pathotype correspond to yellow rust resistance genes; parenthesis indicate interactions either influenced by unrecognised resistance in test-cultivars or a heterozygous state of pathogen isolates. Asterisk means not accessible based on present test varieties (Hovmøller and Justesen 2007).

n number of isolates

Table 2 Rationale for CPO yellow rust resistance grouping of five varieties based on variety and pathogen characteristics: percent diseased leaf area in non-inoculated observation plots, in inoculated nurseries using specific yellow rust isolates (for virulence profile cf. Table 1), varietal resistance characteristics and frequency of relevant *P. striiformis* pathotypes in Denmark

Variety	Year	CPO resistance grouping in following year	Observation plots (non-inoculated) mean across locations			Inoculated nurseries (mean of three assessment dates)					Resistance genes	Frequency of matching virulence in DK population 2001–2007 (average)	APR[b]
			Yellow rust	Septoria	Powdery mildew	111/02 (VII)	16/02 (VI)	70/99 (I)	08/97 (III)	71/93[a]			
Penta	2001	0	0.0	0.8	0.0			0.0	0.1	0.0	Sd, Yr25, +	84	Very high
	2003	0	0.0	1.5	0.0								
	2005	0	0.0	1.1	0.0								
	2006	0	0.0	1.6	0.0								
Ambition	2004	0	0.0	0.8	0.0			0.0	0.0		–	0	Unknown
	2005	0	0.0	0.9	0.1								
	2006	0	0.0	1.1	0.1								
	2007	0	0.0	2.5	2.9								
Robigus	2003	0	0.0	3.6	0.0	0.0		0.0	0.0	0.0			
	2004	1	0.0	5.0	0.1	0.0	0.0			10.4	Yr2, Yr32	0	Low
	2005	1	0.0	5.0	0.0								
	2006	1	0.0	16.0	0.1								
	2007	2	2.3	5.0	3.2	(1.4)							
Bill	2001	1	0.6	9.0	1.0			0.0	2.8	17.5		78	
	2002	1	1.7	8.0	1.6			0.5	0.0	1.8			
	2003	1	0.1	16.0	1.9			0.1			Yr2, Yr3, Yr17	84	Medium
	2004	2	0.0	9.0	3.8	1.4	5.6	2.7					
	2005	2	0.0	15.0	2.4								
Blixen	2001	2	0.8	6.0	2.6			18.9	0.1	1.5	Yr6, Yr9, +	12	Very low
	2002	2	7.0	8.0	3.8			20.6		(7.3)			
	2003	2	0.0	7.0	2.3								
	2004	3	1.6	5.0	2.5	(4.3)	30.7						
	2005	3	6.0	8.0	4.7								

Parentheses refer to incompatible host-pathogen combinations, thus indicating within-trial mixture of isolates.

[a] Pathotype Vyr1, 2, 3, 32, Sd

[b] Level of resistance of adult plant stage, measured by disease severity when exposed to virulent isolate

Table 3 Dynamics of CPO-groupings (0, 1, 2, 3) for 12 varieties 2001–2008 representing different resistance characteristics, and their relative distribution in Denmark based on amounts of certified seed (*t*), in the considered period

Variety	Race-specific resistance	APR[1]	2002	2003	2004	2005	2006	2007	2008	Av. distribution in DK (in %) based on amounts of certified seed
Ambition	–	–				0	0	0	0	1.0
Baltimor	Yr17, +	Very low	3	3	3	3	3			3.5
Bill	Yr2, Yr3, Yr17	Medium	1	1	1	2	2			8.1
Boston	Yr15	Low	0	1	1					1.0
Cardos	Yr6	Very low	2	2	2	3	3			1.2
Deben	Yr2,Yr32+	–	0	0	0	0	0	0		5.8
Penta	Sd, Yr25, +	Very high	0	0	0	0	0	0		0.1
Ritmo	Yr1	Low	2	2	2	2	2	2		12.8
Robigus	Yr2,Yr32	Low			0	1	1	1	2	6.7
Senat	Yr3,Yr32+	–	0	0	0	0				2.4
Smuggler	Yr1, Yr17, +	–			0	1	1	1	1	12.2
Terra	Yr1	High	0	0	0					0.9

[1] Level of resistance at adult plant stage, measured by disease severity when exposed to virulent isolate

decision, also depending on other characteristics than resistance to yellow rust, e.g., yield, market preferences, resistance to other diseases, etc.

A specific disease management recommendation by CPO depends on a range of information, e.g., disease records in the crop, growth stage, crop rotation, CPO resistance grouping, and to some extent weather forecasts. For simplicity, disease assessments were measured by disease incidence (% infected plants), where the basis for an assessment was dependent on growth stage: (1) GS 29–32, whole plant assessment; (2) GS 32–71, three top leaves of main tiller. Note that disease control by fungicides in wheat was only considered relevant between GS 29 and 71 (Anonymous 2007b). The end user (farmer) is asked to supply four types of data to the CPO-model: (1) variety, (2) crop growth stage, (3) previous records (and fungicide treatment) of disease in the crop, (4) disease incidence of yellow rust, leaf rust, powdery mildew, Septoria leaf blotch and tan spot, respectively. An observation above the threshold results in a fungicide recommendation, depending on product and pesticide prices, crop growth stage, presence of additional diseases (below their threshold value) and weather forecast (Table 4). If treatment was not recommended, the farmer will be recommended to make another assessment after 7 days. If the model recommends fungicide treatment against yellow rust, and the variety is susceptible/very susceptible, the model will recommend the treatment to be repeated after 3 weeks.

Past experiences from numerous field trials in >10 years have documented that CPO, in general, has resulted in the highest yield margin over product use and much reduced fungicide inputs as compared to treatments using the recommended standard doses, or treatments recommended by other European decision support systems (Hagelskjær and Jørgensen 2003). Part of these reductions may be ascribed to the CPO-resistance groupings (Jørgensen et al. 2003) who analysed fungicide inputs in 36 winter wheat field trials in three growing seasons and observed an almost 50% reduction in fungicide input in resistant varieties compared to varieties which were susceptible to Septoria leaf blotch or yellow rust.

Although approximately only 1,000 Danish farmers have bought access to CPO (Jørgensen et al. 2007), there are indications that CPO may have influenced the general recommendations by the

Table 4 Yellow rust thresholds in CPO according to resistance grouping, growth stage and disease incidence

CPO resistance grouping	Disease incidence (% infected plants)	
	GS 29–60	GS 61–71
0	>10	>75
1	>1	>50
2 and 3	>1	>10

Danish extension service for using reduced fungicide doses, compared to label standards, thereby contributing to achieving the goals laid down in the various Danish pesticide action plans since 1986 (Jørgensen and Kudsk 2006). Based on farmer interviews (Jørgensen et al. 2007), the main barriers for a wider use of the system may be lack of time for disease assessments, possibly due to increasing farm and field sizes, increasing focus on animal production as compared to crop production, problems in recognizing diseases, which may remain undetected until too late, and access to relatively cheap routine sprays.

Acknowledgement Access to high quality disease assessments from disease observation plots across multiple locations and years by S. Sindberg and colleagues, The Danish Plant Directorate, Tystofte, is greatly acknowledged, as well as valuable comments from L. N. Jørgensen, Flakkebjerg, on the Danish Pesticide Action Plan and the CPO-threshold values.

References

Anonymous (2007a). The official list of varieties. *The Danish Gazette for Plant Varieties, 26*, 42. Retrieved October 2, 2007 from http://www.pdir.dk/Files/Filer/Virksomheder/Froe/Sortsafprovning/Nyt_Gazette/Sortsliste_2007.pdf.

Anonymous (2007b). *Vejledning i bedømmelser i Landsforsøgene.* Århus: Dansk Landbrugsrådgivning, Landscentret.

Bayles, R. A., Flath, K., Hovmøller, M. S., & Vallavielle-Pope, C. (2000). Break-down of the *Yr17* resistance to yellow rust of wheat in northern Europe. *Agronomie, 20*, 805–811.

Bayles, R. A., Hubbard, A. J., & Slater, S. E. (2004). Yellow rust of wheat. Annual report 2003, UKPVSC, 11–20.

Brown, J. K. M. (2003). Little else but parasites. *Science, 299*, 1680–1681.

Bruil, A. A. M., Verhaar, M. A., Vandenbosch, F., Hoogkamer, W., & Zadoks, J. C. (1989). Effect of cultivar mixtures on the wave velocity of expanding yellow stripe rust foci in winter wheat. *Netherlands Journal of Agricultural Science, 37*(1), 75–78.

Chen, X. M. (2005). Epidemiology and control of stripe rust [*Puccinia striiformis* f. sp. *tritici*] on wheat. *Canadian Journal of Plant Pathology, 27*, 314–337.

Christiansen, M. J., Feenstra, B., Skovgaard, I. M., & Andersen, S. B. (2006). Genetic analysis of resistance to yellow rust in hexaploid wheat using a mixture model for multiple crosses. *Theoretical and Applied Genetics, 112*, 581–591.

Enjalbert, J., Duan, X., Leconte, M., Hovmøller, M. S., & de Vallavieille-Pope, C. (2005). Genetic evidence of local adaptation of wheat yellow rust (*Puccinia striiformis* f. sp. *tritici*) within France. *Molecular Ecology, 14*, 2065–2073.

Eriksson, J., & Henning, E. (1896). *Die Getreidenroste.* Stockholm: Norstedt & Söner.

Flor, H. (1971). Current status of the gene-for-gene concept. *Annual Review of Phytopathology, 9*, 275–296.

Hagelskjær, L., & Jørgensen, L. N. (2003). A web-based decision support system for integrated management of cereal pests. *EPPO Bulletin, 33*, 467–471.

Hermansen, J. E. (1968). Studies on the spread and survival of cereal rust and mildew diseases in Denmark. Dr Science thesis, The Royal Veterinary and Agricultural College Copenhagen. *Fresia, 8*(3), 1–206.

Hovmøller, M. S. (2001). Disease severity and pathotype dynamics of *Puccinia striiformis* f. sp. *tritici* in Denmark. *Plant Pathology, 50*, 181–189.

Hovmøller, M. S. (2007). Sources of seedling and adult plant resistance to *P. striiformis* f. sp. *tritici* in European wheats. *Plant Breeding, 126*, 225–233.

Hovmøller, M. S., & Justesen, A. F. (2007). Appearance of atypical *Puccinia striiformis* f. sp. *tritici* phenotypes in north-western Europe. *Australian Journal of Agricultural Sciences, 58*, 518–524.

Johnson, R. (1984). A critical analysis of durable resistance. *Annual Review of Phytopathology, 22*, 309–330.

Johnson, R. (1992). Past, present and future opportunities in breeding for disease resistance, with examples from wheat. *Euphytica, 63*, 3–22.

Johnson, R., Stubbs, R. W., Fuchs, E., & Chamberlain, N. H. (1972). Nomenclature for physiologic races of *Puccinia striiformis* infecting wheat. *Transactions of the British Mycological Society, 58*, 475–480.

Jørgensen, L. N. (Ed.) (2006). Pesticidafprøvning 2006. DJF-rapport Markbrug 129, 136 pp.

Jørgensen, L. N., Hagelskjær, L., & Nielsen, G. C. (2003). Adjusting the fungicide input in winter wheat depending on variety resistance. *Proceedings of BCPC Conference on Crop Science and Technology* (pp. 1115–1120). Glasgow, UK.

Jørgensen, L. N., & Kudsk P. (2006). Twenty years' experience with reduced agrochemical inputs: Effect on farm economics, water quality, biodiversity and environment. (Paper presented at the HGCA conference, Grantham UK).

Jørgensen, L. N., Noe, E., Langvad, A. M., Jensen, J. E., Ørum, J. E., & Rydahl, P. (2007). Decision support systems: Barriers and farmers' need for support. *EPPO Bulletin, 37*, 374–377.

Manners, J. G. (1988). *Puccinia striiformis*, yellow rust (stripe rust) of cereals and grasses. *Advanced Plant Pathology, 6*, 373–387.

McNeal, F. H., Konzak, C. F., Smith, E. P., Tate, W., & Russel, T. S. (1971). A uniform system for recording and processing cereal research data. *US Agricultural Research Service, 42*, 34–121.

Pinnschmidt, H., Hovmøller, M. S., & Østergård, H. (2006). Approaches for field assessment of resistance to leaf pathogens in spring barley varieties. *Plant Breeding, 125*(2), 105–113.

Priestley, R. H., Bayles, R. A., & Thomas, J. E. (1984). Identification of specific resistances against *Puccinia striiformis* (yellow rut) in winter wheat varieties. I. Establishment of a set of type varieties for adult plant tests. *Journal of the National Institute of Agricultural Botany, 16*, 469–476.

Schmidt, K. (2003). Ergebnisse der Meldungen für Pflanzenschutzmittel und Wirkstoffe nach § 19 des Pflanzenschutzgesetzes für die Jahre 1999, 2000 und 2001 in Vergleich zu 1998. *Nachrichtenblatt des Deutschen Pflanzenschutzdienstes, 55*, 121–133.

Stubbs, R. W. (1985). Stripe rust. In A. P. Roelfs, & W. R. Bushnell (Eds.) *The cereal rusts volume 2: Diseases, distribution, epidemiology and control* (pp. 61–101). London: Academic .

Wellings, C. R. (2007). *Puccinia striiformis* in Australia: A review of the incursion, evolution, and adaption of stripe rust in the period 1979–2006. *Australian Journal of Agricultural Research, 58*, 567–575.

Zadoks, J. C. (1961). Yellow rust on wheat: Studies in epidemiology and physiologic specialisation. *Tijdschrift over Planteziekten, 67*, 69–256.

Eur J Plant Pathol (2008) 121:387–397
DOI 10.1007/s10658-007-9252-3

Molecular approaches for characterization and use of natural disease resistance in wheat

Navreet Kaur · Kenneth Street · Michael Mackay ·
Nabila Yahiaoui · Beat Keller

Received: 20 June 2007 / Accepted: 5 November 2007
© KNPV 2007

Abstract Wheat production is threatened by a constantly changing population of pathogen species and races. Given the rapid ability of many pathogens to overcome genetic resistance, the identification and practical implementation of new sources of resistance is essential. Landraces and wild relatives of wheat have played an important role as genetic resources for the improvement of disease resistance. The use of molecular approaches, particularly molecular markers, has allowed better characterization of the genetic diversity in wheat germplasm. In addition, the molecular cloning of major resistance (R) genes has recently been achieved in the large, polyploid wheat genome. For the first time this allows the study and analysis of the genetic variability of wheat R loci at the molecular level and therefore, to screen for allelic variation at such loci in the gene pool. Thus, strategies such as allele mining and ecotilling are now possible for characterization of wheat disease resistance. Here, we discuss the approaches, resources and potential tools to characterize and utilize the naturally occurring resistance diversity in wheat. We also report a first step in allele mining, where we characterize the occurrence of known resistance alleles at the wheat $Pm3$ powdery mildew resistance locus in a set of 1,320 landraces assembled on the basis of eco-geographical criteria. From known $Pm3$ R alleles, only $Pm3b$ was frequently identified (3% of the tested accessions). In the same set of landraces, we found a high frequency of a $Pm3$ haplotype carrying a susceptible allele of $Pm3$. This analysis allowed the identification of a set of resistant lines where new potentially functional alleles would be present. Newly identified resistance alleles will enrich the genetic basis of resistance in breeding programmes and contribute to wheat improvement.

Keywords Allele mining · Genetic diversity · $Pm3$ · Wheat powdery mildew

N. Kaur · N. Yahiaoui · B. Keller (✉)
Institute of Plant Biology, University of Zurich,
Zollikerstrasse 107,
8008 Zürich, Switzerland
e-mail: bkeller@botinst.uzh.ch

K. Street
ICARDA,
P.O. Box 5466, Aleppo, Syria

M. Mackay
Australian Winter Cereals Collection,
4 Marsden Park Road,
Calala, NSW 2340, Australia

Present address:
N. Yahiaoui
UMR Biologie et Génétique des Interactions
Plante-Parasite CIRAD TA A-54/K
Campus International de Baillarguet,
34398 Montpellier cedex 15, France

Introduction

Wheat is globally one of the three most important food crops, the other two being maize and rice. Wheat diseases cause severe yield losses and often reduce grain quality. Some of the most important fungal diseases of wheat include three rust species (stripe, leaf and stem rust), powdery mildew, fusarium, septoria, mycosphaerella and tan spot. To achieve sufficiently high resistance to fungal pathogens is an ongoing challenge for wheat breeding. Various aspects such as understanding pathogen biology, characterization of pathogen avirulence and plant disease resistance genes, and finally the search for new resistance sources, all contribute to the development of wheat with increased resistance to various diseases.

Wheat (*Triticum aestivum*) is an allopolyploid species featuring three distinct homoeologous genomes A, B and D. As the wheat genome is large (16×10^9 bp) and the major fraction consists of repetitive sequences, the molecular cloning of genes, for which only genetic information is available, remains a challenge (Keller et al. 2005). Therefore, the first efforts to characterize loci of disease resistance at the molecular level concentrated on the development of molecular markers linked to these important traits.

The availability of molecular markers linked to specific resistance genes and of information on their genetic location has supported resistance breeding by simplifying the detection of specific genes in breeding material. This makes the selection process faster and more cost effective. In addition, different genes can be combined in a pyramiding strategy for resistance breeding. Despite these efforts, the genetic base of disease resistance in wheat remains dangerously narrow and adaptation of pathogens is always a threat, challenging the resistance of existing elite material. Currently, the emergence and spread of the new virulent stem rust race *Ug99* is considered to be a potential threat to wheat production worldwide. The *Ug99* race has overcome the major stem rust resistance gene *Sr31* (http://www.ars.usda.gov/Main/docs.htm?docid=14649). Previously, there was no report of virulence against *Sr31*, a gene which is widely used in India, China, Europe and South America. This makes it imperative to identify new sources of resistance which are thought to exist in germplasm collections and then to make combinations with existing sources to develop more durable types of resistance.

The introduction of resistance genes from landraces, traditional varieties and wild relatives, e.g. the diploid and tetraploid progenitors of hexaploid wheat, has been successful in broadening resistance (Miranda et al. 2006; Liu et al. 2002; Rong et al. 2000). Over thousands of years, landraces of hexaploid wheat have developed under a variety of different edaphic and climatic environments. This has resulted in the evolution of a large number of ecotypes adapted to specific local environments. Thus, the genetic collections available in gene banks are expected to provide a rich resource to identify new functional genes or alleles of resistance genes. The molecular changes underlying this adaptation are mostly unknown and this diversity is largely unused and uncharacterized at the molecular level. Traditionally, resistance genes in wild relatives of wheat have been introgressed by complex breeding schemes involving irradiation and chromosomal translocations (Baum et al. 1992). This has resulted in the introgression of large chromosomal segments, often carrying negative breeding traits (linkage drag). The molecular isolation of the underlying genes and their use through transgenic technologies will contribute to an efficient future use of resistance genes from wild grasses.

Hybridization and introgression of chromosomal segments, marker-assisted selection and the breeding of synthetic hexaploid wheat (Zhang et al. 2005) are well established methods for broadening the genetic diversity of disease resistance in wheat. Still largely unexplored, a new approach of 'allele mining' has recently become available and shows promise for the more efficient use of genetic diversity (to be discussed in detail later in this paper). The first wheat disease resistance genes have been cloned at the molecular level (Huang et al. 2003; Feuillet et al. 2003; Yahiaoui et al. 2004; Srichumpa et al. 2005; Yahiaoui et al. 2006; Cloutier et al. 2007). The sequence information of these cloned genes facilitates the rapid analysis of the genetic diversity at these loci over a wide range of germplasm and the subsequent identification of new alleles through allele mining. Molecular tools that specifically access the existing genetic diversity at a particular locus provide a promising approach for utilising the diversity maintained in the gene banks globally. In this paper, we discuss the molecular approaches available for detecting and using genetic diversity for improving disease resistance in wheat. We also describe an allele mining strategy for new resistance specificities at the wheat resistance locus *Pm3* against the powdery mildew pathogen (*Blumeria*

graminis f.sp. *tritici*) applied to 1,320 wheat landraces from different geographic origin.

Molecular markers in the characterization of wheat disease resistance diversity

Molecular markers play a significant role in the process of identification and introgression of natural resistance into economically important but susceptible breeding material. The range of molecular markers used for this purpose includes restriction fragment length polymorphism (RFLP), randomly amplified polymorphic DNA (RAPD), amplified fragment length polymorphism (AFLP), sequence tagged sites (STS), simple sequence repeats (SSR)/microsatellites, expressed sequence tags (ESTs) and RGAs (resistance gene analogues). RFLPs were the first molecular markers to be used in wheat in the early 1990s. There are several reports on the essential role of RFLPs in marker-assisted selection, genome mapping as well as characterization and isolation of various disease resistance genes in wheat (Lagudah et al. 2006; Feuillet et al. 2003). Because of lower costs and time requirements, PCR-based markers have a higher potential for applications in the genetic characterization of wheat germplasm than RFLPs. AFLPs combine the simplicity of the RAPDs and robustness of RFLPs. Therefore, they have been frequently used to investigate biodiversity in several crops. Diagnostic markers for different resistance genes in wheat have been developed by using the AFLP technique (William et al. 2003; Adhikari et al. 2003).

STS markers have been highly useful tools for the screening of natural genetic variation as well as for tagging of resistance genes and QTL in wheat (Lagudah et al. 2006; Tyryshkin et al. 2006; Liu et al. 2002). Given the vast molecular information available on various plant disease resistance genes, RGAs have been extensively used. For instance, RGA markers linked to resistance genes *Yr5*, *Pm21* and *Pm31* (Yan et al. 2003; Chen et al. 2006; Xie et al. 2004) have been identified. As RGA sequences are very common in the plant genomes, they can be considered as a type of random marker.

SSR/microsatellite markers are highly polymorphic in wheat and are now widely used in wheat genetics for diversity studies, genotype identification (Prasad et al. 2000), marker-assisted selection, mapping,

identification and tagging of disease resistance genes (Chagué et al. 1999; Miranda et al. 2006; Liu et al. 2002; Adhikari et al. 2003). Recently, microsatellite markers have been found linked to leaf rust resistance gene *Lr34* (Bossolini et al. 2006) supporting the identification of wheat genotypes with *Lr34*. With the increasing availability of molecular maps based on SSR markers, the identification and cloning of important genes is predicted to become more straightforward. Multiallelism, chromosome-specificity, even distribution in the genome and the possibility of high-throughput fingerprinting of large numbers of accessions are the properties that make SSR markers a good choice for detection of genetic polymorphism and diversity among wheat lines. However, microsatellite markers are often not suitable to define homoeologous loci, thus limiting their use in intraspecific and intragenomic studies (Gupta et al. 1999). This also complicates the use of SSR markers for introgression studies involving wild relatives of wheat. Table 1 gives a summary of some molecular markers currently used for tagging different disease resistance genes in wheat.

Map-based cloning of disease resistance genes in wheat

To design precise and targeted molecular tools for diversity analysis based on allele mining, knowledge of the DNA sequence at a particular resistance locus becomes important, requiring the molecular cloning of genes. DNA sequence information is essential to devise rapid and inexpensive PCR strategies to isolate alleles of identified resistance genes from a wide range of cultivars, landraces and related species.

In map-based cloning, saturation of the genomic region of interest is greatly supported by genomic information and markers obtained from the grass model species rice and brachypodium (Griffiths et al. 2006). In addition, there are now a number of BAC libraries available from diploid, tetraploid and hexaploid wheat species (Keller et al. 2005) supporting map-based cloning strategies of wheat genes. Until now, three leaf rust resistance genes (*Lr*) and one allelic series of a powdery mildew (*Pm*) resistance gene have been cloned from wheat: *Lr21* (Huang et al. 2003), *Lr10* (Feuillet et al. 2003), *Lr1* (Cloutier et al. 2007) and the *Pm3* alleles (Yahiaoui et al. 2004; Srichumpa et al. 2005; Yahiaoui et al. 2006).

Table 1 Examples of molecular markers used for characterization and genetic mapping of different disease resistance genes in wheat

Disease	Resistance gene/genomic region	Marker used for identification and/or mapping	Reference
Leaf rust, yellow rust and stem rust	*Lr34/Yr18*	SSR, STS	Bossolini et al. (2006), Lagudah et al. (2006)
	Lr9, Lr19 and *Lr24*	STS	Tyryshkin et al. (2006)
	Lr46	AFLP	William et al. (2003)
	Lr47	RFLP/CAPS	Helguera et al. (2000)
	Yr15	SSR	Chagué et al. (1999)
	Yr5	RGA	Yan et al. (2003)
	Sr30	AFLP, RFLP	Bariana et al. (2001)
	Sr36	SSR	Bariana et al. (2001)
Powdery mildew	*Pm3*	STS	Tommasini et al. (2006)
	Pm21	RGA	Chen et al., (2006)
	Pm26	RFLP	Rong et al. (2000)
	Pm30	SSR	Liu et al. (2002)
	Pm31	RGA	Xie et al. (2004)
	Pm34	SSRs	Miranda et al. (2006)
Septoria leaf blotch	*Stb8*	AFLP and SSRs	Adhikari et al. (2003)

The first cloned wheat disease resistance gene, *Lr21*, was incorporated into bread wheat cv. Thatcher from the diploid wheat ancestor *Ae. tauschii*. A diploid/polyploid shuttle mapping strategy was deployed for map-based cloning of *Lr21* (Huang et al. 2003). There, the genetic analysis was done in hexaploid wheat but the large insert cosmid library was developed from the diploid donor. *Lr21* was chosen for cloning because of its location in a gene rich region and the extensive allelic diversity at this locus in natural populations of *Ae. tauschii* (Huang and Gill 2001).

Feuillet et al. (2003) isolated the leaf rust resistance gene *Lr10*, located on chromosome 1AS in hexaploid wheat by combining subgenome map-based cloning (Stein et al. 2000) and haplotype studies in the genus *Triticum*. The chromosome walking was performed on BAC clones of the diploid wheat *T. monococcum* DV92 (A genome) which had an *Lr10* haplotype, while the genetic map was constructed on the basis of genetic data from a hexaploid wheat population segregating for the *Lr10* resistance.

The wheat powdery mildew resistance gene *Pm3*, a dominant gene on chromosome 1AS, exists in ten different alleles (*Pm3a–Pm3j*) as identified by classical genetic studies. These alleles are predicted to confer resistance to specific races or isolates of the powdery mildew pathogen. Yahiaoui et al. (2004)

used the combined analysis of genomes from wheat species with different ploidy levels for positional cloning of the *Pm3b* allele of *Pm3* in bread wheat. This represented the first molecular isolation of a powdery mildew resistance gene from wheat and a breakthrough for further analysis of diversity and evolution of this important locus. Based on the identification of a specific *Pm3* haplotype and using molecular markers derived from the *Pm3b* locus, additional known *Pm3* alleles (*Pm3a, Pm3b, Pm3c, Pm3d, Pm3e, Pm3f, Pm3g*) were isolated from different wheat lines (Srichumpa et al. 2005; Yahiaoui et al. 2006). Interestingly, it was also found that the three alleles *Pm3h, Pm3i, Pm3j* are actually identical to *Pm3d, Pm3c* and *Pm3b* respectively (Yahiaoui et al. 2006), suggesting that the lines in which the *h* to *j* alleles were identified contained additional resistance genes.

Quantitative resistance is often assumed to be more durable than single-gene resistance. Therefore, the improvement of quantitative resistance through tagging and cloning of QTL is of increasing importance in several wheat research and breeding programmes. A few QTL have been cloned in plants mainly by positional cloning (Salvi and Tuberosa 2005). In wheat, no QTL for disease resistance have yet been cloned but there are many ongoing projects with this goal. These include projects aimed at the isolation of rust resistance

loci such as *Lr34/Yr18* (Bossolini et al. 2006; Lagudah et al. 2006; Spielmeyer et al. 2005) and *Lr46/Yr29* (William et al. 2003; Rosewarne et al. 2006), *Sr2* (Kota et al. 2006) and a major QTL for Fusarium head blight (FHB) resistance (Liu and Anderson 2003). Mardi et al. (2005) reported the tagging of QTL responsible for FHB resistance with SSR markers and suggested that the SSR markers linked to the QTL would facilitate marker-assisted selection for FHB resistance in wheat. QTL tagging and cloning provide tools to the breeder for marker-assisted selection of complex disease resistance traits. It should also help to understand the respective roles of specific resistance loci versus partial resistance genes and the interactions between the genes and the environment.

Genetic resources for improvement of wheat disease resistance

A big advantage of diversity studies in wheat, compared to model plants such as *Arabidopsis*, is the existence of large collections of wild and cultivated diploid, tetraploid and hexaploid species secured in gene banks. However, at the molecular level this diversity remains largely unexplored due to a lack of fast and efficient tools to identify and study potentially useful new alleles. In addition to wheat landraces, wild relatives of wheat have been always explored and exploited as sources of new resistance genes. For example, a number of *R* genes originate from wild wheat relatives: the stem rust resistance gene *Sr39* was transferred from the wild relative *Aegilops speltoides* to bread wheat cv. Thatcher, leaf rust resistance gene *Lr24* from *Agropyron elongatum*, *Lr47* and *Pm32* from *Aegilops speltoides*, *Pm6* from *T. timopheevi* (Allard and Shands 1954), *Pm26* and *Pm30* from *T. turgidum* var. *dicoccoides* and *Pm34* from *A. tauschii* (Miranda et al. 2006; Liu et al. 2002; Rong et al. 2000). Common wheat has also been genetically improved for many decades through the introgression of rye chromatin. The rye chromosome arm 1RS is the most widely incorporated alien variation in the wheat genome. To give an example, wheat cv. Amigo carries the powdery mildew resistance gene *Pm17* on its introgressed 1RS chromosome arm (Forsström and Merker 2001).

As each of these *R* genes usually act only against a subset of the existing pathogen races, combinations of genes as well as the identification of new resistance genes/alleles are essential. Classically, identification of new resistance genes or of new alleles at already known loci is achieved by infection experiments on landraces or wild relatives of wheat followed by crosses necessary to determine if the resistance is due to a single gene and if the gene is a new allele at a known locus or represents a new locus. As resistance may be lost with rapid emergence of new pathogen strains, it becomes a continuous task to identify new resistance genes and to transfer these genes into common wheat if they are present in wild relatives or related species.

Molecular tools for screening the diverse germplasm: allele mining in cereals

Tools to access the existing genetic diversity at specific loci facilitate the rapid analysis of allelic diversity in the gene pool of wheat and its relatives. This in turn allows the molecular isolation of new alleles with potential agronomical relevance and a more efficient and targeted use of genetic resources for research and breeding. The strategy of finding valuable, unknown alleles at a known locus is referred to as 'allele mining.' In allele mining, the sequence of a target gene is used to develop specific markers to amplify, isolate and sequence new alleles at that particular locus. It seems to be a promising, although largely untested method to unlock the diversity in the collections of genetic resources in the world genebanks.

There are reports about the allele mining strategy in several cereal species to isolate alleles of target genes. In barley, an evaluation of cultivated germplasm was carried out to detect the presence of thermostable alleles of β-amylase (*Bmy-Sd2H* and *Bmy-Sd3* alleles) that improve the fermentability during brewing (Malysheva et al. 2004). The study was carried out on 891 accessions originating from different geographic regions worldwide. This led to the identification of 166 accessions with superior alleles, suggesting that the improvement of malting quality in barley could be achieved by introducing these alleles into breeding programmes. Latha et al. (2004) used the rice calmodulin genes and a salt-inducible rice gene for allele mining of stress tolerance genes on identified accessions of rice and related germplasm. They examined the feasibility of

allele mining using PCR primers based on the 5'- and 3'-untranslated regions of genes and found that these primers were sufficiently conserved to be effective over the entire range of germplasm in rice. The new HMW-glutenin alleles encoded by the *Glu-R1* locus of *Secale cereale* (rye) have been analysed and characterized (De Bustos and Jouve 2003) from different rye cultivars and their most closely related wild subspecies. Primers designed from a nucleotide sequence of the allele *Glu-Dly10*, which recognised the upstream and downstream flanking positions of the coding regions of the genes, were used in the study. Thus, allele mining supports the discovery of new alleles of target genes. However, the limitation of this approach in wheat lies in the fact that very few genes of agronomical importance have yet been cloned. This is particularly true for genes involved in disease resistance.

Ecotilling (Comai et al. 2004) represents a specific approach to allele-mining and refers to a high-throughput screening technique for the discovery of polymorphisms in natural populations. It can serve as a cheaper alternative to full DNA sequencing when searching for rare polymorphisms, but similar to the other allele-mining strategies it still requires specific sequence information for the target gene. Ecotilling can also be used for mapping, association analysis, mutational profiling and biodiversity studies. It has been successfully used in *Arabidopsis* (Comai et al. 2004) where 55 haplotypes of five genes have been discovered after screening of more than 150 individuals. The polymorphisms discovered were confirmed by sequencing, and base pair changes, insertions, deletions and variation in microsatellite number were observed.

Focused identification of germplasm strategy (FIGS) and allele-mining for molecular diversity at the *Pm3* locus

To test a strategy of allele mining in wheat using a large set of diverse germplasm, we focused on the *Pm3* resistance locus as there is extensive sequence information available for targeted allele cloning. A subset of bread wheat landraces were selected for the study using the FIGS system (Mackay, Street et al. (unpublished). Also see www.figstraitmine.com). In this case, the eco-geographic profile of 400 acces-

sions, from the USDA-ARS National Small-Grains Collection, with known powdery mildew resistance was identified. This profile was then used as a template to identify environmentally similar collection sites from the FIGS database of nearly 17,000 landraces. Individual accessions were selected using multivariate statistical procedures that determined how eco-geographically similar the collection site of a given accession was to the resistant set template. The FIGS powdery mildew set of accessions now includes 899 landraces from ICARDA (International Centre for Agricultural Research in the Dry Areas, Syria), 295 landraces from AWCC (Australian Winter Cereals Collection) and 126 landraces from VIR (N.I. Vavilov Research Institute of Plant Industry, Russia), making a total of 1,320 landraces. These originate from Turkey (419), Iran (391), Afghanistan (292), Pakistan (133), Armenia (34), Turkmenistan (16), Russia (9), India (6), Azerbaijan (1) and Uzbekistan (1).

Screening and identification of powdery mildew resistant lines

For characterization of the FIGS powdery mildew set we used a combined strategy of screening for genetic diversity with molecular markers and classical pathogenicity tests. The entire FIGS powdery mildew set was screened with a differential set of powdery mildew isolates to select a subset of resistant landraces for molecular analysis. The detached leaf segments from seven day-old plants were placed on phytagar media and subjected to infection with four different isolates of powdery mildew (Fig. 1). The choice of the isolates was based on the pattern of their avirulence/virulence to the known alleles of *Pm3*. The four isolates used were 96224, 98275 and 96244 (avirulent on most known *Pm3* alleles) and 2000.15.Syros (virulent on all the known *Pm3* alleles). The phenotypes were grouped in three categories: resistant (R), intermediate (I) and susceptible lines (S). The experiment was scored 9–10 days after inoculation using a 1 to 100% susceptibility scale, i.e. the leaf area covered with mildew was ranked phenotypically where lines with 100% leaf area covered with mildew were considered fully susceptible and 0% marked complete resistance. This screening led to the selection of 211 resistant or intermediate resistant lines to at least one of the four mildew isolates used in the screen.

Fig. 1 Phenotypic assay of wheat landraces for powdery mildew resistance by infection with powdery mildew isolate 96244. Powdery mildew resistant (R), Intermediate (I) and Susceptible leaves (S) are marked

PCR-based approach for characterization of *Pm3* alleles

Initially we tested the molecular tools available for the detection of the *Pm3* gene in a subset of 295 AWCC landraces. We used an STS marker obtained from haplotype studies at the *Pm3* locus (Yahiaoui et al. 2004; Srichumpa et al. 2005). This *Pm3* haplotype marker amplifies a 946 bp fragment originating from the 5′ non-coding region of *Pm3b* which is diagnostic for the presence of a *Pm3* gene (Fig. 2).

The *Pm3* haplotype was present at an unexpectedly high frequency in the subset of the FIGS powdery mildew set tested. In the 295 AWCC landraces, amplification of the *Pm3* STS marker was found in 257 lines (87.1%). This high percentage prompted us to check this subset for the presence of the already known alleles (*Pm3a–Pm3g*) using *Pm3* allele-specific markers. These markers were developed in our laboratory (Tommasini et al. 2006) based on the specific nucleotide polymorphisms of coding and adjacent non-coding regions of each of the *Pm3* alleles. We found that the *Pm3b* allele was the only known functional *Pm3* allele present in the subset. It was detected in seven lines. This demonstrated that most of the alleles of *Pm3* in the subset do not correspond to known resistance alleles. The infection data obtained from the powdery mildew infection described above showed that only 40 out of 295 lines were resistant or intermediately resistant to at least one of the isolates while the

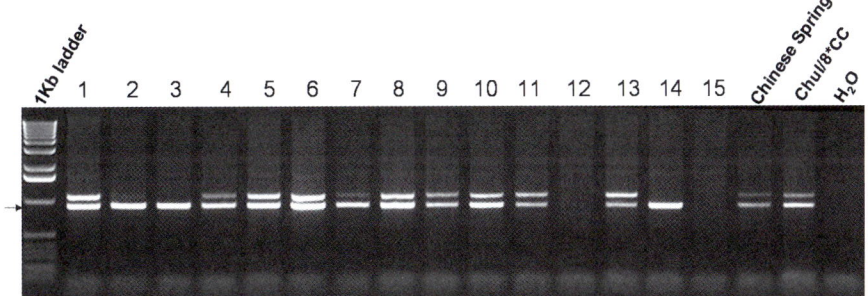

Fig. 2 PCR amplification in the FIGS powdery mildew set of the STS marker specific for the *Pm3* haplotype. The lower band (*see arrow*) corresponding to chromosome 1A is diagnostic for the presence of a *Pm3* like gene. The numbers 1–15 correspond to the tested landraces, Chinese Spring & Chul/8*CC are the positive controls while H$_2$O serves as the negative control

Table 2 *Pm3* allele mining: the frequency of known *Pm3* resistance alleles found in the FIGS powdery mildew set of 1,320 lines

Specific PCR	Number of lines tested for *Pm3* haplotype and known *Pm3* alleles	Number of lines possessing *Pm3* haplotype and *Pm3* known alleles	Landrace	Origin
Pm3 haplotype	211	145		
Pm3a, Pm3d, Pm3e, Pmf, Pm3g	145	0		
Pm3b	145	15	AUS9943, AUS9948, AUS10003, AUS10033, AUS13239, AUS13297, AUS13306, AUS13307, AUS13311, AUS14504, AUS14532, AUS14840, VIR 45538, VIR 49005, VIR 49006	Afghanistan
		6	IG 122348, IG 122354, IG 122361, IG 122373, IG 122502, VIR 38613	Iran
		2	VIR 16766, VIR 31595	Azerbaijan
		6	VIR 23918, VIR 23922, VIR 34986, VIR 35021, VIR 35030, VIR 34984	Russia
		1	VIR 35203	Turkey
Pm3c	145	3	IG 122491, IG 122372, IG 122346	Iran
		1	VIR 46301	Azerbaijan

Detection was done with the *Pm3* haplotype-specific and *Pm3* allele-specific primers

other 255 lines were susceptible to the tested isolates. This indicated that susceptible alleles of *Pm3* were present in at least 86.4% of the lines and are therefore expected to be widespread among the landraces. This percentage is possibly even higher, given the fact that resistance to powdery mildew might not be due to a gene at the *Pm3* locus but may be caused by any of the known or still uncharacterized resistance genes in the germplasm. Therefore, in the particular case of *Pm3* allele mining, the strategy of screening the lines with different mildew isolates prior to sequencing was

chosen. However, for other genes and traits sequencing the complete set of germplasm without prior phenotypic analysis might be a valid alternative strategy.

The 211 intermediate or resistant lines selected during the infection screen were subjected to molecular analysis for the *Pm3* locus. Out of the 211 lines (from 1,320), 145 showed the presence of a *Pm3* haplotype. The search for the seven known *Pm3* resistance alleles in the 145 lines revealed the presence of *Pm3b* and *Pm3c* in 30 and four lines, respectively. Thus, *Pm3b* was the most frequent *Pm3*

Fig. 3 Geographic origin of the 30 *Pm3b* lines detected in the FIGS powdery mildew set of 1320 land-races. The collection sites are indicated by *red triangles*

allele in the landrace set. It was identified in landraces originating from Afghanistan (15), Iran (6), Russia (6), Azerbaijan (2) and Turkey (1), while the four landraces with *Pm3c* allele originated from Iran (3) and Azerbaijan (1) (see Table 2 for a summary of results and Fig. 3 for geographic distribution of *Pm3b* lines). The first identification of the *Pm3b* allele was in a landrace from Uzbekistan (http://www.ars-grin.gov/npgs/index.html), which is consistent with its frequency and actual geographical distribution, particularly in Afghanistan, a neighbouring country to Uzbekistan (Fig. 3).

To summarize, the strategy for the identification of new *Pm3* alleles described in this paper included phenotypic screening of the FIGS powdery mildew set of landraces for powdery mildew resistance, molecular analysis of *Pm3* haplotype composition and determination of known *Pm3* resistance alleles. This resulted in 111 candidate lines (9% of total set) to specifically target for further characterization of the gene present at the *Pm3* locus. These candidate lines (a) are resistant or intermediately resistant to at least one of the isolates tested (b) possess the *Pm3* haplotype and (c) lack any of the known *Pm3* alleles (145 − 34 = 111). It is predicted that new *Pm3* resistance alleles will be found among these lines, although the presence of susceptible *Pm3* alleles cannot be ruled out (based on the results of AWCC subset, presented earlier in this paper). The resistant phenotype in such cases may be attributed to the presence of other *Pm* genes.

The approach described above is one of the first large-scale attempts of a systematic resistance allele-mining from wheat landraces using molecular tools derived from the target gene sequence. Putative new *Pm3* resistance genes will be tested functionally by using a transient transformation assay (Yahiaoui et al. 2004) and other approaches such as virus-induced gene silencing (VIGS, Scofield et al. 2005). This should reveal whether the newly isolated genes are actually active resistance alleles and whether they confer new resistance specificities to the plant. Finally, the newly identified *Pm3* alleles could be transferred by classical genetic crosses to powdery mildew-susceptible cultivars or alternatively be transformed into susceptible varieties as single genes. In addition, they could be combined as *R* gene cassettes to achieve a disease control which is possibly more durable. Besides these more applied aspects in wheat breeding, the analysis of allelic diversity and accu-

mulation of diverse allelic sequences will contribute to a better characterization of the mechanisms involved in resistance gene evolution.

The identification of new functional *Pm3* alleles from diverse germplasm will also contribute to the molecular understanding of *R* gene function. The comparison of sequences from new alleles can clarify the molecular basis of *Pm3* specificity, e.g. by studying chimeric genes created by domain swap experiments with domains from the newly identified sequences.

Concluding remarks

A more efficient exploitation of the genetic diversity in gene banks is essential for meeting the challenges that wheat breeding is facing in the decades to come. However, the use of this diversity is hampered by the sheer number of accessions available and the limited resources which are at hand for phenotypic characterization of all these lines. Therefore, it is necessary to (a) develop strategies to assemble focused sets of material for specific traits based on rational criteria for selection of the lines but also (b) to identify genes underlying agronomically important traits and (c) establish the molecular tools for rapid characterization of new alleles.

Acknowledgements The reported research was supported by the Swiss National Science Foundation grant 3100-105620.We also gratefully acknowledge the gene banks of AWCC (Australia), ICARDA (Syria) and VIR (Russia) for providing us with the FIGS powdery mildew set of landraces.

References

Adhikari, T. B., Anderson, J. M., & Goodwin, S. B. (2003). Identification and molecular mapping of a gene in wheat conferring resistance to *Mycosphaerella graminicola*. *Phytopathology, 93*, 1158–1164.

Allard, R. W., & Shands, R. G. (1954). Inheritance of resistance to stem rust and powdery mildew in cytologically stable spring wheat derived from *Triticum timopheevi*. *Phytopathology, 44*, 266–274.

Bariana, H. S., Hayden, M. J., Ahmed, N. U., Bell, J. A., Sharp, P. J., & McIntosh, R. A. (2001). Mapping of durable adult plant and seedling resistances to stripe rust and stem rust diseases in wheat. *Australian Journal of Agricultural Research, 52*, 1247–1255.

Baum, M., Lagudah, E. S., & Appels, R. (1992). Wide crosses in cereals. *Annual Review of Plant Physiology and Plant Molecular Biology, 43*, 117–143.

Bossolini, E., Krattinger, S. G., & Keller, B. (2006). Development of simple sequence repeat markers specific for the *Lr34* resistance region of wheat using sequence information from rice and *Aegilops tauschii*. *Theoretical and Applied Genetics, 113*, 1049–1062.

Chagué, V., Fahima, T., Dahan, A., Sun, G. L., Korol, A. B., Ronin, Y. I., et al. (1999). Isolation of microsatellite and RAPD markers flanking Yr15 gene of wheat using NILs and bulked segregant analysis. *Genome, 42*, 1050–1056.

Chen, Y. P., Wang, H. Z., Cao, A. Z., Wang, C. M., & Chen, P. D. (2006). Cloning of a resistance gene analog from wheat and development of a codominant PCR marker for *Pm21*. *Journal of Integrative Plant Biology, 48*, 715–721.

Cloutier, S., McCallum, B. D., Loutre, C., Banks, T. W., Wicker, T., Feuillet, C., et al. (2007). Leaf rust resistance gene *Lr1*, isolated from bread wheat (*Triticum aestivum* L.) is a member of the large psr567 gene family. *Plant Molecular Biology, 65*, 93–106.

Comai, L., Young, K., Till, B. J., Reynolds, S. H., Greene, E. A., Codomo, C. A., et al. (2004). Efficient discovery of DNA polymorphisms in natural populations by Ecotilling. *The Plant Journal, 37*, 778–786.

De Bustos, A., & Jouve, N. (2003). Characterisation and analysis of new HMW-glutenin alleles encoded by the Glu-R1 locus of *Secale cereale*. *Theoretical and Applied Genetics, 107*, 74–83.

Feuillet, C., Travella, S., Stein, N., Albar, L., Nublat, A., & Keller, B. (2003). Map-based isolation of the leaf rust disease resistance gene *Lr10* from the hexaploid wheat (*Triticum aestivum* L.) genome. *Proceedings of the National Academy of Sciences of the United States of America, 100*, 15253–15258.

Forsström, P., & Merker, A. (2001). Sources of wheat powdery mildew resistance from wheat-rye and wheat-*Leymus* hybrids. *Hereditas, 134*, 115–119.

Griffiths, S., Sharp, R., Foote, T. N., Bertin, I., Wanous, M., Reader, S., et al. (2006). Molecular characterization of *Ph1* as a major chromosome pairing locus in polyploid wheat. *Nature, 439*, 749–752.

Gupta, P. K., Varshney, R. K., Sharma, P. C., & Ramesh, B. (1999). Molecular markers and their applications in wheat breeding. *Plant Breeding, 118*, 369–390.

Helguera, M., Khan, I. A., & Dubcovsky, J. (2000). Development of PCR markers for the wheat leaf rust resistance gene *Lr47*. *Theoretical and Applied Genetics, 100*, 1137–1143.

Huang, L., Brooks, S. A., Li, W., Fellers, J. P., Trick, H. N., & Gill, B. S. (2003). Map-based cloning of leaf rust resistance gene *Lr21* from the large and polyploid genome of bread wheat. *Genetics, 164*, 655–664.

Huang, L., & Gill, B. S. (2001). An RGA-like marker detects all known *Lr21* leaf rust resistance gene family members in *Aegilops tauschii* and wheat. *Theoretical and Applied Genetics, 103*, 1007–1013.

Keller, B., Feuillet, C., & Yahiaoui, N. (2005). Map-based isolation of disease resistance genes from bread wheat: cloning in a supersize genome. *Genetical Research, 85*, 93–100.

Kota, R., Spielmeyer, W., McIntosh, R. A., & Lagudah, E. S. (2006). Fine genetic mapping fails to dissociate durable stem rust resistance gene *Sr2* from pseudo-black chaff in common wheat (*Triticum aestivum* L.). *Theoretical and Applied Genetics, 112*, 492–499.

Lagudah, E. S., McFadden, H., Singh, R. P., Huerta-Espino, J., Bariana, H. S., & Spielmeyer, W. (2006). Molecular genetic characterization of the *Lr34/Yr18* slow rusting resistance gene region in wheat. *Theoretical and Applied Genetics, 114*, 21–30.

Latha, R., Rubia, L., Bennett, J., & Swaminathan, M. S. (2004). Allele mining for stress tolerance genes in *Oryza* species and related germplasm. *Molecular Biotechnology, 27*, 101–108.

Liu, S., & Anderson, J. A. (2003). Targeted molecular mapping of a major wheat QTL for Fusarium head blight resistance using wheat ESTs and synteny with rice. *Genome, 46*, 817–823.

Liu, Z., Sun, Q., Ni, Z., Nevo, E., & Yang, T. (2002). Molecular characterization of a novel powdery mildew resistance gene *Pm30* in wheat originating from wild emmer. *Euphytica, 123*, 21–29.

Malysheva, L., Ganal, M. W., & Röder, M. S. (2004). Evaluation of cultivated barley (*Hordeum vulgare*) germplasm for the presence of thermostable alleles of β-amylase. *Plant Breeding, 123*, 128–131.

Mardi, M., Buerstmayr, H., Ghareyazie, B., Lemmens, M., Mohammadi, S. A., Nolz, R., et al. (2005). QTL analysis of resistance to Fusarium head blight in wheat using a 'Wangshuibai'-derived population. *Plant Breeding, 124*, 329–333.

Miranda, L. M., Murphy, J. P., Marshall, D., & Leath, S. (2006). *Pm34*: A new powdery mildew resistance gene transferred from *Aegilops tauschii* Coss. to common wheat (*Triticum aestivum* L.). *Theoretical and Applied Genetics, 113*, 1497–1504.

Prasad, M., Varshney, R. K., Roy, J. K., Balyan, H. S., & Gupta, P. K. (2000). The use of microsatellites for detecting DNA polymorphism, genotype identification and genetic diversity in wheat. *Theoretical and Applied Genetics, 100*, 584–592.

Rong, J. K., Millet, E., Manisterski, J., & Feldman, M. (2000). A new powdery mildew resistance gene: Introgression from wild emmer into common wheat and RFLP-based mapping. *Euphytica, 115*, 121–126.

Rosewarne, G. M., Singh, R. P., Huerta-Espino, J., William, H. M., Bouchet, S., Cloutier, S., et al. (2006). Leaf tip necrosis, molecular markers and β1-proteasome subunits associated with the slow rusting resistance genes *Lr46/Yr29*. *Theoretical and Applied Genetics, 112*, 500–508.

Salvi, S., & Tuberosa, R. (2005). To clone or not to clone plant QTLs: Present and future challenges. *Trends in Plant Science, 10*, 297–304.

Scofield, S. R., Huang, L., Brandt, A. S., & Gill, B. S. (2005). Development of a virus-induced gene-silencing system for hexaploid wheat and its use in functional analysis of the *Lr21*-mediated leaf rust resistance pathway. *Plant Physiology, 138*, 2165–2173.

Spielmeyer, W., McIntosh, R. A., Kolmer, J., & Lagudah, E. S. (2005). Powdery mildew resistance and *Lr34/Yr18* genes for durable resistance to leaf and stripe rust co-segregate at a locus on the short arm of chromosome 7D on wheat. *Theoretical and Applied Genetics, 111*, 731–735.

Srichumpa, P., Brunner, S., Keller, B., & Yahiaoui, N. (2005). Allelic series of four powdery mildew resistance genes at the *Pm3* locus in hexaploid bread wheat. *Plant Physiology, 139*, 885–895.

Stein, N., Feuillet, C., Wicker, T., Schlagenhauf, E., & Keller, B. (2000). Subgenome chromosome walking in wheat: A 450-kb physical contig in *Triticum monococcum* L. spans the *Lr10* resistance locus in hexaploid wheat (*Triticum aestivum* L.). Proceedings of the National Academy of Sciences of the United States of America, 97, 13436–13441.

Tommasini, L., Yahiaoui, N., Srichumpa, P., & Keller, B. (2006). Development of functional markers specific for seven *Pm3* resistance alleles and their validation in the bread wheat gene pool. *Theoretical and Applied Genetics, 114,* 165–175.

Tyryshkin, L. G., Gul'tyaeva, E. I., Alpat'eva, N. V., & Kramer, I. (2006). Identification of effective leaf-rust resistance genes in wheat (*Triticum aestivum*) using STS markers. *Russian Journal of Genetics, 42,* 662–666.

William, M., Singh, R. P., Huerta-Espino, J., Islas, S. O., & Hoisington, D. (2003). Molecular marker mapping of leaf rust resistance gene *Lr46* and its association with stripe rust resistance gene *Yr29* in wheat. *Phytopathology, 93,* 153–159.

Xie, C., Sun, Q., Ni, Z., Yang, T., Nevo, E., & Fahima, T. (2004). Identification of resistance gene analogue markers closely linked to wheat powdery mildew resistance gene *Pm31*. *Plant Breeding, 124,* 198–200.

Yahiaoui, N., Brunner, S., & Keller, B. (2006). Rapid generation of new powdery mildew resistance genes after wheat domestication. *The Plant Journal, 47,* 85–98.

Yahiaoui, N., Srichumpa, P., Dudler, R., & Keller, B. (2004). Genome analysis at different ploidy levels allows cloning of the powdery mildew resistance gene *Pm3b* from hexaploid wheat. *The Plant Journal, 37,* 528–538.

Yan, G. P., Chen, X. M., Line, R. F., & Wellings, C. R. (2003). Resistance gene-analog polymorphism markers co-segregating with the *Yr5* gene for resistance to wheat stripe rust. *Theoretical and Applied Genetics, 106,* 636–643.

Zhang, P., Dreisigacker, S., Melchinger, A. E., Reif, J. C., Kazi, A. M., VanGinkel, M., et al. (2005). Quantifying novel sequence variation and selective advantage in synthetic hexaploid wheats and their backcross-derived lines using SSR markers. *Molecular Breeding, 15,* 1–10.

Eur J Plant Pathol (2008) 121:399–409
DOI 10.1007/s10658-008-9273-6

Integration of breeding and technology into diversification strategies for disease control in modern agriculture

Maria R. Finckh

Received: 26 July 2007 / Accepted: 17 January 2008
© KNPV 2008

Abstract While diversity for resistance has been recognised for more than 60 years as a key factor in disease management, and diversification strategies such as cultivar mixtures and multilines are described and advocated in almost every plant pathology text-book, the general view in modern agriculture is that diversity would be too difficult and expensive to implement. In addition, difficulties in marketing the produce are emphasised. The question thus arises if and how such strategies can be designed to find a place in modern agriculture. Considering the general ecological benefits of diversification and the possible economical benefits for growers and society, several possible approaches to the solution of actual and perceived problems in modern agriculture are dis-cussed. An important route towards achieving diver-sity would be to integrate it into the breeding process. Selection criteria would include inducibility of resis-tance and competitive ability, in order to produce diversified varieties able to adapt both to unpredict-able environmental conditions (especially climatic) and to changing pest and pathogen populations through co-evolution. Evolutionary breeding methods such as composite crosses and modern landraces and some of the legal problems associated with these approaches are discussed. Technical solutions are integral to the future use of diversification strategies and range from more or less simple adjustments to machinery for planting and harvesting to devices designed for separation of the harvested products.

Keywords Mixtures · Intercropping ·
Composite crosses · Evolutionary breeding ·
Co-evolution · Plant varietal protection

Introduction

Since the advent of modern plant breeding in the early twentieth century, the trends in agriculture have been towards genetic uniformity within crops. This has greatly enhanced possibilities for mechanisation in agriculture. However, this has resulted simultaneously in major losses in agricultural biodiversity (Fowler and Mooney 1990). Much of the resilience of agricultural ecosystems is due to complex interactions among species and genotypes at all levels of the system (inter- and intra-specific) and there are many examples throughout the nineteenth and twentieth century which demonstrate how a lack of diversity for resistance within crops renders crops and whole agricultural systems vulnerable to pest and disease attacks (e.g. Harlan 1972; Ullstrup 1972; Trenbath 1977; Juska et al. 1997). By the middle of the twentieth century breeders and pathologists realised that while "plant

M. R. Finckh (✉)
Faculty of Organic Agricultural Sciences,
Ecological Plant Protection Group, University of Kassel,
Nordbahnhofstr. 1a,
37213 Witzenhausen, Germany
e-mail: mfinckh@uni-kassel.de

diseases are shifting enemies" (Stakman 1947), reacting to changes in host resistance with continuous adaptation human interference with crop genetic make-up may pose dangers. These are best described with the following quote: "...by the wide use of hybrid corn we are depriving this important crop of its power of taking care of itself and by continued crossing and variation continually adjusting itself to the equally variable parasites which attack it" (Stevens 1942).

In reaction to the recognition of the dangers of genetic uniformity, diversification concepts have been developed by pathologists and breeders since the 1950s (see Wolfe and Finckh 1997 for review). Diversification strategies in time and space are being used in many agricultural systems to achieve acceptable levels of important diseases, pests, and weeds together with high product quality and stable yields (Finckh and Wolfe 2006).

There are a great number of possibilities for diversifying cropping systems, ranging from diversification of resistance genes within monoculture systems (multilines, cultivar mixtures) to species mixtures and very complex perennial polyculture (Finckh and Wolfe 2006). Polyculture systems in particular are usually designed to achieve many ecological benefits such as control of erosion, weeds, pests and diseases together with improved soil fertility (Altieri 1995; Altieri et al. 1996). While the focus of this paper is on crop diseases, implications for pests, weeds and other problems will also be considered to show how system modifications can be introduced which ultimately reduce costs, increase production and reduce stress on the environment.

The purpose of this paper is to give an overview of the effects of diversity on diseases and the wider ecological benefits of diversity. While much of the benefits of diversity are seen in intercropped systems, the focus here is largely on intraspecific diversity or intercropping in modern agriculture. Subsequently, the integration of breeding for diversity-including some of the legal implications-and technological approaches needed to achieve greater diversity within the agricultural system are discussed.

Effects of diversity on diseases

While monoculture usually refers to the continuous use of a single crop species over a large area, it is important to clearly differentiate the different types of monoculture that exist in agriculture with respect to pests and pathogens (Finckh and Wolfe 2006). Monocultures also exist at the level of species, variety or gene. For example, within a species monoculture, farmers can diversify among different varieties which exhibit different disease resistance genotypes. However, many of the modern varieties of a given species often possess the same gene(s) for resistance to a particular pathogen, resulting in a monoculture with respect to that resistance (resistance gene monoculture). A current example in western and central Europe is the use of the *mlo*-resistance to barley powdery mildew (caused by *Blumeria graminis*) in about 60% of the area presently grown with spring barley (Schwarzbach, personal communication). Monoculture of susceptibility has led to many catastrophic losses in, for example, potatoes, coffee, grapes, elms, or chestnuts during the nineteenth and twentieth century. The losses usually occurred in the wake of accidental intercontinental pest- and pathogen migrations. However, even without such migration events, many plant disease epidemics occur as a consequence of the cultivation of genetically uniform crops over large areas and thus, plant disease epidemics may be 'normal agricultural accidents' (Juska et al. 1997) due to the breeding for uniformity and the use of too few differentially resistant varieties in space and time.

Diversity that limits pathogen and pest expansion and that is designed to make use of knowledge about host-pest/pathogen interactions to direct pathogen evolution has been termed functional diversity (Schmidt 1978). The mechanisms leading to disease reduction in diversified systems will only be summarised here as there are several detailed recent reviews of the subject (Finckh and Wolfe 2006, Finckh et al. 2000; Mundt 2002).

On a mechanistic level, an increased distance between plants with the same susceptibility and the barrier function of differentially susceptible plants in between play a major role. In addition, microclimatic effects due to differences in crop architecture may become important, especially when plants are grown in alternating rows or strips. For example, in south western China, single rows of highly susceptible tall rice cultivars are interspersed in between several rows of dwarf hybrids (Fig. 1). In addition to the distance effects, reduced lodging of the tall plants and reduced

Fig. 1 Rice cultivar mixtures in China. Single rows of tall traditional varieties are interspersed within fields of high yielding hybrid cultivars (source: own photograph)

relative humidity in the well-aerated tall rice lead to reductions in rice neck blast severity (caused by *Pyricularia grisea*) from 40% to 50% to 4% to 5% (Zhu et al. 2005). But even in more intricate cultivar mixtures of wheat, e.g., differing in height, microclimatic conditions in the crop canopy may be drier than in uniform stands.

Overall, in species mixtures, the nutrient use efficiency is usually increased due to niche differentiation of the different crops (Tilman et al. 2001). While I am not aware of direct measurements of such effects in intraspecific mixtures, this may be true for these too, provided there are differences in nutrient uptake patterns and root spatial distribution. Differences in the microbial communities associated with the rhizosphere influencing plant health (termed plant probiotic microorganisms, PPM) are variety-specific (Picard and Bosco 2007) and it is likely that PPMs may also affect the nutrient availability and uptake (Picard et al. 2008). In addition, increased or decreased competition among plants due to differential infections may lead to the favouring of more resistant genotypes and allow for compensation of yield losses (Finckh and Mundt 1996; Finckh et al. 2000). Therefore, yield stability (i.e. consistent high yield over a range of environments) is commonly greater in mixtures than in pure stands. In large data sets, the treatment variance (mean square error) has been shown to be a good measure of yield stability (e.g. Allard 1961; Finckh et al. 2000).

On the pathogen side, different pathogens and pathotypes of a given pathogen may interfere by competing with each other for available host space (e.g. Lannou et al. 2005), resistance induction (e.g. Chin et al. 1984; Calonnec et al. 1996; Lannou et al. 2005), and possibly other unknown mechanisms. As host diversity favours diversity among pathogens, it will increase the chances of avirulent pathotypes occurring on a given host genotype. This, in return will enhance induced resistance in the system. This was shown impressively for yellow rust of wheat (Calonnec et al. 1996; Finckh et al. 2000).

Scale effects of diversification strategies are important but often not considered: as the area planted to a given host genotype increases, so does the amount of inoculum produced by this genotype and consequently the contribution of regional dispersal to epidemic development. Conversely, as the area sown to mixtures increases and the overall disease severity decreases, regional dispersal will also be reduced. While it is almost impossible to demonstrate this experimentally, results from a computer modelling study by Mundt and Brophy (1988) demonstrated this effect (Table 1). Also, when used on more than 300,000 ha in the former German Democratic Republic, powdery mildew of barley was reduced by 80% in barley cultivar mixtures within five years (Wolfe 1992).

If cultivar mixtures or multilines are used over time, there will also be evolutionary responses of the

Table 1 Effect of scale of diversification on area under the disease progress curve (AUDC) in computer simulated epidemics

Number of fields	Total Area (ha)	AUDC	
		Non-diversified	Diversified
4	10	92	70 (0.75)
16	38	127	75 (0.59)
64	154	172	83 (0.48)
256	614	218	91 (0.41)
1,024	2,458	261	89 (0.38)

In the study, either all fields were susceptible (non-diversified) or 75% of the fields resistant (diversified). Numbers in parentheses represent the relative disease in the diversified compared to the non-diversified situation (data from Mundt and Brophy 1988)

pathogens which could lead to the development of so-called super-races, i.e. pathotypes able to attack all genotypes in the mixtures, if the same type of host diversity is used continuously (see Mundt 2002; Wolfe and Finckh 1997; Finckh et al. 2000 for reviews). Thus, care has to be taken to either change the host resistances used over time or to allow for evolutionary responses on the host side (see below).

The described mechanisms have been shown to work well in reducing incidence and severity not only for more or less specialised wind and also some rain splash-dispersed foliar pathogens of cereals but also for other crops such as coffee (see Finckh and Wolfe 2006 for review). For potato late blight (caused by *Phytophthora infestans*), however, only moderate to insignificant effects of cultivar mixtures or plantings in alternating rows and strips of differentially susceptible varieties have been reported (Andrivon et al. 2003; Garrett and Mundt 2000; Phillips et al. 2005; Stolz et al. 2003) with greater mixture effects under moderate natural inoculum pressure than under high natural inoculum pressure (Garrett et al. 2001; Pilet et al. 2006). Part of the explanation for these results may be that, at least in Europe, most race-specific resistances have been overcome and we have observed that in mixtures and alternating row plantings of potato varieties differing in partial resistance, the more susceptible varieties were less diseased than in pure stands but the more resistant varieties were more diseased (Stolz et al. 2003 and own unpublished data). However, intercropping potatoes with strips of non-hosts led to significant disease reductions in field experiments over three years (Finckh et al. 2005;

Bouws and Finckh 2008). The reductions were due to reduced initial inoculum in smaller plots, reduced infection rates, changes in microclimatic conditions and, most importantly, loss of inoculum from strips planted perpendicular to the prevailing wind direction (Bouws and Finckh 2008). Late blight epidemics are usually started either through seed-borne infections or by wind-borne inoculum. In smaller fields, the absolute number of seed-borne infections per field as well as the probability of inoculation by an inoculum cloud is lower than in large fields. This should lead, on average, to a later epidemic start in smaller fields. Experimentally, this can only be tested without artificial inoculation. In the intercropping experiments of Bouws and Finckh (2008) the variable start of the epidemics was clearly visible and a detailed geostatistical analysis of the data showed the effects of plot size and plot location in the field (upwind/ downwind, neighbour spring wheat or grass-clover; Finckh et al., unpublished data).

Comparison of different management methods such as strips and whole fields in a meaningful way, has to be done on-farm and in different fields. Late blight epidemics in pure stands of potatoes grown in strips 12 m wide and 50 m long planted perpendicular to the prevailing wind direction, were compared with epidemics in regular potato fields on four commercial organic farms. Copper applications were tested as an additional factor. Overall, disease severity was highly variable among the different sites and field locations within sites with no obvious effect of cropping pattern. However, when copper was applied, the area under the disease progress curve in whole fields was reduced by an average of 23% but by 34% in the strips. It is likely that this was due to a greater loss of inoculum out of the strips than of the field (Finckh et al., unpublished data). In a theoretical modelling approach, Skelsey et al. (2005) predicted about 40% reductions in what the authors called 'final disease severity' on single potato rows surrounded by resistant plants in fields surrounded by non-hosts. However, as the simulations were stopped long before any spores could reach field edges, final disease severity in the model was not even 50% in the pure stands. In contrast, the complete epidemics were compared in our field experiments. It would be highly interesting to include wind direction, variable amounts of randomly distributed inoculum, and genotype units consisting of several rows into this model.

While late blight was reduced in strip-cropped potatoes, we also found that the outer rows of the potatoes in the strips often suffered from competition by neighbouring cereals (but not grass-clover) making it necessary to find a balance between epidemiological benefits and possible negative effects on the yield of the edge rows. This example shows that, despite positive effects of diversification strategies on diseases, these effects may not always translate into yield benefits. Care has to be taken that crops and species are truly amenable to mixing. While in variety mixtures, potatoes may still yield better overall than in pure stands (Phillips et al. 2005), in cassava (*Manihot esculenta*), significantly reduced yields were observed in cultivar and species mixtures as compared with pure stands, leading to the recommendation to plant small fields of pure stands separated by different crops for erosion control, e.g. rather than using intricate mixtures (Daellenbach et al. 2005).

Where intercropping is not an option, because of competition or the lack of appropriate partners for a system, mulches may offer solutions. For example, virus infestation levels of organic potatoes were significantly reduced by applying straw mulch in between the rows which reduced host finding by aphids (Saucke and Döring 2004). In addition, erosion was reduced by > 90% and after harvest, soil mineral nitrogen was prevented from leaching (Döring et al. 2005).

Diversification in time through crop rotation may deprive pathogens of their hosts for one to several seasons, thereby reducing inoculum of specialised pathogens substantially. In addition, certain green manure and other crops may have direct negative effects on pathogens and weeds due to allelopathy. For example, certain *Brassica* species release volatile compounds that have direct inhibitory effects on the growth of many pathogens (e.g. Gimsing and Kirkegaard 2006; Kasuya et al. 2006; Mayton et al. 1996). Other

crops such as hairy vetch (*Vicia villosa*; Zhou and Everts 2004) as well as various oat species and varieties, have been found to suppress several soil-borne pathogens including nematodes (Elmer and LaMondia 1999; Vilich 1993).

Besides pathogen starvation, different soil tillage practices applied to different crops in the rotation help reduce weed seed banks, and usually, soil microbial activity and soil fertility are increased by crop rotation. Thus, yields of winter wheat could only be maximised when grown after a pre-crop other than wheat even when fungicides were applied (Odoerfer et al. 1994; Table 2).

Integrating breeding and technological approaches to achieve diversity within the agricultural system

Despite the described positive effects of diversity, modern agriculture is based on monocultures, which are therefore targeted by current breeding programmes. Consequently, the ideotype of modern wheat has been defined as being relatively non-competitive to allow for dense pure stands (Hamblin and Donald 1974). It is for this reason, that most of the currently available crop varieties are not necessarily amenable to mixed cropping and there is a need to develop genetic resources adapted to diversified growing systems. In addition, managerial, legal, and technological problems have to be tackled.

Genetic resources for mixed cropping

Diversity can be achieved at the level of species, varieties or genes within species, variety mixtures as well as multilines in use worldwide (Finckh and Wolfe 2006). In contemporary usage of cultivar mixtures, the components have not been selected for performance in mixtures. It is unlikely, therefore, that

Table 2 Effects of different pre-crops and fungicide sprays on leaf diseases of wheat caused by *Drechslera tritici-repentis* (DTR) and *Septoria tritici* and on yields of wheat

Data from Odoerfer et al. 1994

Pre-crop	% Diseased leaf area with DTR and *S. tritici*		Yield (t ha^{-1})	
	No fungicide	Fungicide	No fungicide	Fungicide
Winter wheat	43	4	6.77	7.67
Faba beans	5	0	8.76	10.02
Red clover	7	1	8.39	9.38
Winter rape	5	0	7.98	8.69
Maize	13	0	7.43	8.26

they will perform as well as lines selected for mixture use, although even in pedigree line breeding programmes, the F1 and F2 generations are usually grown as populations. However, for example, beans that had evolved within a composite cross were more amenable to mixed cropping than beans selected early in the breeding process as pure lines (Allard 1961). Some trial results suggest that tests of two-way mixtures can provide useful indications of specific and general mixing ability for predicting performance of more complex mixtures of wheat (Knott and Mundt 1990; Lopez and Mundt 2000; Mille et al. 2006), barley (Gacek et al. 1996) and even potatoes (Phillips et al. 2005) in terms of both yield and disease restriction.

While the mixture strategies contribute substantially to plant protection, they all are based on at least a periodical remixing of the host populations. In this way, while the pathogen populations adapt constantly through natural selection, there is no possibility for reciprocal evolution on the host side, i.e. co-evolution does not happen. Before the advent of modern plant breeding, locally evolved and usually genetically diverse crop landraces were grown, and where landraces are still in use, these are often diverse for resistances to pests and pathogens. Thus, rice landraces in Bhutan differed in their resistance to rice blast (caused by *Pyricularia grisea*) among environments due to varying selective pressures (Thinlay 1998; Thinlay et al. 2000). Similarly, high diversity for resistance has been found in barley landraces in the Middle East (van Leur et al. 1989) and in phaseolus beans in Rwanda (Trutmann et al. 1993). Where landraces are still being used by growers, different approaches have been used to maintain their diversity and adaptability and some research is geared towards showing the role of diverse landraces in plant protection (see Jarvis et al. 2007 for review). Approaches being made use of in attempts to increase population resistance and yield potential while maintaining diversity include farmer participatory breeding (Ceccarelli et al. 2000), top-crosses (Yadav et al. 2000), partial replacement of local varieties through high yielding and/ or resistant varieties (Trutmann and Pyndji 1994) or population selection out of landraces (Finckh 2003).

Because of the absence of landraces in modern agriculture and also their low attractiveness in terms of yield and quality in comparison to modern varieties, Murphy et al. (2004) called for new approaches to the

development and selection of what they termed 'modern landraces'. These are bulk populations developed from superior germplasm and further subjected to local selection, or even farmer participatory improvement, using simple selection schemes that enable farmers to "breed crop varieties and landraces that will help improve the sustainability and profitability of their farm" (Murphy et al. 2004). Such an approach to the development of high yielding, but also highly diverse and adaptable plant populations, was already taken in California during the first half of the twentieth century (Harlan and Martini 1929) and later termed 'evolutionary plant breeding' (Suneson 1956). Through composite crosses (CC), i.e. the crossing in all possible combinations of several to many parental lines (Jain and Qualset 1975), genetically diverse populations were created that were later subjected to mass selection under different environmental conditions, a process mimicking the development of landraces while integrating the advantages of modern high yielding and high quality varieties.

Composite crosses have been made for barley, oats, wheat, phaseolus beans (see Phillips and Wolfe 2005 for review), and faba beans (Ghaouti et al. 2005). For barley, the CC approach has been shown to be extremely powerful: innumerable elite varieties released in the twentieth century trace their origin to the CC populations produced in California in the 1920s while the CC populations have remained genetically variable even after 50 generations of propagation at single sites (Allard 1988). When exposed to new environments including diseases that had not been important in the original environment, the CC populations proved to be readily adaptable to these (e.g., Danquah and Barrett 2002a, b; Hensleigh et al. 1992; Webster et al. 1986). Similarly, French work on dynamic management of wheat CCs shows that adaptive changes within the populations in response to local selection pressures occur, while simultaneously maintaining genetic variation and thus adaptability (Goldringer et al. 2006).

Several varieties that were not based on pure lines were selected from the barley CCs with varying success (Suneson 1956; Jain and Qualset 1975). More recently, wheat breeders at Washington State University (Dawson et al. 2006) and in Europe (Wolfe et al. 2006) have taken a similar approach for low-input wheat. Thus, such approaches to diversification in the

breeding process may be taken in areas where landraces do not play an important role any more. This would contribute to an overall diversification of the agricultural system and, at the same time to dynamic development and conservation of genetic resources. The advantage of the modern landrace approach is that selection for the ability to be grown in diversified systems is favoured. In addition, increased intra-varietal diversity may become more and more important in a future of rapidly changing and unpredictable climatic conditions.

While intra-varietal diversity might be important in the longer-term evolutionary context, it will only be of interest for practical use if yield levels and stability are comparable to pure lines or variety mixtures. For wheat, early work has shown increased yields in the F2 over physical mixtures (Qualset 1968). In an initial comparison of three wheat composite crosses with mixtures of their parental lines in four different sites, the composite crosses had higher yields in 11 out of 12 cases in 2005 (Wolfe et al. 2006). In 2006, results were similar but not in 2007 (Wolfe and Jones, personal communication). Polycross progenies of winter and spring faba beans tested in four sites over three years yielded significantly higher than the inbred parental lines, while the variances of the polycrosses were significantly lower, indicating superior yielding ability and stability (Ghaouti et al. 2005; Ghaouti and Link 2007).

Little formal research has been conducted into breeding for multiple cropping and mixtures (for review see Francis 1990). However, farmers often have long-standing experience in growing mixtures and may be an important resource for breeders. Thus, the success of heterogeneous bean varieties in Rwanda was greater when selected by local farmers than when selected by breeders (Sperling et al. 1993). Since 2004, a European COST project on 'Sustainable low-input cereal production: required varietal characteristics and crop diversity' (see www.COST860.DK) is addressing issues of breeding and diversification, bringing together scientists from more than 20 countries. There is also growing interest in organic, sustainable, and participatory breeding approaches as documented by several recent conferences (see website above) including the creation of a new section on sustainable and organic breeding within EUCARPIA (European Association for Plant Breeding Research) on November 7, 2007.

Legal considerations

If variety and species mixtures are to be sold as seed, the exact amount of each mixture component has to be stated by law. This precludes the production of mixtures from mixtures for sale, as the component frequencies in such seed mixtures will differ from the original frequencies. Producing pure lines and mixing these is, however, less efficient than simply producing mixed seed, making mixture seed often more expensive.

Registration of diversified varieties presents legal difficulties both in Europe and in the USA because of the Union for the Protection of New Varieties (UPOV) guidelines or EU rules (Regulation 2100/94/EC) and the Plant Varietal Protection Act (PVPA) in the USA, which require that a variety must be uniform genetically and in appearance and be readily distinguishable from other varieties in order to be accepted for registration. Over the past 50 years, innumerable varieties and landraces have disappeared from the market because they did not fulfill the legal requirements. For example, it has been estimated that approximately 75% of all vegetable varieties in Europe have disappeared within 10 years since the inception of UPOV in 1961 (Mooney 1979). Indeed, great concern has been voiced world-wide about the precipitous genetic erosion in agriculture due to legislative measures, general breeding methods and genetic engineering technology (e.g. Fowler and Mooney 1990; Kloppenburg 2004). Besides some local national provisions for the maintenance and limited circulation of landraces (i.e. already existing diverse populations), there is no provision for the release of newly bred diversified varieties such as composite crosses, or top crosses, since they do not comply with current law.

Technological and managerial questions

There may be technological limits to the mixture concept where mixtures are not desirable or special quality requirements have to be met. However, many products are the result of mixing. Thus, maltsters usually mix different varieties together for malting to achieve a desired malting quality and the malting quality of a batch of barley is dependent much more on the conditions under which it was grown than on the individual variety. Furthermore, variety mixtures of the same quality class can provide a more stable quality across environments than the single compo-

nent varieties (Baumer 1983). Similarly, in the former German Democratic Republic, breeders and brewers cooperated closely to achieve high quality malting barley cultivar mixtures (Wolfe 1992). For wheat, millers mix together different batches of grain to obtain the quality requirements of processors. This can also be achieved with cultivar mixtures and where farmers and bakers directly cooperate, e.g. it is common practice in the organic sector (personal observation). While in general, it is difficult to market mixtures in Europe, no such difficulty exists in the USA, as documented by the large area grown with wheat cultivar mixtures (Finckh and Wolfe 2006).

While coffee is a particularly sensitive crop from the point of view of quality some of the best coffee worldwide is produced in Colombia from more than 350,000 ha of mixtures that have been selected to be variable in terms of rust resistance but uniform for quality characteristics (Moreno-Ruiz and Castillo-Zapata 1990).

The Chinese example with rice in strips (Fig. 1, Zhu et al. 2005) is successful because the tall traditional varieties fetch a premium price on the market and they are harvested manually before the hybrids. Where the agricultural system is fully mechanised, however, strips of the width of commonly used machines should be chosen.

Introducing different crops into highly specialised farming operations usually requires additional know-how and machinery. Many of the beneficial break crops such as grass-clover are only useful if animals are present on the farm to make use of them. While green manure crops are an option on stockless farms, care has to be taken to prevent leaching of nutrients. Either cooperation between animal producers and stockless farmers or the integration of biomass production for biofuels may offer solutions.

Farmers often argue that the extra labour required when changing field sizes and arrangements will not pay. Net returns were estimated for the four on-farm potato strip intercropping experiments described above in comparison to normal fields. It was estimated that up to six more person hours ha^{-1} and season will be required for the management of strips including all operations (four copper applications) from planting to harvest. In three of the four farms, slightly higher net returns were realised from strip-cropped potatoes, twice in unsprayed strips and once in sprayed strips (own unpublished results).

Farmers and agricultural engineers continuously develop and adapt machinery for new applications. Row or strip intercropping is often practiced with the help of adapted machinery, especially in vegetable cultures, where machines are usually smaller, and strip or row-application of fertilisers, compost or agrochemicals is already common practice. Machinery for direct drilling and for undersowing does exist and mechanical separation of seed mixtures where different seed sizes are involved (e.g. faba bean-cereal mixtures) is practiced by some farmers (personal observation). Modern optics might offer solutions for post-harvest separation of fruits or vegetables.

Conclusions

Adaptability and buffering capacity are the product of functional diversity and the ability of crops to evolve. Considering current environmental problems such as erosion and global warming and the reduced predictability of the local climate, there are a number of breeding goals that need to be taken care of at the same time. Besides yield and quality, the ability to adjust to changing environments is especially important. Any properties that increase erosion control and/or allow for reduced inputs such as competitiveness against weeds, better soil cover, pest and disease resistance, as well as nutrient use efficiency will increase in importance.

While the positive effects of functional diversity for resistance but also for other important traits providing ecosystems services such as erosion control, reduced nutrient leaching and genetic resource conservation, are well known, there are many apparent and true technological and legal impediments to a more widespread use of diversity in agriculture. While technical solutions can and will be found relatively easily, possibly the most important impediment to the use of functional diversity is the lack of diversified varieties and the associated legal problems.

Clearly, breeders must be rewarded for their work and one way is through royalties. However, the current legal situation in all member countries of the Union for the Protection of New Varieties (UPOV) prevents the inclusion of functional genetic diversity for disease, pest, and other abiotic stress resistances into population varieties. In addition, it does not allow for the deliberate production of population varieties

where pure lines are the standard. There is a need to find new solutions that will allow breeders to be compensated for their efforts without constraining the potential for improving yield stability and the durability of disease and pest resistance in practical agriculture through intra-crop diversity. There is also a need for breeding crops adapted to mixed cropping, and technological solutions need to be found to allow farmers to efficiently increase systems diversity in agriculture.

References

Allard, R. W. (1961). Relationship between genetic diversity and consistency of performance in different environments. *Crop Science, 1*, 127–133.

Allard, R. W. (1988). Genetic changes associated with the evolution of adaptedness in cultivated plants and their wild progenitors. *Journal of Heredity, 79*, 225–238.

Altieri, M. A. (1995). *Agroecology. The science of sustainable agriculture*. Boulder, Colorado, London: Westview Press, IT Publications.

Altieri, M A., Nicholls, C. I., & Wolfe, M. S. (1996). Biodiversity - a central concept in organic agriculture: Restraining pests and diseases. In T. V. Ostergaard (Ed.), *Fundamentals of Organic Agriculture, vol. 1* (pp. 91–112). IFOAM, Ökozentrum Imsbach, D-66636 Tholey-Theley.

Andrivon, D., Lucas, J. M. A., & Ellisseche, D. (2003). Development of natural late blight epidemics in pure and mixed plots of potato cultivars with different levels of partial resistance. *Plant Pathology, 52*, 586–594.

Baumer, M. (1983). Neue Ergebnisse mit Sortenmischungen bei Sommergerste. *Top Agrar, 2*, 82–86.

Bouws, H., & Finckh, M. R. (2008). Effects of strip-intercropping of potatoes with non-hosts on late blight severity and tuber yield in organic production. *Plant Pathology*, in press.

Calonnec, A., Goyeau, H., & Devallavieillepope, C. (1996). Effects of induced resistance on infection efficiency and sporulation of *Puccinia striiformis* on seedlings in varietal mixtures and on field epidemics in pure stands. *European Journal of Plant Pathology, 102*, 733–741.

Ceccarelli, S., Grando, R., Tutwiler, R., Baha, J., Martini, A. M., Salahieh, H., et al. (2000). A methodological study on participatory plant breeding. I. Selection phase. *Euphytica, 111*, 91–104.

Chin, K. M., Wolfe, M. S., & Minchin, P. N. (1984). Host-mediated interactions between pathogen genotypes. *Plant Pathology, 33*, 161–171.

Daellenbach, G. C., Kerridge, P. C., Wolfe, M. S., Frossard, E., & Finckh, M. R. (2005). Plant productivity in cassava-based mixed cropping systems in Colombian hillside farms. *Agriculture, Ecosystems and Environment, 105*, 595–614.

Danquah, E. Y., & Barrett, J. A. (2002a). Grain yield in composite cross five of barley: Effects of natural selection. *Journal of Agricultural Sciences, 138*, 171–176.

Danquah, E. Y., & Barrett, J. A. (2002b). Evidence of natural selection for disease resistance in Composite Cross Five (CCV) of barley. *Genetica, 115*, 195–203.

Dawson, J., Murphy, K., Piaskowski, J., Arteburn, M., Lyon, S., Balow, K., et al. (2006). Increasing the diversity of wheat cultivars in Washington State (US). In H. Ostergaard, & L. Fontaine (Eds.) *Proceedings of the COST SUSVAR workshop on cereal crop diversity: implications for production and products, 13–14 June 2006, La Besse, France* (pp. 63–67). Paris, France: ITAB (Institut Technique de l'Agriculture Biologique).

Döring, T. F., Brandt, M., Heß, J., Finckh, M. R., & Saucke, H. (2005). Effects of straw muluch on soil nitrate dynamics, weeds, yield and soil erosion in organically grown potatoes. *Field Crops Research, 94*, 238–249.

Elmer, W. H., & LaMondia, J. L. (1999). Influence of ammonium sulfate and rotation crops on strawberry black root rot. *Plant Disease, 83*, 119–123.

Finckh, M. R. (2003). Ecological benefits of Diversification. In T. W. Mew, et al. (Ed.) *Rice science: innovations and impact for livelihood* (pp. 549–564). Los Banos, Philippines: Internatonal Rice Research Institute.

Finckh, M. R., Bouws-Beuermann, H., Piepho, H.-P., & Büchse, A. (2005). Effects of field geometry, neighbour culture and exposition on the spatial and temporal spread of *Phytophthora infestans* in organic farming. 16th Triennial Conference of the European Association of Potato Research, 17–22.7.2005, Bilbao. Abstracts of Papers and Posters II, pp. 429–431.

Finckh, M. R., Gacek, E. S., Goyeau, H., Lannou, C., Merz, U., Mundt, C. C., et al. (2000). Cereal variety and species mixtures in practice, with emphasis on disease resistance. *Agronomie, 20*, 813–837.

Finckh, M. R., & Mundt, C. C. (1996). Temporal dynamics of plant competition in genetically diverse wheat populations in the presence and absence of stripe rust. *Journal of Applied Ecology, 33*, 1041–1052.

Finckh, M. R. & Wolfe, M. S. (2006). Diversification strategies. In B. M. Cooke et al. (Eds.), *The Epidemiology of Plant Disease*. (pp. 269–308). Berlin: Springer.

Fowler, C., & Mooney, P. R. (1990). *Shattering: Food politics, and the loss of genetic diversity*. Tucson, AR: University of Arizona Press.

Francis, C. A. (1990). Breeding hybrids and varieties for sustainable systems. In C. A. Francis, et al. (Ed.) *Sustainable agriculture in temperate zones* (pp. 24–54). New York: Wiley.

Gacek, E. S., Czembor, H. J., & Nadziak, J. (1996). Disease restriction, grain yield and its stability in winter barley cultivar mixtures. In *Proceedings of the Third Workshop on Integrated Control of Cereal Mildews Across Europe. Kappel a. Albis, Switzerland, 5–9 Nov. 1994* (pp. 185–190). Brussels, Belgium: Office for Official Publications of the EC.

Garrett, K. A., & Mundt, C. C. (2000). Host diversity can reduce potato late blight severity for focal and general patterns of primary inoculum. *Phytopathology, 90*, 1307–1312.

Garrett, K. A., Nelson, R. J., Mundt, C. C., Chacon, G., Jaramillo, R. E., & Forbes, G. A. (2001). The effects of host diversity and other management components on

epidemics of potato late blight in the humid highland tropics. *Phytopathology, 91*, 993–1000.

Ghaouti, L. & Link, W. (2007). Appropriate breeding approach and type of cultivar in breeding faba bean for organic farming. In *Plant breeding for organic and sustainable low-input agriculture: dealing with genotype-environment interactions. Book of Abstracts. EUCARPIA Symposium, 7–9 November 2007, Wageningen, NL* (p. 55) Wageningen, NL: Wageningen University, Plant Breeding Group.

Ghaouti, L., Vogt-Kaute, W., & Link, W. (2005). Entwicklung ökologischer Regionalsorten bei Ackerbohnen [Development of region-specific organic cultivars in faba bean]. In *8. Wissenschaftstagung Ökologischer Landbau–Ende der Nische, Kassel 01.0.–04.3.2005*. (pp. 61–62).

Gimsing, A. L., & Kirkegaard, J. A. (2006). Glucosinolate and isothiocyanate concentration in soil following incorporation of *Brassica* biofumigants. *Soil Biology and Biochemistry, 38*, 2255–2264.

Goldringer, I., Prouin, C., Rousset, M., Galic, N., & Bonnin, I. (2006). Rapid differentiation of experimental populations of wheat for heading-time in response to local climatic conditions. *Annals of Botany, 98*, 805–817.

Hamblin, J., & Donald, C. M. (1974). The relationship between plant form, competitive ability and grain yield in a barley cross. *Euphytica, 23*, 535–542.

Harlan, J. R. (1972). Genetics of disaster. *Journal of Environmental Quality, 1*, 212–215.

Harlan, H. V., & Martini, M. L. (1929). A composite hybrid mixture. *Journal of the American Society Agronomy, 21*, 487–490.

Hensleigh, P. F., Blake, T. K., & Welty, L. E. (1992). Natural selection on winter barley composite cross XXVI affects winter survival and associated traits. *Crop Science, 32*, 57–62.

Jain, S. K., & Qualset, C. O. (1975). New developments in the evaluation and theory of bulk populations. In *Third International Barley Genetics Symposium* (pp. 739–749). München: Karl Thiemig.

Jarvis, D. I., Brown, A. H. D., Imbruce, V., Ochoa, J., Sadiki, M., Karamura, E., et al. (2007). Managing crop disease in traditional agroecosystems:the benefits and hazards of genetic diversity. In D. I. Jarvis, et al. (Ed.) *Managing biodiversity in agricultural ecosystems* (pp. 292–319). New York: Columbia University Press.

Juska, A., Busch, L., & Tanaka, K. (1997). The blackleg epidemic in Canadian rapeseed as a "normal agricultural accident". *Ecological Applications, 7*, 1350–1356.

Kasuya, M., Olivier, A. R., Ota, Y., Tojo, M., Honjo, H., & Fukui, R. (2006). Induction of soil suppressiveness against *Rhizoctonia solani* by incorporation of dried plant residues into soil. *Phytopathology, 96*, 1372–1379.

Kloppenburg, J. R. (2004). *First the seed. The political economy of plant biotechnology*. Madison, WI: The University of Wisconsin Press.

Knott, E. A., & Mundt, C. C. (1990). Mixing ability analysis of wheat cultivar mixtures under diseased and non-diseased conditions. *Theoretical and Applied Genetics, 80*, 313–320.

Lannou, C., Hubert, P., & Gimeno, C. (2005). Competition and interactions among stripe rust pathotypes in wheat-cultivar mixtures. *Plant Pathology, 54*, 699–712.

Lopez, C. G., & Mundt, C. C. (2000). Using mixing ability analysis from two-way cultivar mixtures to predict the performance of cultivars in complex mixtures. *Field Crops Research, 68*, 121–132.

Mayton, H. S., Oliviar, C., Vaughn, S. F., & Loria, R. (1996). Correlation of fungicidal activity of *Brassica* species with allyl isothiocyanate production in macerated leaf tissue. *Phytopathology, 86*, 267–271.

Mille, B., Belhaj Fraj, M., Monod, H., & de Vallavieille-Pope, C. (2006). Assessing four-way mixtures of winter bread wheat for disease resistance, yield, and grain quality using the performance of their two-way and individual cultivar components. *European Journal of Plant Pathology, 114*, 163–173.

Mooney, P. R. (1979). Seeds of the Earth. A public or private resource. Published by Inter Pares (Ottawa) for the Canadian Council for Intern. Co-operation and the Intern. Coalition for Development Action (London). Ottawa: Ottawa Mutual Press Limited.

Moreno-Ruiz, G., & Castillo-Zapata, J. (1990). The variety Colombia: A variety of coffee with resistance to rust (*Hemileia vastatrix* Berk. & Br.). *Cenicafe Chinchiná-Caldas-Colombia Technical Bulletin, 9*, 1–27.

Mundt, C. C. (2002). Use of multiline cultivars and cultivar mixtures for disease management. *Annual Review of Phytopathology, 40*, 381–410.

Mundt, C. C., & Brophy, L. S. (1988). Influence of host genotype units on the effectiveness of host mixtures for disease control: A modeling approach. *Phytopathology, 78*, 1087–1094.

Murphy, K., Lammer, D., Lyon, S., Brady, C., & Jones, S. S. (2004). Breeding for organic and low-input farming systems: An evolutionary-participatory breeding method for inbred cereal grains. *Renewable Agriculture and Food Systems, 20*, 48–55.

Odoerfer, A., Obst, A., & Pommer, G. (1994). The effects of different leaf crops in a long lasting monoculture with winter wheat. 2. Disease development and effects of phytosanitary measures. *Agribiology Research, 47*, 56–66.

Phillips, S. L., Shaw, M. W., & Wolfe, M. S. (2005). The effect of potato variety mixtures on epidemics of late blight in relation to plot size and level of resistance. *Annals of Applied Biology, 147*, 247–252.

Phillips, S. L., & Wolfe, M. S. (2005). Evolutionary plant breeding for low input systems. *Journal of Agricultural Science, 143*, 245–254.

Picard, C., Baruffa, E., & Bosco, M. (2008). Enrichment and diversity of plant-probiotic microorganisms in the rhizosphere of hybrid maize during four growth cycles. *Soil Biology and Biochemistry, 40*, 106–115.

Picard, C. & Bosco, M. (2007). Genotypic and phenotypic diversity in populations of plant-probiotic *Pseudomonas* spp. colonizing roots. *Naturwissenschaften, 95*, 1–16.

Pilet, F., Chacon, G., Forbes, G. A., & Andrivon, D. (2006). Protection of susceptible potato cultivars against late blight in mixtures increases with decreasing disease pressure. *Phytopathology, 96*, 777–783.

Qualset, C. O. (1968). Population structure and performance in wheat. In K. W. Finlay & K. W. Shepherd (Eds.) *Proceedings of the third international wheat genetics symposium* (pp. 397–402). Canberra: Butterworth.

Saucke, H., & Döring, T. F. (2004). Potato virus Y reduction by straw mulch in organic potatoes. *Annals of Applied Biology, 144*, 347–355.

Schmidt, R. A. (1978). Diseases in forest ecosystems: The importance of functional diversity. In J. G. Horsfall, & E. B. Cowling (Eds.) *Plant disease: An advanced treatise, vol 2* (pp. 287–315). New York: Academic.

Skelsey, P., Rossing, W. A. H., Kessel, G. J. T., Powell, J., & van der Werf, W. (2005). Influence of host diversity on development of epidemics: An evaluation and elaboration of mixture theory. *Phytopathology, 95*, 328–338.

Sperling, L., Loevinsohn, M. E., & Ntabomvura, B. (1993). Rethinking the farmer's role in plant breeding: Local bean experts and on-station selection in Rwanda. *Experimental Agriculture, 29*, 509–519.

Stakman, E. C. (1947). Plant diseases are shifting enemies. *American Scientist, 35*, 321–350.

Stevens, N. E. (1942). How plant breeding programs complicate plant disease problems. *Science, 95*, 313–316.

Stolz, H., Bruns, C., & Finckh, M. R. (2003). Einfluß genetischer Vielfalt auf den Befall mit *Phytophthora infestans* und auf die Ertragsbildung in Kartoffelbeständen. In *Beiträge zur 7. Wissenschaftstagung zum Ökologischen Landbau. Ökologischer Landbau der Zukunft, 24.26.2.2003, Wien* (pp. 569–570). Vienna: Universität für Bodenkultur.

Suneson, C. A. (1956). An evolutionary plant breeding method. *Agronomy Journal, 48*, 188–191.

Thinlay. (1998). Rice blast, caused by *Magnaporthe grisea*. In Bhutan and development of strategies for resistance breeding and management. Zürich, Switzerland: Dissertation ETH No. 12777, Swiss Federal Institute of Technology.

Thinlay, Zeigler, R. S., & Finckh, M. R. (2000). Pathogenic variability of *Pyricularia grisea* from the high- and mid-elevation zones of Bhutan. *Phytopathology, 90*, 621–628.

Tilman, D., Reich, P. B., Knops, J., Wedin, D., Mielke, T., & Lehman, C. L. (2001). Diversity and productivity in a long-term grassland experiment. *Science, 294*, 843–845.

Trenbath, B. R. (1977). Interactions among diverse hosts and diverse parasites. *Annals of the New York Academy of Sciences, 287*, 124–150.

Trutmann, P., & Pyndji, M. M. (1994). Partial replacement of local common bean mixtures by high yielding angular leaf spot resistant varieties to conserve local genetic diversity while increasing yield. *Annals of Applied Biology, 125*, 45–52.

Trutmann, P., Voss, J., & Fairhead, J. (1993). Management of common bean diseases by farmers in the Central African Highlands. *International Journal of Pest Management, 39*, 334–342.

Ullstrup, A. J. (1972). The impacts of the Southern com leaf blight epidemics of 1970–1971. *Annual Review of Phytopathology, 10*, 37–50.

van Leur, J. A. G., Ceccarelli, S., & Grando, S. (1989). Diversity for disease resistance in barley landraces from Syria and Jordan. *Plant Breeding, 103*, 324–335.

Vilich, V. (1993). Crop rotation with pure stands and mixtures of barley and wheat to control stem and root rot diseases. *Crop Protection, 12*, 373–379.

Webster, R. K., Saghai-Maroof, M. A., & Allard, R. W. (1986). Evolutionary response of Barley Composite Cross II to *Rhynchosporium secalis* analyzed by pathogenic complexity and by gene-by-race relationships. *Phytopathology, 76*, 661–668.

Wolfe, M. S. (1992). Barley diseases: Maintaining the value of our varieties. In L. Munk (Ed.) *Barley genetics VI* (pp. 1055–1067). Copenhagen: Munksgaard.

Wolfe, M. S., & Finckh, M. R. (1997). Diversity of host resistance within the crop: effects on host, pathogen and disease. In H. Hartleb, et al. (Ed.) *Plant resistance to fungal diseases* (pp. 378–400). Jena: G. Fischer Verlag.

Wolfe, M. S., Hinchscliffe, K. E., Clarke, S. M., Jones, H., Haigh, Z., Snape, J., et al. (2006). Evolutionary breeding of wheat. In H. Ostergaard, & L. Fontaine (Eds.) *Proceedings of the COST SUSVAR workshop on cereal crop diversity: Implications for production and products, 13–14 June 2006, La Besse, France* (pp. 77–80). Paris, France: ITAB (Institut Technique de l'Agriculture Biologique).

Yadav, O. P., Weltzien, R. E., Bidinger, F. R., & Mahalakshmi, V. (2000). Heterosis in landrace- based topcross hybrids of pearl millet across arid environments. *Euphytica, 112*, 295.

Zhou, X. G., & Everts, K. L. (2004). Suppression of Fusarium wilt of watermelon by soil amendment with hairy vetch. *Plant Disease, 88*, 1357–1365.

Zhu, Y., Fang, H., Wang, Y., Fan, J. X., Yang, S., Mew, T. W., & Mundt, C. C. (2005). Panicle blast and canopy moisture in rice cultivar mixtures. *Phytopathology, 95*, 433–438.